Inferential Statistics

NIPA® GENX ELECTRONIC RESOURCES & SOLUTIONS P. LTD.
New Delhi-110 034

www.nipaers.com
Browse, Search, Read & Buy...
Print Books
eChapters
eBooks
Publishers
Forthcoming
Subject Catalogues
New Titles
Online Resources on:
✓ Current Affairs
✓ Reasoning, Logic & Aptitude
✓ Competitive Examinations
✓ English Language Lab
✓ Writing & Pronunciation Tools
✓ Personality Development
✓ Online Programmes for Professionals
✓ Interview Preparation
Pay Using
PhonePe
Google Pay
paytm
PayPal
UPI
CC Avenue

Inferential Statistics

K.S. Kushwaha
M.Sc., P.S.C.C., Ph.D. (Statistics)
Recipient of Dr. Radhakrishnan Award (1992)
Associate Professor (Statistics)
Department of Mathematics & Statistics
Jawaharlal Nehru Krishi Vishwavidyalaya
Jabalpur-482004, M.P. (India)

NIPA® GENX ELECTRONIC RESOURCES & SOLUTIONS P. LTD.
New Delhi-110 034

NIPA® GENX ELECTRONIC RESOURCES & SOLUTIONS P. LTD.

101,103, Vikas Surya Plaza, CU Block
L.S.C. Market, Pitam Pura, New Delhi-110 034
Ph : +91 11 27341616, 27341717, 27341718
E-mail: newindiapublishingagency@gmail.com
Web: www.nipabooks.com

For customer assistance, please contact
Phone: + 91-11-27 34 17 17
Fax: + 91-11- 27 34 16 16
E-Mail: feedbacks@nipabooks.com

ISBN: 978-81-19215-16-4

Composed and Designed by NIPA®.

Preface

The book entitled "Inferential Statistics" has been designed for all P.G. and Ph.D. students of "Pure Statistics" and Agricultural Statistics and for those who have to appear in competitive examinations of I.S.S., I.E.S., state's PS.C and IAS. This book is also useful for faculties of "Department of Statistics" of Indian Universities. The book is the outcome of 30 years of teaching experience of U.G., P.G. and Ph.D. students of different disciplines of Agriculture, Agril. Engg. and Agril Statistics in J.N.K.V.V., Jabalpur. The content of the book covers the syllabus on the topic "Statistical Inference."

The book contains 9 chapters out of which Chapters (1-2) deals with the "Point Estimation", (3) Interval estimation, (4-5) Test of hypothesis, (6) Sequential statistics, (7) Decision theory, (8) Linear estimation and (9) Theory of ordered statistics. In each chapter three type of questions. "True/False, Fill in the Blanks and Multiple Choice Questions & Answers" alongwith "Key Answers" have been provided.

During the preperation of this manuscript, the author has incorporated the fruitful academic advices provided by Dr. H.P. Singh, Prof & Head, School of Studies in Statistics, Vikram University, Ujjain and Dr. B.B. Khare, Prof. (Statistics), Department of Statistics, B.H.U., Varanasi. It is expected to have a good popularity due to its usefulness amongst students and its users.

The author is very much thankful and extends his honourable honour of high order to Prof. N.N. Pathak, the Vice-chancellor (JNKVV, Jabalpur) for his kind permission to hand over the manuscript to "New India Publishing Agency" New Delhi to publish it in the present shape of a book.

The author acknowledges very strongly to his wife Smt. Veena Kushwaha, daughter Miss Akansha Kushwaha (Asstt. Engineer, Ministry of Water Resources, Central Ground Water Board) and son Mr. Utkarsh Kushwaha who cared a lot during the preparation of this manuscript.

Last but not the least, the author is very much thankful to the entire staff of "New India Publishing Agency" New Delhi for taking keen interest in publishing the manuscript in its present shape.

Although, the author has tried his best in designing the subject matter very nicely and systematically but that can be further improved based on the suggestions, received from its readers and users. Therefore, the fruitful academic suggestion received from all sides will be welcomed and will be highly appreciated by the author.

10th January, 2016

K.S. Kushwaha
Jabalpur

Contents

Preface .... *v*

1. Point Estimation 1

1.1 Write True/False 1

1.1.1 Cramer, Rao Inequality 6

1.1.2 Minimum Variance Bound Estimators 10

1.1.3 Blackwellization and Minimum Variance Unbiased Estimator (M.V.U.E.) 12

1.1.4 Sufficient Statistic and Completeeness of Sufficient Statistic 14

1.1.5 Distributions Possessiong Sufficient Statistic 15

1.1.6 Sufficient Statistic for Several Parameters 16

1.1.7 Ordered Statistic 20

1.2 Fill in the Blanks 37

1.2.1 Cramer, Rao Inequality 42

1.2.2 Minimum Variance Bound Estimators 45

1.2.3 Blackwellization and M.V.U.E. 47

1.2.4 Sufficient Statistic and Completeness of Sufficient Statistic 49

1.2.5 Distributions Possessing Sufficient Statistic 49

1.2.6 Sufficient Statistic for Several Parameters 50

1.2.7 Ordered Statistic 53

1.3 Multiple Choice Questions and Answers 67

1.3.1 Cramer, Rao Inequality 76

1.3.2 Minimum Variance Bound Estimators 84

1.3.3 Blackwellization and M.V.U.E. 88

1.3.4 Sufficient Statistic and Completeness of Sufficient

Statistic 91

1.3.5 Distributions Possessing Sufficient Statistic 92

1.3.6 Sufficient Statistic for Several Parameters 94

1.3.7 Ordered Statistic 102

1.4 Key Answers 126

2. Methods of Estimation 135

2.1 Write True/False 135

2.2 Fill in the Blanks 150

2.3 Multiple Choice Questions and Answers 163

2.4 Key Answers 186

3. Interval Estimation 191

3.1 Write True/False 191

3.1.1 Theory of Confidence Sets 197

3.2 Fill in the Blanks 203

3.3 Multiple Choice Questions and Answers 212

3.4 Key Answers 234

4. Test of Hypothesis 239

4.1 Write True/False 239

4.1.1 Preliminaries 239

4.1.2 Neyman Pearson Lemma 243

4.2 Fill in the Blanks 248

4.2.1 Preliminaries 248

4.2.2 Neyman - Pearson Lemma 251

4.3 Multiple Choice Questions and Answers 255

4.3.1 Preliminaries 255

4.3.2 Neyman-Pearson Lemma 261

4.4 Key Answers 269

5. More on Testing of Hypothesis 273

5.1 Write True/False 273

5.2 Fill in the Blanks 283

5.3 Multiple Choice Questions and Answers 289
5.4 Key Answers 301

6. Sequential Analysis 305
6.1 Write True/False 305
6.2 Fill in the Blanks 317
6.3 Multiple Choice Questions and Answers 326
6.4 Key Answers 342

7. Decision Theory 347
7.1 Write True/False 347
7.2 Fill in the Blanks 353
7.3 Multiple Choice Questions and Answers 359
7.4 Key Answers 369

8. Linear Estimation 373
8.1 Write True/False 373
8.2 Fill in the Blanks 383
8.3 Multiple Choice Questions and Answers 391
8.4 Key Answers 403

9. Distribution of Order Statistics 407
9.1 Write True/False 407
9.2 Fill in the Blanks 410
9.3 Multiple Choice Questions and Answers 412
9.4 Key Answers 416

10. Bibliography 419

Chapter - 1

Point Estimation

1.1 Write True/False

Q.1 The process of knowing all the relevant information from the known sample to the unknown population, is known as "Statistical Inference."

Q.2 An enquier may be completely unknown to some feature of the population but he may want to guess completely about that feature based on a random sample drawn from the population. This type of problem is known as the "Problem of Estimation."

Q.3 The estimation of population parameter by a single value (an estimate of parameter) based on a random sample drawn from the population, is known as "Point Estimation."

Q.4 A process of getting an interval within which the parameter is expected to lie, is known as "Interval Estimation."

Q.5 An estimator having the properties of consistancy, sufficiency and efficiency (by Prof. R.A. Fisher) is known as "Best Estimator."

Q.6 Any statistic (estimator) whose mathematical expectation is equal to the population parameter, say θ, is known as unbiased statistic (estimator) of the parameter.

Q.7 An estmator whose mathematical expectation is not equal to the population parameter, is known as "Biased Estimator."

Q.8 Let t_n $(x_1, x_2, x_3, \ldots\ldots\ldots\ldots, x_n)$ be an estimator defined on a random sample $(x_1, x_2, \ldots\ldots\ldots\ldots, x_n)$ of size n drawn from a population having its parametric value, say θ, then t_n is said to be an unbiased estimator of θ if

$E(t_n) = \theta$

Q.9 An estimator t_n defined in $\theta(8)$ is said to be a biased estimator of the parameter θ, if :

$E(t_n) \neq \theta$

Q.10 The amount of bias of an estimator $t_n(\underline{x})$ can be worked out using the expression.

$B(t_n) = E(t_n) - \theta$

Q.11 An estimator $t_n(x)$ is said to be positively biased estimator of the parameter θ if :

$E(t_n) > \theta$

Q.12 An estimator $t_n(x)$ of the parameter θ is said to be negatively biased if :

E(t\) < θ

Q.13 An estimator $t_n(x)$ is said to be asymptotically unbiased for the parameter θ if :

$$\lim_{n \to \infty} E(t_n) = \theta$$

Q.14 An estimator $t_n = \frac{1}{n}\sum_{i}^{n}(x_1 - \bar{x}_n)^2$ is asymptotically unbiased for the population variance σ^2 because $\lim_{n \to \infty} E(t_n) = \lim_{n \to \infty} \frac{n-1}{n}\sigma^2 = \sigma^2$

Q.15 The sample mean $\bar{x}_n = \frac{1}{n}\sum_{i}^{n} x_i$ is an absolutely unbiased estimator of the population mean μ_x because :

$$E(\bar{x}_n) = \mu_x$$

Q.16 Let $(x_1, x_2, \ldots\ldots\ldots\ldots, x_n)$ be a random sample of sixe n drawn from a normal population N(μ, 1) than the estimator $t_n = \frac{1}{n}\sum_{i}^{n} x_i^2$ is an unbiased estimator of $(\mu^2 + 1)$, since :

$$E(t_n) = (\mu^2 + 1)$$

Q.17 Let $(x_1, x_2, \ldots\ldots\ldots\ldots, x_n)$ be random observations (sample) taken on a Bernoulli variable α such that :

$P(x = 1) = \theta$

$P(x-0) = 1-\theta$

then $t^* = \dfrac{t(t-1}{n(n-1)}$ is an unbiased estimator for θ^2 where $t = \sum_{i=1}^{n} x_i$

Q.18 Unbiasedness property of an estimator $t_n(x)$ simply shows that the estimate on the average becomes equal to the parameter θ.

Q.19 Unbiasedness property of an estimator $t_n(x)$ does not say anything about the difference between the estimated value and true value of the parameter θ.

Q.20 The difference between the estimated value and true value of the parameter θ is described by the "Consistency" property of the estimator.

Q.21 An estimator $t_n(x)$ of the parameter θ is said to be consistent for θ if :

$$\lim_{n\to\infty} P\,[\,|\,t_n - \theta\,| \le \epsilon] > 1-\delta.$$

for (ϵ, δ) being infinitely small positive quantities.

Q.22 An estimator t_n is said to be a consistent estimator for the parameter θ if it converges in probabilty to θ.

Q.23 An estimator t_n is said to be consistent for the parameter θ if :

$\lim_{n\to\infty} P\,[\,|\,t_n - \theta\,| \le \epsilon] \to 1$, for ϵ being infinitely small positive value.

Q.24 A consistent estimator may not be an unbiased estimator of the parameter say, θ.

Q.25 An unbiased estimator may not be a consistent estimator of the parameter, say θ.

Q.26 A consistent estimator with finite mean value may tend to be unbiased to the parameter θ, in large samples.

Q.27 If there exists one consistent estimator t_n for the parameter θ then an infinite number of consistent estimators can be constructed.

Q.28 If t_n is an unbiased estimator of parameter θ_n with $V(t_n) = \sigma_n^2$ then t_n will be consistent estimator for θ if :

$$\theta_n \to \theta \;\&\; \sigma_n^2 \to 0 \text{ for } n \to \infty.$$

Q.29 The sample median x is a consistent estimator for population mean μ if $\tilde{x} \sim N(\mu, \sigma^2)$.

Q.30 In case of normal distribution $N(m, \sigma^2)$, the sample variance s^{*2} =

$\dfrac{1}{n}\sum_{i}^{n}(x_1 - \overline{x})^2$ is a consistent estimator for population variance σ^2.

Q.31 In cauchy distribution $f(x, \theta) = \frac{1}{\pi} \times \frac{1}{1+(x-\theta)^2}$, $-\infty < x < \infty$, the sample mean $\bar{x}_n$ is not consistent estimator for parameter θ.

Q.32 In Cauchy distribution $f(x, \theta) = \frac{1}{\pi} \frac{1}{1+(x-\theta)^2}$ $-\infty < x < \infty$, the sample median $\tilde{x}$ is a consistent estimator for parameter θ.

Q.33 In Cauchy distribution $f(x, \theta) = f(x, \theta) = \frac{1}{\pi} \frac{1}{1+(x-\theta)^2}$ $-\infty < x < \infty$, the mean $E(x) = \bar{x}$ does not exist.

Q.34 In poisson distribution $f(x, \lambda)$, the consistent estimator for $\frac{1}{\lambda}$ is given as : $t_n = \frac{1}{\bar{x}}$ where $\bar{x} = \frac{1}{n}\sum_{i}^{n} x_i$.

Q.35 Let t be a consistent estimator for any parameter θ, then a consistent estimator for e^{θ} will be given by e^{t}.

Q.36 An estimating function of the sample valueus $(x_1, x_2, \ldots\ldots\ldots, x_n)$ to estimate the population parameter say θ, is known as as estimator.

Q.37 A sample statistic whose sampling distribution concentrates as closely as possible near to the true value of the parameter θ, may be considered as an estimator for θ.

Q.38 In the class of consistent estimators for a parameter θ, if there exists one estimator t^* whose sampling variance is less than that of any such estimator in the class, then t^* is known as "most efficient" estimator.

Q.39 The most efficient estimator t^* for the parameter θ, provides the criteria for measuring the "efficiency" of such other estimators.

Q.40 Let $(t_n, t_{n'})$ be two consistent estimators for the parameter θ such that :

$$t_n \sim N(\theta, \frac{\sigma^2}{n})$$

$$t_{n'} \sim N(\theta, \frac{\sigma'^2}{n})$$

then t_n is said to be more efficient than $t_{n'}$ if $\sigma^2 - \sigma'^2$ fpr $\not\forall_n$.

Q.41 Let e denotes the efficiency of an estimator t_n then e is defined as :

$$e = \frac{Sampling\ variance\ of\ most\ efficient\ estimator}{Sampling\ variance\ of\ any\ given\ estimator.......} < 1.0$$

Q.42 If e denotes the efficiency of two unbiased estimators (T_1 and T_2) with equal sampling variance then we have the relation as :

corelation $(T_1, T_2) \geq (e - 1)$.

Q.43 Let (t_1, t_2) be two most efficient estimators for the parameter θ then $\int t_1, t_2 = 1.0$ i.e. t_1 and t_2 are linearly related.

Q.44 In case of normal distribution N($N(\mu, \sigma^2)$, the sample mean $\overline{x}_n$ is more efficient than the sample median $\widetilde{x}_n$ for the parameter μ.

Q.45 All the efficient estimators of the same parameter say θ, are equivalent.

Q.46 Uniformly minimum variance estimator (U.M.V.E.) of a parameter say θ, is unique.

Q.47 An estimator t_i ($\underset{\sim}{x}$) of parameter θ_i, is said to be uniformly minimum variance estimator (U.M.B.R) of θi, if.

$V(t_i(x)) <$ V (hi (x))

for any other estimator hi ($\underset{\sim}{x}$) of θ_i for all $\theta_i \in \Omega\ \theta$. where θ is the parameter space.

Q.48 Let (T_1, T_2) be two consistent/unbiased estimators of a parameter θ out of which T_1 is more efficient then T_2 with efficiency e then :

$$\int t_1, t_2 = \text{corr.}\ (T_1, T_2) = \sqrt{e}$$

Q.49 Let (T_1, T_2) be two unbiased estimators of parameter ψ (θ) with efficiency (e_1, e_2) respectively with correlation (T_1, T_2) as $\int$ then there exists a following relation.

$$\sqrt{e_1 e_2} - \sqrt{(1-e_1)(1-e_2)} \leq \int \leq \sqrt{e_1 e_2} + \sqrt{(1-e_1)(1-e_2)}$$

Q.50 Let t be the minimum variance unbiased estimator (M.V.U.E.) of a parameter θ and t^* be on other estimator of θ with efficiency 'e' then 'e' is defined as :

$e\ (t^*/t) = V(t)/V(t^*) < 1.0$

Q.51 In Cauchy distribution, let $t_n(x)$ denotes the maximum likelihood estimator (m.l.e.) of a parameter θ then t_n ($\underset{\sim}{x}$) is most efficient estimator with :

$$V(t_n) = \frac{2}{n}$$

Q.52 In Cauchy distribution we have variance (sample median x) $= \frac{\pi^2}{4n}$

Q.53 In Cauchy distribution, the efficiency of sample median x is defined as :

$$e(\tilde{x}) = \frac{2}{n} \Big/ \left(\frac{\pi^2}{4n}\right) = \frac{8}{\pi^2} = 0.8 \text{ approximately.}$$

Q.54 In normal distribution $N(\mu, \sigma^2)$, let the sample mean deviation

$$s^* = \frac{1}{n}\sum_{i}^{n} | x_i - \mu |_{,}^{2}$$

then $V(\$^*) = \left(\frac{\pi - 2}{\pi}\right)\frac{\sigma^2}{n} < V(\text{sample mean } \overline{x}_n)$

Q.55 An uniformly minimum variance unbiased estimator (U.M.V.U.E.) of a parameter say θ, may not exist even though an unbiased estimator does exist.

Q.56 The correlation between a U.M.V.U.E. and any other unbiased estimator is always positive.

Q.57 Let (T_1, T_2) be U.M.V.U.E. of two parametric functions say $(g_1(\theta), g_2(\theta))$ respectively then an U.M.V.U.E. (T_0) of the parametric function $a_1 g_1(\theta) + a_2 g_2(\theta)$ will be given by :

$T = a_1 T_1 + a_2 T_2$

where (a_1, a_2) are fixed constants.

Q.58 An U.M.V.U.E. may not attain the Cramer Rao lower bound.

Q.59 Let $T_i (i = 1, 2, ..., k)$ be k independent unbiased estimators of a parametric function $\phi(\theta)$, and $V(T_i) = \sigma_i^2 < \sigma^2 \; \forall$ i, then an unbiased estimator T of $\phi(\theta)$ will be given as :

$$T = \frac{1}{k}\sum_{i}^{k} T_i$$

1.1.1 Cramer, Rao Inequality

Q.60 If $t_n(x)$ is an unbiased estimator of a parametric function $\psi(\theta)$ then C.R. inequality states that :

$$V(t_n) \geq \frac{\left(\frac{\partial}{\partial_\theta}\psi(\theta)\right)^2}{E\left(\frac{\partial}{\partial_\theta}\log L\right)^2} = \frac{\{\psi'(\theta)\}^2}{I_\theta}$$

where I_θ is the information on θ supplied by the random sample.

Q.61 One of the regularity conditions for C.R. inequality to hold good is that :

$$\frac{\partial}{\partial_\theta} f(x, \theta) \text{ exists } \forall\ \theta \in \Omega_\theta.$$

Q.62 One of the regularity conditions for C.R. inequality to hold good is that : The range of integration is independent of the parameter θ so that

$\int \prod_i^n f(x_i, \theta)\, dx$ can be differentiated under integral sign :

Q.63 For C.R. inequality to hold good, the following criterio written as :

$$E\left\{\frac{\partial}{\partial_\theta}\log f(x, \theta)\right\}^2 > 0\ \forall\ \theta \in \Omega_\theta$$

is true.

Q.64 For C.R. inequality to hold good, the function written as :

$\int t_n(x) \prod_i^n f(x_i, \theta)\, dx$ is differentiable under integral sign.

Q.65 Inc ase of Poisson distribution $f(x, \lambda) = \frac{e^{-\lambda}\lambda^x}{x1}$, $x = 0, 1\ 2$..............,

a statistic $T_n = \sum_i^n x_i$ is complete and sufficient for parameter λ.

Q.66 In case of Poisson distribution $f(x, \lambda)$, the M.V.U.E. of $e^{-\lambda}$ is given by a statistic T defined as :

$$T = (1 - \frac{1}{n})^{T_n} = \left(\frac{n-1}{n}\right)^{\sum_i^n x_i}$$

with $V(T) = e^{-2\lambda}(e^{\lambda/n} - 1)$

Q.67 In θ66, the C.R. lower bound for T is equal to $e^{-2\lambda}.(n/\lambda)$.

Q.68 If an unbiased estimator T_n of parameter θ has the C.R. lower bound as its sampling variance, then Tn must be sufficient statistic for θ.

Q.69 The C.R. lower bound attends equality sign if :

$$\frac{\partial}{\partial_\theta} \log \prod_i^n f(x_1, \theta) = \kappa(\theta, n)(T_n - \psi(\theta)) \text{ holds good.}$$

Q.70 If there exists an estimator $T_n(x)$ such that the condition written as in Q.69, holds good then Tn is sufficient and is U.M.V.U.E. of ψ(θ).

Q.71 In case of Q.69 the sampling variance of the estimator $T_n(x)$ is given as :

$$V(T_n(x)) = \frac{\psi^1(\theta)}{\kappa(\theta, n)}$$

Q.72 Cramer Rao inequality gives lower bound to the sampling variance of an unbiased estimator $T_n(x)$ of parameter θ.

Q.73 Information I_θ on parameter θ is provided by the expression written as :

$$I_\theta = \nu\left(\frac{1}{L}\frac{\partial L}{\partial \theta}\right) = n\, E\left[\frac{\partial}{\partial \theta} log\ f\right]^2$$

Q.74 Let $T_n(x)$ be an unbiased estimator of parametric function ψ(θ) then following relation exists :

$$E(t_n - \theta)^2 \geq [\theta - \psi(\theta)]^2 + \frac{(\psi'(\theta))^2}{I_\theta}$$

Q.75 Let ψ(θ) = θ + b(θ) and $E(t_n) = \psi(\theta)$ then we have the relation :

$V(t_n) \geq [1 + b'(\theta)] \mid I_\theta$

Q.76 Let $E(\theta - \theta)^2$ denotes the mean squared deviation of an estimator θ of the parameter θ and b(θ) denotes the bias of θ then the following relation exists :

$$E(\theta - \theta)^2 \geq \frac{[1 + b'(\theta)]}{I_\theta} + (b(\theta))^2$$

Q.77 If $\frac{\partial}{\partial\theta} \log L = \frac{1}{\lambda}(T_n - \psi(\theta))$ then. C.R. inequality states that T_n is M.V.U.E. of $\psi(\theta)$ with $V(T_n) = \lambda$.

Q.78 Let x~ N(μ, σ^2) then C.R. inequality states that M.V.U.E. of μ is given by the sample mean $\bar{x}_n$ with minimum variance $(\bar{x}_n) = \frac{\sigma^2}{n}$ when σ^2 is known.

Q.79 In case of Poisson distribution $f(x, \lambda) = \frac{e^{-\lambda}\lambda^x}{\lfloor x}$, a M.V.U.E. of the parameter λ is given by the sample mean $\bar{x}n$ with its sampling variance (λ/n).

Q.80 In Q.79, the following expression exists :

$\frac{\partial}{\partial\lambda} \log L = \frac{n}{\lambda}(\bar{x}_n - \lambda)$ where L denotes the likelihood function of random sample $(x_1, x_2, \ldots\ldots\ldots\ldots\ldots, x_n)$.

Q.81 Let X be a random variable following binomial distributes β(X, n, p), we can write :

$$\frac{\partial}{\partial_p} \log L = \frac{\left(\frac{r}{n} - p\right)}{p(1-p)/n}$$

Q.82 In Q.81, the M.V.U.E. of p is given by $T_n = \frac{r}{n} = \bar{x}_n$ with sampling variance $V(T_n) = p(1-p)/n = pq/n$

Q.83 In case of Cauchy distribution $f(x, \theta) = \frac{1}{\pi} - \frac{1}{1+(x-\theta)^2} - \infty < x < \infty$ the M.V.U.E. does note exist.

Q.84 Let x~ N(μ, σ^2). then following result exists :

$$\frac{\partial}{\partial\sigma^2} \log L = \frac{n}{2\sigma^4}\left[\frac{1}{n}\sum^{n}(x_i - \mu)^2 - \sigma^2\right] = \frac{n}{2\sigma^4}\left(s^{*2} - \sigma^2\right)$$ where s* = $\frac{1}{n}\sum^{n} |x_i - \mu|$ is sample mean deviation.

Q.85 In Q.84, the M.V.U.E. of population variance σ^2 is given by s^{*2} with its minimum variance $V(s^{*2}) = \frac{2\sigma^4}{n}$

Q.86 In case of Poisson distribution $f(x, \lambda)$ the M.V.U.E. of λ^2 does not exist.

Q.87 Consider the p.d.f. of type III population as :

$$P(x, p, \theta) = \frac{1}{\Gamma p\ \theta^p} - \underline{x^{p-1}} \cdot e^{\frac{-x}{\theta}}, \quad p > 0,\ 0 \leq x < \infty.$$

The M.V.B.E. of θ for fixed p is given as $t_n = \frac{\bar{x}_n}{p}$ with

variance $(t_n) = (t_n) = \frac{\theta^2}{np}$.

1.1.2 Minimum Variance Bound Estimators

Q.88 A condition for the existence of M.V.B.E. for ψ(θ) is writeen as :

$$\frac{\partial}{\partial\theta} \log L = A(\theta)\ \{t - \psi(\theta)\}$$

where A(θ) is independent of sample values $(x_1, x_2, \ldots\ldots\ldots\ldots, x_n)$, but may be the function of parameter θ.

Q.89 If the frequency density function can not be put in the form as in Q.88, then there may still be an estimator t^* of ψ(θ) which has uniformally smaller variance than that of any other estimator. Such estimator to known as "Minimum Variance Estimator (M.V.E.)."

Q.90 The least attainable variance of an estimator say t', may be greater than that of M.V.B.E. of the parameter ψ(θ).

Q.91 If the regularity conditions leading to the existence of M.V.B.E. do not hold, then least attainable variance may be less than that of M.V.B. estimator.

Q.92 In binomial distribution b(x, n, p), an unbiased estimator of p(1 – p) = pq is given by :

$$t = \frac{r\ (n-r)}{n\ (n-1)} \quad \text{where } r = \sum_{i=1}^{n} x_i.$$

Q.93 If a M.V.B.E. exists, it will exist for one specified function say $\psi(\theta)$ of the parameter θ but for other function of θ.

Q.94 In case of normal distribution $N(\mu, \sigma^2)$, where μ being known, the M.V.B.E. of σ^2 exits.

Q.95 In Q.94, M.V.B.E for σ does not exist.

Q.96 Consider the logistic distribution :

$$f(x,\theta) = e\frac{-(x-\theta)}{1+e^{-(x-\theta)}}$$

The M.V.B.E. for θ is given as $t_n = \frac{3}{n}$

Q.97 In Q.96, the exact sampling variance of the sample mean $\overline{x}$ is given as :

$$V(\overline{x}) = \frac{\pi}{3n}$$

Q.98 In case of Q.96, the exact sampling variance of the sample median x is given as :

$$v(\tilde{x}) = \frac{4}{n}$$

Q.99 Consider the normal distribution $N(X, \mu, \sigma^2)$. Then for known σ^2, the M.V.B.E. of μ^2 is given as :

$$\hat{\mu}^2 = \delta = \left(\overline{x}^2 - \frac{\sigma^2}{n}\right)$$

Q.100 In Q.99, the sufficient statistic for μ^2 is the sample mean $\overline{x}_n$.

Q.101 In case of normal distribution $N(\mu, \sigma^2)$, the statistic $t = \sum_{i}^{n} a_i x_i$ being an unbiased estimate of population mean μ, the satisfying condition is $\sum_{n}^{1} a_i x_i$ = 1.0 for a set of unknown constants (a_i, i = 1, 2, ..., n)

Q.102 A minimum variance bound estimator (M.V.B.E.) of a parametric function $\psi(\theta)$ exists only if a sufficient statistic for $\psi(\theta)$ exists.

Q.103 If M.V.B.E. of function $\psi(\theta)$ exists then the statistic for $\psi(\theta)$ may or may not exist.

Q.104 Let 't' be a sufficient statistic for parameter θ then any one to one function of 't' will also be sufficient statistic.

Q.105 If a M.V.B.E. for a parameter θ exists then it is given by the method of "Maximum Likelihood Estinator."

Q.106 Consider the geometric distribution :

$f(x, \theta) = \theta\,(1-\theta)^{x-1},\ 0 < \theta < 1$

Then an unbiased estimator of $\frac{1}{\theta}$, is given by sample mean.

$$t = \frac{1}{n}\sum_{i}^{n} x_i = \bar{x}.$$

Q.107 Consider the distribution :

$$f(x, \theta) = \frac{1}{\overline{|p}}\,(x-\theta)^{b-1}\,e^{-(x-\theta)},\ \begin{matrix} 0 \le x < \infty \\ p>2 \end{matrix}$$

Then M.V.B.E. for the parameter θ holds and is given by $t = \left(\frac{p-2}{2}\right)$

Q.108 For the distribution given in Q.107, the M.V.B.E. for the θ is not attainable because no signle sufficient statistic exists for θ.

Q.109 If the zero frequency for the Poisson distribution does not exist then the distribution is known as "Truncated Poisson Distribution."

Q.110 In case of truncated Poission distribution $\frac{e^{-\theta}.\,\theta^{x}}{x1}$, from a sample observations (x = 1, 2, 3, ...), the only unbiased estimate of $(1-e^{-\theta})$ takes the value 0 (zero). if x is odd.

Q.111 In Q.110, when x is even then unbiased estimate of $(1-e^{-\theta})$ is equal to 2.

Q.112 In case of Poisson distribution, the parameter λ can not be estimated by the "Method of moment."

1.1.3 Blackwellization and Minimum Variance Unbiased Estimator (M.V.U.E.)

Q.113 The C.R. inequality provides us a means of judging whether or not a given unbiased estimator is also. M.V.U.E.

Q.114 Rao-Blackwell theorem enables us to obtain a M.V.U.E. from any other unbiased estimator by using a sufficient statistic say 't' for parameter θ.

Q.115 To apply Rao-Blackwell theorem, the only condition to be satisfied is that the sufficient statistic 't' should have a complete family of distribution.

Q.116 The implication of "Rao-Blackwell Theorem" is that we are given an unbiased estimator 'U' then we can improve upon 'U' by forming a new statistic ϕ (T)' based upon 'U' and a sufficient statistic 'T'.

Q.117 The process of finding an improved estimator (from point of view of unbiasedness and minimum variance) starting from an unbiased estimator is known as "Blackwellisation" after D. Blackwell.

Q.118 In case of Rao-Blackwell theorem, the new estimator $\phi(T)$ will not only be a better estimator than U but it will also be best estimator if 'T' has a complete family of distribution.

Q.119 Cramer-Rao inequality method is too much restrictive due to the regularity conditions to be satisfied.

Q.120 This is the fact that "If an estimator is not M.V.B. estimator it does not mean that "it is also not a M.V.U. estimator.

Q.121 D. Blackwellisation theorem does not suffer from the deficites mentioned in Q.119, and Q.120.

Q.122 Let X and Y be two r.v.'s such that $E(Y) = \mu$, and $v(Y) = \sigma^2 y$. Let $E(y/x) = \phi(x)$ then Rao-Blackwell theorem states that :

$E(\phi(x)) = \mu$ and $v(\phi(x)) \le \sigma y^2$

Q.123 The statistic $\phi(x)$ defined in Q.122, is an unbiased statistic (estimator) with minimum variance for the parameter μ.

Q.124 In Q.122, the random variable x has been treated as a sufficient statistic for the parameter μ.

Q.125 If a M.V.B.E. for a parameter θ, exists then it is the function of sufficient statistic.

Q.126 Consider the binomial distribution $f(x,n,p)$. Then according to Rao-Blackwell theorem, a M.V.U.E. of parametor p is given. as $t = \frac{r}{n}$.

Q.127 Consider Poisson distribution $p(x, \lambda) = e^{-\lambda} . \frac{\lambda^x}{x!}$. Then a sufficient statistic for parameters λ is given as $t = \sum_{i}^{n} xi$.

Q.128 In Q.127, the sufficient statistic for $e^{-\lambda}$ is given as $t = \sum_{i}^{n} xi$.

Q.129 In Q.127, Rao-Blackwell theorem states that a M.V.U.E. of $e^{-\lambda}$ is given by a statistic $t = \left(1 - \frac{1}{n}\right)^{r}$ for r = $\sum_{i}^{n} x_i$.

Q.130 In Q.127, a sufficient statistic T for λ^2 is given as :

$$T = \left(\bar{x}^2 - \frac{\bar{x}}{n}\right)$$

Q.131 In Q.127, the M.V.B.E. for λ^2 does not exist.

Q.132 Consider the distribution - $f(x, \theta) = \frac{1}{\theta} \cdot e^{\frac{-x}{\theta}}$, $0 < x < \infty$, Then according to C.R. inequality, an unbiased estimator of θ is given as $t = \frac{1}{n} \sum_{i}^{n} x = \bar{x}$.

Q.133 In Q.132, the sampling variance of the estimator 't' is given as v(t) $= \frac{\theta^2}{n}$.

Q.134 If t is a M.V.U.E. and 'U' is another unbiased estimator of the parameter θ, then

correlation $(t, U - t) = 0$.

1.1.4 Sufficient Statistic and Completeness of Sufficient Statistic

Q.135 Let $f(x, \theta)$ be the p d f of r v. x with parameter θ. A statistic $T(x)$ is sufficient for the parameter θ if :

$T(x)$ gives all relevant information about parameter θ. (By. Prof. R.A. Fisher).

Q.136 If in Q.135, $y_2(x)$, $y_3(x)$, ..., $y_n(x)$ are $(n - 1)$ other statistics then

$$J = \left|\frac{\partial(T, y_2, y_{333}, \ldots\ldots\ldots\ldots, y_n)}{\partial\,(x_1, x_2, \ldots\ldots\ldots\ldots\ldots\ldots x_n}\right| \neq 0$$

Q.137 In Q.135, the conditional distribution of $(y_2, y_3, \ldots\ldots\ldots\ldots\ldots, y_n)$ given $T(x)$, is independent of parameter θ.

Q.138 Let $T(x)$ be a sufficient statistic with pd.f denoted by $G(T, \theta)$. If the range of x is independent of the parameter θ then "Fisher Neyman Factorisation Theorem" states that $T(x)$ is sufficient for θ if :

$$\prod_{i=1}^{n} f(x_i, \theta) = G(T, \theta)\ \mathrm{H}(x)$$

where H(x) is the function of ($\underset{\sim}{x}$) and is independent of θ.

Q.139 If (T_1, T_2) are two sufficient statistics for the parameter θ then there exists a functional relationship between T_1 & T_2.

Q.140 A sufficient statistic T for the parameter θ is unique.

Q.141 A sufficient statistic T is said to be complete, if no function of it has zero expectation unless it is zero itself atmost every where.

Q.142 A statistic $t(x)$ is said to be sufficient and complete if :

E (q (t(x))) ⇔ p [q (t($\underset{\sim}{x}$) = 0] = 1.

1.1.5 Distributions Possessing Sufficient Statistic

Q.143 The statisticians Dormois, Pitman and Koopman have given the following precise form of exponential distribution as :

$f(x, \theta) = e^{A(\theta)\, B(x) + C(x) + D(\theta)}$

which is the condition for the existence of M.V.B. estimator for some function of parameter θ in case of the range of variate is independent of θ.

Q.144 If the range of variate x is independent of parameter θ. then

$$\frac{\partial}{\partial\theta} \log f(x, \theta) = p(\theta)\, k(x) + q(\theta) / n$$

is the sufficient condition for the existance of sufficient statistic for parameter θ where (p, q) being the arbitrary constants.

Q.145 If the lower terminal of the range of r.v x, depends on the parameter θ then frequency function is of the form $f(x, \theta),\ a(\theta) \le x \le b$

and can be factorised as :

$f(x, \theta) = g(x)./h(\theta)$ ....(A)

Then the lowest order statistic $x_{(1)}$ of $x_{(1)} < x_{(2)} <$ .... $< x_{(n)}$ is sufficient statistic for θ.

Q.146 If the range of variate is like $a \le x \le b(\theta)$, where $b(\theta)$ is the monotone function in θ then highest order statistic $x_{(n)}$ of $x_{(1)} < x_{(2)} < x_{(3)} <$ .... $< x(_{n})$. is the sufficient statistic for paramerter θ. if the condition A in

Q.145 is statisfied.

Q.147 If both the terminal of range of r.v. x depend upon parameter θ i.e. it is of the form $a(\theta) \leq \times \leq b(\theta)$ and

$$L(x/\theta) = \prod_{i}^{n} f(x_i\ \theta)\ U(x_{(1)} - \theta)\ V(b(\theta) - x_{(n)}).$$

Then extreme observations $x_{(1)}$ & $x_{(n)}$ are pair of sufficient statistic for θ. if:

$f(x, \theta) = g(x)/h(\theta).$

Q.148 Let $f(x,\theta)$ be factored as $f(x,\theta) = g(x)/h(\theta)$. Given $f(x,\theta)$, $a(\theta) \leq \times \leq b$, where $a(\theta)$ is monotone function in θ then $x_{(1)}$ is single sufficient and complete statistic for θ.

Q.149 In Q.148, the range of r.v. × is as $a \leq \times \leq b(\theta)$ where $b(\theta)$ is monotone function in θ then $x_{(n)}$ is single sufficient and complete statistic for θ.

Q.150 Given $f(x,\theta)$, $a(\theta) \leq \times \leq b(\theta)$ where $a(\theta)$ & $b(\theta)$ are monotone function in θ (Both increasing or both decreasing function of θ) then no single sufficient statistic exists but $x_{(1)}$ & $x_{(n)}$ are jointly sufficient and complete statistics for θ.

Q.151 Given $f(x,\theta)$, $a(\theta) \leq \times \leq b(\theta)$ and if $a(\theta)$ is decreasing and $b(\theta)$ is an increasing function in θ then

$T = max\ \{a^{-1}\ (x_{(1)},\ b^{-1}\ (x_{(n)})\ \}$

is sufficient and complete statistic for θ.

Q.152 In Q.151, if $a(\theta)$ is an increasing function and $b(\theta)$ is decreasing function in θ, then

$T = mix\ \{a^{-1}\ (x_{(1)},\ b^{-1}\ (x_{(n)})\ \}$

is sufficiant and complete statistic for θ.

1.1.6 Sufficient Statistic for Several Parameters

Q.153 Let us define each observations x_i as a vector variate, t as s component vector of statistics and θ as k component vector of parameters. If we can write :

L(x/θ) = g(t/θ) h (x)

Then, we call the components of 't' as a set of jointly sufficient statistic for parameter θ.

Q.154 In Q.153, if t has s components, we may have s as $s \leq k$ or $s \geq k$. If s = 1, we may call 't' as single sufficient statistic for θ.

Q.155 In Q.154, if we put t = x, we say that the observations x themselves always constitute a set of sufficient statistic for θ with $s = n$.

Q.156 In Q.153, if any particular component of 't' say $t^{(1)}$ is individually sufficient for θ_1, then individual sufficiency of all the $t(i)$ when all others are known, does not imply joint sufficiency.

Q.157 The statistic $s^2 = \frac{1}{n}\sum_{i}^{n}(x_i - \bar{x})^2$ alone is not sufficient for σ^2 when μ is unknown.

Q.158 The statistic $s^2 = \frac{1}{n}\sum_{i}^{n}(x_i - \mu)^2$ is sufficient statistic for σ^2 is μ is known.

Q.159 If a single sufficient statistic 't' exists for parameter θ, the M.L.E. θ of θ is a function of 't' alone.

Q.160 The M.L.E. θ is unique if L.F. has a unique maxima.

Q.161 All the optimum properties of single sufficient statistic are conferred upon M.L. estimators which are one to one function of them.

Q.162 A sufficient statistic 'T' is said to be minimal sufficient statistic for parameter θ, if it is the function of every other sufficient statistic.

Q.163 A sufficient statistic $T(x)$ is minimal sufficient if the data represented by $\underset{\sim}{x}$ can not be condenced beyond T without sacrificing sufficiency.

Q.164 The statistic $T(\underset{\sim}{x})$ is said to be complete if for any measurable function $\psi(T)$ the :

$$E_\theta[\psi(T)] = 0 \ \forall \ \theta \in \Omega\,\theta.$$

implies that $\psi(T) = 0$, almost everywhere $\forall$ θ.

Q.165 In Q.164, the concept of completeness utilizes both the concepts of "Points Estimation" and "Testing of Hypothesis."

Q.166 The family { P_θ (T) / θ ε Ω θ } of distributions is said to be complete if for any measurable function $\psi(T)$,

$$\int_\theta y(T) \,.\, Pq(T(x))\, d_\theta = 0 \ \forall \ \theta \in \Omega_\theta.$$

which implies that ψ (T) 0 $\forall \ \theta \in \Omega_\theta$

Q.167 A sample statistic $T(\underset{\sim}{x})$ is said to be boundedly complete if there exists any measurable function $\psi(T)$ such that :

$|\psi(T)| < M$ for some M

and $E_\theta\, \psi(T) = 0 \;\forall\; \theta \in \Omega_\theta$.

which implies that $\psi(T) = 0 \;\forall\; \theta \in \Omega_\theta$ almost every where.

Q.168 A family $\{P(T, \theta), \theta \in \Omega_\theta\}$ of distributions is said to be boundedly complete if there exists a measureable function $\psi(T)$ of sufficient statistic T(x) such that :

$|\psi(T)| < M$ for some M

and $\int_\theta \psi(T)\; P(T, \theta), d_\theta = 0 \;\forall\; \theta \in \Omega_\theta$.

which implies that $\psi(T) = 0 \;\forall\; \theta \in \Omega_\theta$.

Q.169 If $T(x)$ is a complete statistic then it is necessarily a boundedly complete but converse is not true.

Q.170 Let (T_1, T_2) be jointly sufficiant statistics for two parameters (θ_1, θ_2) respectively and further let T_1 be sufficient for θ_1 when θ_2 is known. Then conditional distribution of T_2 for a given value of T_1 is independent of θ_1.

Q.171 A sufficient estimator $T(x)$ for the parameter θ is most efficient estimator of most efficient estimator exists.

Q.172 The method of minimum variance will provide a solution only if there exists a sufficient statistic $T(x)$ for parameter θ.

Q.173 If $(T_1, T_2, \ldots\ldots\ldots\ldots, T_r)$ is the set of jointly sufficient statitics for the parameter θ, then any set of one to one functions or transformation of $(T_1, T_2, \ldots\ldots\ldots\ldots, T_r)$ is also jointly sufficient statistic for θ.

Q.174 Let $x_1, x_2, \ldots\ldots\ldots\ldots, x_n$ be i id.v's from a continuous distribution $f(x, \theta)$ then the set of order statistic $\{x_{(1)}, x_{(2)}, \ldots\ldots\ldots\ldots, x_{(n)}\}$ is sufficient for θ.

Q.175 The original sample $(x_1, x_2, \ldots\ldots\ldots\ldots, x_n)$ drawn from a population $f(x, \theta)$, is always sufficient statistic for parameter θ.

Q.176 Let $f(x, \theta)$ belong to one parametric exponential family of distributions. Let $T(x_i)$ be sufficient statistic for θ. Then $\sum_i^n T(x_i)$ is complete and minimal sufficient statistic for θ.

Q.177 If $f(x, \theta_1, \theta_2, \ldots\ldots\ldots\ldots, \theta_k)$ denotes the regular case of k parameters exponential family of distributions. Then $\sum_{i}^{n} T_1(x_i), \sum_{i}^{n} T_2(x_i), \ldots\ldots\ldots\ldots,$

$\sum_{i-1}^{n} T_k(x_i)$ are complete and minimal sufficient statistics for $\theta_1, \theta_2, \ldots\ldots\ldots\ldots,$ θ_k respectively.

Q.178 Bounded completeness implies the minimal sufficiency but minimum sufficiency does not imply the bounded completeness of the statistic.

Q.179 There can not be more than one boundedly complete sufficient statistic for a parameter θ.

Q.180 Let $x_1, x_2, \ldots\ldots\ldots\ldots x_n$ be iid random variable with p.d.f. $f(x, \theta)$. If $S = S(\underset{\sim}{x})$ is a complete and sufficient statistic for θ, and if T = T(S) is a function of S and is an unbiased estimator of $\psi(\theta)$ then T(S) is U.M.V.U.E. of ψ (θ).

Q.181 Let $(x_i, y_i, i = 1, 2, \ldots\ldots\ldots\ldots, n)$ be independently normaly distributed with same mean μ and variances (σ_1^2, σ_2^2) respectively. Then $T = \{\bar{x}, s_x^2, \bar{y}, s_y^2\}$ is minimal sufficient statistic but not complete statistic.

Q.182 Let (X_1, X_2) be two random variables distributed as $N(\theta, 1)$. Let a new variable Y be defined as Y = (X_1 +X_2) then Y is sufficient statistic for parameter θ.

Q.183 Consider normal distribution i.e. X ~ $N(\mu, \sigma^2)$. Then if σ^2 is known then sample mean $\bar{x}$ is sufficient statistic for μ.

Q.184 If in Q.183, μ is known then $s^2 = \frac{1}{n}\sum_{i}^{n}(x_i - \bar{x})^2$ is not sufficient statistic for σ^2.

Q.185 In Q.184, the statistic $s^{*2} = \frac{1}{n}\sum_{i=1}^{n}(x_i - \mu)^2$ is sufficient statistic for σ^2.

Q.186 Consider normal distribution $N(0, \theta)$. Then the statistic $T_{(x)} = \frac{\sum^{n} x^2}{n}$ is an unbiased statistic for the parameter θ. having $V(T) = \frac{2\theta^2}{n}$.

Q.187 In Q.186, the statistic $T(x)$ defined as :

$$T(x) = \sum_{i}^{n} x_i^2$$

is sufficient statistic for parameter θ.

Q.188 Consider the following p.d.f. :

$$f(x, \theta) = \theta \,.\, (1-\theta)^x,\; x = 0, 1, 2, \ldots\ldots\ldots\ldots,\; 0 < \theta < 1$$

Then the statistic $T(x) = \sum_{i}^{n} x_i$ is sufficient for the parameter θ.

Q.189 Consider a random sample $(x_1, x_2, \ldots\ldots\ldots\ldots, x_n)$ drawn from exponential distribution.

$$f(x, \theta) = e^{-(x-\theta)},\; 0 < \theta < \infty$$

Then the statistic $T(x)$ = minimum $(x_1, x_2, \ldots\ldots\ldots\ldots, x_n)$ is sufficient for parameter θ.

1.1.7. Ordered Statistic

Q.190 Let the random sample $(x_1, x_2, \ldots\ldots\ldots, x_n)$ be arranged in ascendig order of magnitude as :

$y_1 < y_2 < \ldots\ldots\ldots < y_n$. Then y (r), (r = 1, 2, ..., $_n$) is called rth order statistic.

Q.191 The order statistic $y(r)$ $(r = 1, 2, \ldots\ldots\ldots, n)$ are the function of random sample $(x_1, x_2, \ldots\ldots, x_n)$.

Q.192 The ordered statistic is particularly used in "Non-parametric methods."

Q.193 Let a statistic M be defined as :

$$M = \begin{cases} x_{\left(\frac{n+1}{2}\right)} & \text{if } n \text{ is odd} \\ \dfrac{x_{(n)} + x_{(n+1)}}{2} & \text{if } n \text{ is even} \end{cases}$$

Then M is known as the median of the sample values where n is the sample size.

Q.194 Let R_m be defined as :

$$R_m = {}^{x}(n-m+1) - {}^{x}(m)$$

Then R_m is known as quasi-range of the sample values $x_1, x_2, \ldots\ldots\ldots\ldots x_n$.

Q.195 In Q.194, if we put m = 1 than we have

$R_1 = {}^{x}(n) - {}^{x}(1)$.

and R_1 is known as the range of the sample values.

Q.196 Let R_m^* be defined as :

$$R_m^* = \frac{1}{2}[{}^{x}(n-m+1) + {}^{x}(m)]$$

Then R_m^* is known as m^{th} mid range.

Q.197 If in Q.196, we put m = 1, then we have

$$R_1^* = \frac{1}{2}[{}^{x}(n) + {}^{x}(1)]$$

and R_1^* is known as mid range and is the measure of central tendency.

Q.198 Let ${}^{x}(r)$, $(r = 1, 2, \ldots\ldots\ldots\ldots, n)$ be the rth order statistic from a continuous distribution function F_x, then the transformation

$U(r) = F(x_{(r)})$

provides a r.v. which is the order statistic from a uniform distribution on (0, 1) under probability integral transformation.

Q.199 Even if $x_1, x_2, \ldots\ldots\ldots\ldots, x_n$ are independent r,v.'s then order statistics generated from them are not independent r.v.'s.

Q.200 Let $(x_1, x_2, \ldots\ldots\ldots\ldots, x_n)$ be i i d. random variables from a continous distribution. Then the set of ordered statistics say $(y_1 < y_2 < y_3 < \ldots\ldots < yn)$ is both complete and sufficient statistic.

Q.201 Let x be a continuous or discrete r.v. and let $(x_1 < x_2 < \ldots\ldots < x_n)$ be a set of ordered statistic such that :

$E(X_{(r)}) = \mu_{(r)}$.

Then $\mu_{(r)}$ exists provided μ exists but converse is not necessarily true.

Q.202 In Q.201, if μ does not exist then $\mu_{(r)}$ may exist for certain '*r*' but not for all values of *r*.

Q.203 In case of Cauchy distribution :

$$f(x, \theta) = \frac{1}{\pi}\frac{1}{(1+x^2)}$$

μ does not exist but $\mu_{(r)}$ exists.

Q.204 Let $y_1 < y_2 < y_3 < \ldots\ldots\ldots\ldots < y_n$ be a set of ordered statistics. Then joint p.d.f of (y_n, y_1) is given by

$$f(y_1, y_n) = n(n-1)\{F(y_n) - F(y_1)\}^{n-2} f(y_1) f(y_n)$$

where $F(y_n), F(y_1), f(y_1)$ & $f(y_n)$ have usual meaning in the theory of distribution.

Q.205 In Q.204, The p.d.f. of largest variate y_n is given as :

$$f(y_n) = n[F(y_n)]^{n-1} f(y_n), \quad a \le y_n \le b$$

Q.206 In Q. 204, the p.d.f. of any arbitrary variate y_k is given as :

$$f(y_k) = \frac{\overline{|(n+1)}}{\overline{|k}\ \overline{|n-k+1}} [F(y_k)]^{k-1} [1 - F(y_k)]^{n-k} f(y_k)$$

Q.207 In case of Q.204, the p.d.f. of two arbitrary ordered statistics say

$(y_{k_1}$ and $y_{k_1+k_2})$ is written as :

$$(y_{k_1}, y_{k_1+k_2}) = \overline{|(n-1)}\ \{F(y_{k_1})\}^{k_1-1}\ [F(y_{k_1+k_2}) - F(y_{k_1})^{k_2-1}$$

$$\times [1 - F(Y_{k_1+k_2})^{n-k_1-k_2} f(y_{k_1}) f(y_{k_1+K_2})$$

Q.208 Consider the rectangular distribution :

$$f(x, \theta) = \begin{cases} \frac{1}{\theta}\ for\ 0 < x < \theta \\ 0,\ otherwise \end{cases}$$

Let $x_1 < x_2 < x_3 < \ldots\ldots\ldots\ldots < x_n$ be an ordered statistic then $t = \max\ [x_1, x_2, \ldots\ldots\ldots\ldots\ldots\ x_n\}$ is a sufficient statistic for parameter θ.

Q.209 In Q.208, consider a statistic $\lambda(x)$ defined as :

$$\lambda(x) = \left(\frac{n+1}{n}\right) t$$

Then $\lambda(x)$ a U.M.V.U.E. for the parameter θ.

Q.210 Consider the binomial distribution :

$$f(x, n, p) = n_{c_x}\ p^x q^{n-x}, \quad \begin{array}{l} x = 0, 1, 2, \ldots\ldots\ldots\ldots, n \\ 0 < p < 1, 0 < q < 1 \\ p + q = 1 \end{array}$$

Then $t = \sum_{i}^{n} x_i = r$ is complete and sufficient statistic for the parameter p.

Q.211 Consider Poisson distribution

$$f(x, \lambda) = \frac{e^{-\lambda} \lambda^x}{x!}, \quad x = 0, 1, 2, \ldots\ldots\ldots\ldots, \quad \lambda > 0$$

Then the statistic $t = \sum_{i}^{n} x_i$ is complete and sufficient for the parameter λ.

Q.212 Consider the negative binomial distribution

$$f(x, p, r) = (r + x - 1)_x \; p^r \; (1 - p)^x \qquad x = 0, 1, 2, \ldots\ldots\ldots$$

$$r > 0$$

$$o \leq p \leq 1$$

Then if r is known then $t = \sum_{i=1}^{n} x_i$ is complete and sufficient statistic for parameter p.

Q.213 In Q.212, the statistic t* defined as :

$$t^* = \left(\frac{n-1}{n-1+t}\right)$$

is U.M.V.U.E. for the parameter p for known r.

Q.214 Consider the following uniform discrete distribution defined as :

$$f(x, k) = \frac{1}{k}, \; x = a, \; a + h, \; a + 2h, \ldots\ldots\ldots\ldots\ldots a + (k-1)\,h$$

Then the population mean μ is defined as :

$$\mu = a + \frac{h(k-1)}{2}$$

Q.215 In Q.214, the population variance σ^2 is defined as :

$$\sigma^2 = \frac{h^2(k^2 - 1)}{12}$$

Q.216 The distribution defined in Q.214, is highly platy Kourtic.

Q.217 In Q.214, let $x_1 < x_2 < \ldots\ldots\ldots < x_n$ be a set of ordered sample values.

Then $t = x_n$ is complete and sufficient statistic for the parameter 'k'.

Q.218 In Q.214, we define a statistic t_1 as :

$$t_1 = \frac{x_n^{n+1} - (x_{n-1})^{n+1}}{x_n^n - (x_{n-1})}$$

Then t_1 is an U.M.V.U.E. for the parameter 'k'.

Q.219 In Q.214, we define another statistic t_2 as :

$$t_2 = \left(\frac{x_n + 1}{2}\right)$$

Then t_2 is an U.M.V.U.E. for $\left(\frac{k+1}{2}\right)$.

Q.220 Consider the hyper geometric distribution defined as :

$$f(x, \theta) = \frac{\binom{\theta}{x}\binom{N-\theta}{n-x}}{\binom{N}{n}}, \quad \begin{matrix} x = 0, 1, 2, 3, \ldots\ldots\ldots. \\ \theta = 1, 2, 3, \ldots\ldots\ldots., N \end{matrix}$$

Then the distribution $f(x, \theta)$ is complete w.r. to parameter θ.

Q.221 In Q.220 the sample $x = (x_1, x_2, x_3, \ldots\ldots\ldots\ldots\ldots, x_n)$ is complete and sufficient for the parameter θ.

Q.222 In Q.220, let T(x) is a complete and sufficient statistic for parameter θ then

$$t = \frac{N}{n} . T(x).$$

is U.M.V.U.E. for θ.

Q.223 Consider the following power series distribution :

$$f(x, \nu, \lambda) = \frac{a(x) . \lambda^x}{h(\nu, \lambda)}, x = \nu, \nu + 1, \ldots\ldots\ldots$$

Consider the ordered sample $x_1 < x_2 < \ldots\ldots\ldots < x_n$.

Let us define $T_1 = x_1$, and $T_2 = \sum_i^n x_i$, then (T_1, T_2) is jointly complete and sufficient statistic for the parameters (ν, λ) respectively.

Q.224 Consider the normal distribution :

$$f(x, \mu, \sigma^2) = \frac{1}{\sqrt{2\pi\sigma^2}} e^{-\frac{1}{2}\left(\frac{x-\mu}{\sigma}\right)^2}, \quad \begin{array}{l} -\infty < x < \infty \\ 0 < \sigma^2 < \infty \end{array}$$

When σ^2 is known then $f(x, \mu, \sigma^2)$ is complete w.r. to μ.

Q.225 In Q.224, when μ is known then $f(x, \mu, \sigma^2)$ is not even boundedly complete w.r. to σ^2.

Q.226 Let $x_i (i = 1, 2, ..., n)$ be a random sample drawn from normal distribution $N(\mu\ \sigma^2)$. When σ^2 is known then $t = \sum_i^n x_i$ is complete and sufficient statistic for μ.

Q.227 In Q.226, the statistic $\left(\sum_i^n x_i, \sum_i^n x_i^2\right)$ are jointly complete and sufficient for (μ, σ^2) respectively.

Q.228 In Q.226, when μ is known then $t = \sum_i^n (x_i - \mu)^2$ is complete and sufficient statistic for parameter σ^2.

Q.229 In Q.226, when μ is unknown then

$$s^2 = \frac{1}{n-1} \sum_i^n (x_i - \bar{x})^2$$

is an U.M.V.U.E. for population variance σ^2.

Q.230 In Q.226, let μ be known then

$$\$^{*2} = \frac{1}{n} \sum^n (x_i - \mu)^2$$

is an U.M.V.U.E. for parameter σ^2.

Q.231 In Q.226, let σ^2 be knwon then

$$t = \frac{\left(\sum^{n} x_i\right)^2 - n\,\sigma^2}{n^2}$$

is an U.M.V.U.E. for μ^2.

Q.232 Consider the log normal distribution given as :

$$f(x, \mu, \sigma^2) = \frac{1}{\sqrt{2\pi}\,\sigma} e - \frac{1}{2\sigma^2} (\log x - \mu)^2 \qquad \begin{matrix} x > 0 \\ \sigma^2 > 0 \end{matrix}$$

If σ^2 is known then $f(x, \mu, \sigma^2)$ is complete and sufficient w.r. to μ.

Q.233 In Q.232, if σ^2 is known then :

$$T_1 = \sum_{i}^{n} \log x_1$$

is complete and sufficient for the parameter μ.

Q.234 In Q.232, the statistic t defined as :

$$t = \left\{ \sum_{i=1}^{n} \log x_i,\ \sum_{i}^{n} (\log x_i)^2 \right\}$$

is jointly complete and sufficient for (μ, σ^2) respectively.

Q.235 In Q.232, if μ is known then statistic t defined as :

$$t = \sum_{i}^{n} (\log x_i - \mu)^2$$

is sufficient for parameter σ^2.

Q.236 In q.232, the statistic t defined as :

$$t = \frac{1}{n} \sum_{i}^{n} \log x_i$$

is an U.M.V.U.E. for the parameter μ.

Q.237 In Q.232 if μ is known then the statistic T' defined as :

$$T^* = \frac{1}{n} \sum_{i}^{n} (\log x_i - \mu)^2$$

is an U.M.V.U.E. for the population variance σ^2.

Q.238 In Q.232, if μ is unknown then the statistic T' defined as :

$$T' = \frac{n \sum^{n} (\log x_i)^2 - \left(\sum^{n} \log x_i \right)^2}{n\,(n-1)}$$

is an U.M.V.U.E. for the parameter σ^2.

Q.239 Consider the power series distribution defined as :

$$f(x, \theta, \alpha) = \frac{\theta \,.\, x^{\theta-1}}{\alpha^{\theta}}, \quad \begin{matrix} 0 < x < \alpha \\ \theta > 0 \end{matrix}$$

The $f(x, \theta, \alpha)$ is complete w.r. to θ when α is known.

Q.240 In Q.239, when θ is known then $f(x, \theta, \alpha)$ is complete w.r. to α.

Q.241 In Q.239, when α is known then the statistic $T(x)$ defined as :

$$T(x) = \sum_{i-1}^{n} \log x_i$$

is complete and sufficient for the parameter θ.

Q.242 Let $x_1 < x_2 < \ldots\ldots\ldots\ldots < x_n$ be an ordered sample drawn from power series distribution. Then for θ being known, the largest sample value x_n is complete and sufficient statistic for α.

Q.243 Consider Cauchy distribution defined as :

$$f(x, \mu, \kappa) = \frac{\kappa}{\pi\,[k^2 + (x-\mu)^2]} \quad \begin{matrix} -\infty < x < \infty \\ -\infty < \mu < \infty \\ \kappa > 0 \end{matrix}$$

when (κ = 1) then $f(x, \mu, \kappa)$ is complete w.r. to μ.

Q.244 In Q.243, when μ = 0 then $f(x, \mu, \kappa)$ is not even boundedly complete w.r. to κ^2.

Q.245 In Q.243, when μ = 0, then the statistic T defined as :

$$T = \sum_{i}^{n} x_i^2$$

is complete for κ^2.

Q.246 In Q.243, for $\kappa = 1$, $f(x, \mu, \kappa)$ reduces to the form :

$$f(x, \mu) = \frac{1}{\pi[1+(x-\mu)^2]}, \quad \begin{matrix} -\infty < x < \infty \\ -\infty < \mu < \infty \end{matrix}$$

Q.247 In Q.246, the m.l.e. of μ is obtained by using the "method of scoring."

Q.248 In Q.243, the sample median x is the consistent estimator of population median μ.

Q.249 Let x_1, x_2 be a sample of size 2 drawn from Cauchy distribution then the statistic T defined as :

$$T = \frac{x_1 + x_2}{2}$$

is a Pitman closure estimator for μ than x_1.

Q.250 The Cauachy distribution defined in Q(246) is symmetrical with mode and median equal to μ.

Q.251 Consider Gamma distribution defined as :

$$f(x, \theta) = \frac{\theta^x \, x^{m-1} \, e^{-\theta x}}{\Gamma m}, \quad \begin{matrix} x > 0 \\ \theta > 0 \\ m > 0 \end{matrix}$$

when m = 1, then $f(x, \theta)$ is known as exponential distribution.

Q.252 In Q.251, when m is known then the statistic $T(x)$ defined as :

$$T(x) = \sum_{i-1}^{n} x_i$$ is complete and sufficient for the parameter θ.

Q.253 In Q.251, when m is known then $f(x, \theta)$ is complete w.r. to the parameter θ.

Q.254 In Q.251, when m is known then the statistic $T(x)$ defined as :

$$T(x) = \frac{\sum_{i}^{n} x_i}{mn}$$

is an U.M.V.U.E. for $\frac{1}{\theta}$

Q.255 Consider Beta distribution defined as :

$$f(x, p, q) = \frac{\Gamma(p+q)\, x^{p-1} \,.\, (1-x)^{q-1}}{\Gamma(p)\, \Gamma(q)} \qquad \begin{matrix} 0 < x < 1 \\ p > o \\ q > o \end{matrix}$$

The distribution $f(x, p, q)$ is complete w.r. to p when q is known and vice-versa.

Q.256 In Q.255, when q is known then the statistic $T(x)$ define as :

$$T(x) = \sum_{i}^{n} \log x_i$$

is complete and sufficient for parameter 'p'.

Q.257 In Q.255, when p is known then the statistic $T^*(x)$ defined as :

$$T^*(x) = \sum_{i-1}^{n} \log(1 - x_i)$$

is complete and sufficient for the parameter 'q'.

Q.258 In Q.255, for q = 1, and n > 1, the statistic $T(x)$ defined as :

$$T(x) = \frac{1}{\frac{1}{n-1} \sum_{i}^{n} \log x_i}$$

is a U.M.V.U.E. for parameter p.

Q.259 In Q.255, for p = 1 and n > 1, the statistic $T^*(x)$ defined as :

$$T^*(x) = \frac{1}{\frac{1}{n-1} \sum_{i-1}^{n} \log(1 - x_i)}$$

is an U.M.V.U.E. for the parameter 'q'.

Q.260 Consider the continuous uniform distribution defined as :

$$f(x, \theta) = \frac{1}{\theta}, \quad 0 < x < \theta, \ \theta > 0$$

Then $f(x, \theta)$ is complete w.r. to parameter θ.

Q.261 In Q.260, the statistic $T(x)$ defined as :

$T_{(x)} = max\ (x_1, x_2, \ldots\ldots\ldots\ldots\ldots\ldots, xn) = x(n).$
is complete and sufficient for the parameter θ.

Q.262 In Q.260, the statistic $T^*(x)$ defined as :

$$T^*(x) = \left(\frac{n+1}{n}\right) x_{(n)}$$

is an U.M.V.U.E. of θ.

Q.263 In Q.260, the statistic Tr defined as :

$$Tr = \left(\frac{n+1}{n}\right) x_{(n)}^{r}$$

is an U.M.V.U.E. for θ^r.

Q.264 Consider the uniform distribution defined as :

$$f(x, \theta_1, \theta_2) = \frac{1}{\theta_2 - \theta_1}, \quad \theta 1 < x < \theta_2$$

then $f(x, \theta_1, \theta_2)$ is complete w.r. to θ_1 when θ_2 is known and vice-versa.

Q.265 If in Q.264, $(x_1 < x_2 < x_3 < \ldots\ldots\ldots\ldots\ldots\ldots < x_n)$ is set of ordered sample and θ_2 is known then statistic T defined as :

$$T = \frac{(n+1)\ x_1 - \theta_2}{n}$$

is an U.M.V.U.E. for θ_1.

Q.266 In Q.264, if θ_1 is known then for ordered sample $x_1 < x_2 < \ldots\ldots\ldots < x_n$, the statistic T* defined as :

$$T^* = \frac{(n+1)\ x_n - \theta_1}{n}$$

is an U.M.V.U.E. for θ_2.

Q.267 Consider the following p.d.f. defined as :

$$f(x, \theta_1, \theta_2) = \frac{1}{\theta_2}, \quad \begin{array}{c} \theta_1 - \frac{\theta_2}{2} \leq x \leq \theta_1 + \frac{\theta_2}{2} \\ -\infty < \theta_1 < 0, \quad \theta_2 > 0 \end{array}$$

then $f(x, \theta_1, \theta_2)$ is not even boundedly complete w.r. to θ_2 when θ_1 is known.

Q.268 In Q.267, for a ordered sample $x_1 < x_2 < x_3 < \ldots\ldots\ldots\ldots < xn$. the statistic T defined as :

$$T = \frac{1}{2}(x_1 + x_n)$$

is an U.M.V.U.E. for parameter θ_1.

Q.269 In Q.267, the statistic T* deformed as :

$$T^* = \left(\frac{n+1}{n-1}\right)(x_1 + x_n)$$

is an U.M.V.U.E. for parameter θ_2.

Q.270 Consider the p.d.f., $f(x, \theta)$ defined as :

$$f(x,\theta) = 1, \quad \begin{matrix} (\theta - \frac{1}{2}) < x < (\theta + \frac{1}{2}) \\ -\infty < \theta < \infty \end{matrix}$$

Then for ordered sample $x_1 < x_2 < \ldots\ldots\ldots\ldots < x_n$, the statistic $T = (x_1, x_n)$ is sufficient but not even boundedly complete statistic for parameter θ.

Q.271 In Q.270, the statistic :

$$T^* = \frac{1}{2}(x_1 + x_n)$$

is M.V.U. linear estimator for parameter θ.

Q.272 Consider the p.d.f. given as :

$$f(x,\theta) = 1, \;\; \theta < x < \theta + 1, \; -\infty < \theta < \infty$$

Then $f(x, \theta)$ is not even boundedly complete w.r. to θ.

Q.273 In Q.272, the statistic T defined as :

$$T = (x_1, \; x_n)$$

is jointly sufficient for the parameter θ.

Q.274 In Q. 272, the statistic T* defined as :

$$T^* = \frac{(x_1 + x_n - 1)}{2}$$

is M.V.U. linear estimator for the parameter θ.

Q.275 Consider the p.d.f. defined as :

$$f(x,\theta) = \frac{1}{2\theta}, \;\; -\theta < x < \theta, \;\; \theta > 0$$

then $f(x, \theta)$ is not even boundedly complete w.r. to θ.

Q.276 In Q.275, the statistic T defined as

$T = m_{ax}(-x_1, x_n)$

is complete and sufficient for parameter θ.

Q.277 In Q. 275, the statistic T* defined as :

$$T^* = \left(\frac{m+1}{n}\right) m_{ax}(-x_1, x_n)$$

is an U.M.V.U.E. for parameter θ.

Q.278 Consider Pareto distribution defined as :

$$f(x, \alpha, \beta) = \frac{\alpha \beta^\alpha}{x(\alpha + 1)}, x > 1, \beta > 0, \alpha > 1$$

Then $f(x, \alpha, \beta)$ is complete w.r. to α when β is known and vice-versa.

Q.279 In Q.278, the statistic T_1 defined as :

$$T_1 = \sum_{i}^{n} \log x_i$$

is complete and sufficient for α when β is known.

Q.280 In Q.278, the statistic T_2 defined as :

$T_2 = x_n^{\cdot}$ is complete and sufficient for β when α is known where x_n is the maximum of ordered sample $x_1 < x_2 < \ldots\ldots\ldots\ldots < x_n$.

Q.281 Consider the p.d.f., $f(x, \theta)$ defined as :

$$f(x,\theta) = \frac{1}{2\theta}, \quad -\theta < x < \theta, 0 < \theta < \infty$$

Let the sample values are arranged in an ascending order of Magnitude as :

$-\theta < x_1 < x_2, < \ldots\ldots\ldots\ldots\ldots < x_n < \theta$

Then the statistic T defined as

$T = \max(x_1, x_2, \ldots\ldots\ldots\ldots\ldots, x_n) = x_n$ is sufficient for the parameter θ.

Q.282 In Q.281, the sufficient statistic 'T' is complete for the parameter θ.

Q.283 Consider the p.d.f. $f(x, \theta)$ defined as :

$$f(x, \theta) = \frac{1}{\theta}, \kappa\, \theta \le \times \le (\kappa + 1)\, \theta, \kappa > 0$$

Then the statistic T defined as $T = \{x_1, x_n\}$ is the pair of joint sufficient statistic for parameter θ.

Q.284 In Q.283, no signle sufficiet statistic for parameter θ. exists.

Q.285 In Q.283, Let θ be the m.l.e. of parameter θ. Then we have $\theta = \frac{x_n}{k+1}$

Thus we see that θ is the function of x_n alone where as the pair $\{x_1, x_n)$ are jointly sufficient statistic for θ.

Q.286 In Q.285, due to symmetry, the $\{x_1, x_n)$ have the same variance, say ν, then

$$\nu\ (\theta) = \frac{\nu}{(\kappa+1)^2}$$

Q.287 In Q.285, let us take :

$$\theta^* = \frac{x_1}{\kappa},\quad \nu\ (\theta^*) = \frac{\nu}{\kappa^2}$$

Since (x_1, x_n) are asymptotically, independently distributed, the function T* defined as :

$$T^* = a\ \theta + (1-a)\ \theta^*$$

will be consistant estimator for the parameter θ.

Q.288 In Q.287, the variance (T^*) will be given as :

$$V\ (T^*) = V\left[\frac{a^2}{(\kappa+1)^2} + \frac{(1-a)^2}{\kappa^2}\right]$$

Q.289 Consider the p.d.f. of normal distribution $N(0, \sigma^2)$ as :

$$f(x, 0, \sigma^2) = \frac{1}{\sqrt{2\pi\sigma}} e^{-\frac{1}{2}\frac{x^2}{\sigma^2}}$$

Then *m* l.e. and sufficient statistic for σ is given by :

$$T = \sigma = \sqrt{\frac{1}{n}\sum^{n} x_i^2}$$

Q.290 Consider the following p.d.f. of double exponential distribution

$$f(x, \theta) = \frac{1}{2} e^{-|x-\theta|}, \quad -\infty < \theta < \infty$$

then no single sufficient statistic for the parameter θ exists.

Q.291 In Q.290, the set of ordered sample $(x_1 < x_2, < .... < x_n)$ is sufficient statistic for the parameter θ.

Q.292 In Q.290 the m l e of θ say θ is given by :

$$\theta = \text{Median } \{x_1, x_2, ..., x_n\}$$

Q.293 Let x be a r.v. following the p.d.f. as :

$$f(x, \theta) = \frac{1}{\theta} e^{-\frac{x}{\theta}}, \quad 0 < x < \infty, 0 < \theta < \infty$$

Then $T = \sum_{i}^{n} x_i$ or $\overline{x}$ is the sufficient statistic for θ.

Q.294 In Q.293, the estimator $t(x)$ defined as :

$$t(x) = \frac{n-1}{n\,\overline{x}}$$

is unbiased for the parameter θ.

Q.295 Consider the following two parametric exponential p.d.f. as :

$$f(x, \theta_1, \theta_2) = \frac{1}{\theta_2} e - \frac{(x-\theta_1)}{\theta_2}, \quad \begin{matrix} \theta_1 < x < \infty \\ -\infty < \theta_1 < \infty \\ 0 < \theta_2 < \infty \end{matrix}$$

Let $x_1 < x_2 < .... < x_n$ be the ordered sample then $(x_1, \sum_{i}^{n} x_i)$ is a pair of jointly sufficient statistic for the parameter (θ_1, θ_2) respectively.

Q.296 Consider the following family of gamma distribution

$$f(x, \theta) = \frac{1}{2^{\theta}\,\overline{|}(\theta)} e \frac{-x}{2} \lambda^{\theta-1}, \quad \begin{matrix} x > 0 \\ 0 < \theta \infty \end{matrix}$$

Then the distribution $f(x, \theta)$ is complete for parameter θ.

Q.297 Let 't' be the best estimator for $\lambda(\theta)$ then for any other estimator t', we have

$$P_\theta [\,|t-\lambda(\theta)|<c\,] \geq P_\theta [\,|t'-\lambda(\theta)|<c\,] \;\forall\; \theta \in \Omega\; \theta.$$

Q.298 Let us consider the following p.d.f. as :

$$f(x,\alpha)=e^{-(x-\alpha)},\quad \alpha \leq x < \infty$$

Then min $\{x_1 < x_2 < \dots\dots\dots < x_n\}$ is sufficient statistic for lower terminal of 'α'.

Q.299 Consider the following p.d.f. written as :

$$f(x,p)=\frac{1}{\overline{|p}}(x-\alpha)^{p-1}\, e^{-(x-\alpha)},\quad \alpha \leq x < \infty$$

Then for $n \geq 2$, there is no single sufficient statistic for α.

Q.300 Consider the following two parametric distribution function

$$f(x.\alpha,\beta)=\frac{1}{\beta-\alpha},\quad \alpha \leq x \leq \beta.$$

Let $x_1 < x_2 < \dots\dots\dots\dots\dots < x_n$ be an ordered sample, then for given β, the Min $\{x_1, x_2, \dots\dots\dots, x_n\}$ is sufficient statistic for α.

Q.301 In Q.300, for given α, the max $\{x_1, x_2, \dots\dots\dots, x_n\}$ is sufficient statistic for β.

Q.302 In Q.300, the pair of observations (x_l, x_n) is jointly sufficient statistic for (α, β) respectively.

Q.303 Let t be a complete statistic for parameter θ. then any other statistic say t* depending on random sample x_1, x_2, xn through 't' is also complete for θ.

Q.304 Consider the Bernoulli distribution with parameter p as :

$p\,(x=1)=p \qquad 0<p<1$

$p\,(x=0)=(1-p)$

Then the statistic $t_n=\sum_{i}^{n} x_i$ is sufficient statistic for p.

Q.305 In standardized bivariate normal distribution, consider a pair of sufficiant statistic, say t defined as :

$$t=\left(\sum_{i}^{n} x_i^2+\sum_{i}^{n} y_i^2,\ \sum_{i}^{n} x_i\, y_i\right)$$

Then t is sufficient statistic for population correlation coefficient ρ.

Q.306 Consider the following p.d.f. defined as :

$$f(x, \theta) = 1, \left(\theta - \frac{1}{2}\right) \leq x \leq \left(\theta + \frac{1}{2}\right)$$

Then for any ordered sample $x_1 < x_2 < \ldots\ldots\ldots < x_n$, the statistic say t = (x_1, x_n) is jointly sufficient for parameter θ.

Q.307 In Q.306, the likelyhood function (L.F.) is maximum at any value in the interval $\left(x_n - \frac{1}{2}, x_1 + \frac{1}{2}\right)$

Q.308 In Q.306, the mid point of the interval $\left(x_n - \frac{1}{2}, x_1 + \frac{1}{2}\right)$ is an unbiased estimate of the parameter θ.

Q.309 In Q.306, the method of moment gives the sample mean $\overline{x}$ as an estimate of θ.

Q.310 Consider the power series distribution defined as :

$$f(x, \theta) = \frac{a_x \ \theta^x}{(f(x))^n} \qquad x = c, c+1, c+2, \ldots\ldots\ldots$$

Then from a random sample $x_1, x_2, \ldots\ldots\ldots\ldots, x_n$. the statistic

$t = \sum_{i}^{n} x_i$ is complete sufficient for parameter θ.

Q.311 In Q.310, the statistic 't' has the distribution of the same form defined as :

$$g(t, \theta) = \frac{b_t \ . \ \theta^t}{(f(\theta)\ )^n}, \text{ for } t = \text{h, h + 1, h + 2, ………}$$

Q.312 Consider binomial distribution as :

$$B(x, n, p) = n_{c_x} \ p^x \ q^{n-x}, \qquad \begin{array}{l} x = 0, 1, 2, \ldots\ldots\ldots, n) \\ 0 < p < 1 \\ 0 < q < 1 \\ p + q = 1 \end{array}$$

when $p = q = \frac{1}{2}$, the value of n is least for $B(x, n.b)$ to be approximated by normal distribution.

1.2 Fill in the Blanks

Q.1 The process of knowing all the relevant information from the sample to the unknown population, is known as "..........................."

Q.2 An attempt to guess completely about the unknown feature of a population from a random sample drawn from the population, is known as the "...............".

Q.3 The estimation of population parameter, say θ by a single value (an estimate) from a random sample drawn from the population is known as "..................".

Q.4 A process of getting an interval within which the population parameter, θ is expected to lie, is known as "..................".

Q.5 An estimator having the properties of consistency, sufficiency, unbiasedness and efficiency, is known as "......................".

Q.6 Any estimator whose mathematical expectation is exactly equal to the population paramerer say θ, is known as an ".................." of the parameter θ.

Q.7 An estimator whose mathematical expectation is not equal to the population parameter, is known as : "......................".

Q.8 Let t_n (x) be an estimator of the parameter θ, then $t_n(x)$ is said to be unbiased if

E (...........) = ...(1)

Q.9 An estimator $t_n(x)$ of a parameter θ is said to be a biased estimtor if

E (..........) $\neq$ ....(2)

Q.10 The amount of bias of an estimator t_n (x) for θ can be worked out using the expression

B $(t_n\ (x))$ = ...(3)

Q.11 An estimator t_n (x) is said to be positively biased estimator of the parameter θ if

E $(t_n\ (x))$ ....(4)

Q.12 An estimator $t_n(x)$ is said to be negatively biased estimator of the parameter θ if

E $(t_n\ (x))$ ....(5)

Q.13 An estimator $t_n(x)$ is said to be asymptotically unbiased for the parameter θ if

$\lim_{n\to\infty}$ .... $= \theta$ ...(6)

Q.14 An estimator $s^{*2} = \frac{1}{n}\sum_{i}^{n}(x_i - \bar{x}_n)^2$ is asymptotically unbiased for the population variance $\sigma^2 x$ because

$$\lim_{n\to\infty} E(\$^{*2}) = \lim_{n\to\infty}(.......)\,\sigma_x^2 = \sigma_x^2 \qquad ...(7)$$

Q.15 The sample mean $\bar{x}_n = \frac{1}{n}\sum_{i}^{n} x_i$ is an absolutely unbiased estimator of the population mean μ_{x} because

E (..........) = ...(8)

Q.16 Let $(x_1, x_2, ..., x_n)$ be a random sample drawn from normal population $N(\mu, 1)$ and $t_n = \sum_{i}^{n} x_i^2$ is an estimator then we have :

$E(t_n) = (..........)$. ...(8)

Q.17 Let $(x_1, x_2, ..., x_n)$ be a random sample taken on a Bernouilli variable x such that :

$P(x = 1) = \theta$

$P(x = 0) = (1 - \theta)$

then an unbiased estimator for the parameter θ^2 is given as :

$$t^* = \frac{t(......)}{n(n-1)} \qquad ...(9)$$

where $t = \sum_{i}^{n} x_i$

Q.18 The difference between the estimated value and true value of the parameter θ is decribed by the "...........". property of the estimator.

Q.19 An estimator tn(x) of the parameter θ is said to be consistant for θ if.

$$\lim_{n\to\infty} P[\,|........| \le \in] > (1-\delta) \qquad ...(10$$

Q.20 An estimator t_n is said to be a consistent estimator for the parameter θ, if it converges in (.............). to θ.

Q.21 An estimator t_n is said to be consistent for the parameter θ if :

$$\lim_{n\to\infty} P[\,|\,........|\leq \in] \to 1 \qquad ...(11)$$

where $\in$ is a infinitely small positive value.

Q.22 A consistent estimator may or may not be an "............." estimator of the parameter.

Q.23 An unbiased estimator may not be a . "..............." estimator of the parameter.

Q.24 A consistent estimator with finite mean value may tend to be ".............." estimator to the parameter θ in large sample.

Q.25 If $t_n(x)$ is an unbiased estimator of the parameter θ_n with $V(t_n) = \sigma^2_n$ and $t_n(x)$ will be consistent estimator for θ if :

$\theta_n \to$ .... and $\sigma^2_n \to$ .... for $n \to \infty$

Q.26 A sample median x is a "............" estimator for population mean μ in normal distribution $N(\mu, \sigma^2)$.

Q.27 In case of normal distribution $N(\mu, \sigma^2)$, the sample variance $s^{*2} = \frac{1}{n}\sum_{i-1}^{n}(x_i - \bar{x}_n)^2$ is a " .... " estimator for population variance σ^2.

Q.28 In Cauchy distribution $f(x, \theta) = \frac{1}{\pi}\frac{1}{(1+(x-\theta)^2)}$, $-\infty < x < \infty$ the sample mean $\bar{x}_n$ is not a " ...." estimator for parameter θ.

Q.29 In Cauchy distribution :

$$f(x, \theta) = \frac{1}{\pi}\frac{1}{(1+(x-\theta)^2)}, -\infty < x < \infty$$

the sample meadian x is a "..................." estimator for the parameter θ.

Q.30 In Cauchy distribution :

$$f(x, \theta) = \frac{1}{\pi}\frac{1}{(1+(x-\theta)^2)}, -\infty < x < \infty$$

the sample mean x "....................." exist.

Q.31 In Poisson distribution $f(x, \lambda)$, the " ...." estimator for $\frac{1}{\lambda}$ is given as $t_n = \frac{1}{\bar{x}_n}$ where $\bar{x}_n$ is the sample mean.

Q.32 Let t be a consistent estimator for any parameter θ then a consistent estimator for e^{θ} will be given as ....

Q.33 An estimating function of sample values ($x_1, x_2, \dots\dots\dots x_n$) to estimate the parameter θ is known as an "......................".

Q.34 A sample statistic whose sampling distribution concentrates as closely as possible to the true value θ_0, may be considered as an "............." for θ_0.

Q.35 In the class of consistent estimators for a parameter θ, if there exists an estimator t^* whose sampling variance is less than that of any such estimator in the class, then t^* is known as most "....................." estimator.

Q.36 A most efficient estimator t^* for the parameter θ provides the criteria to measure the "............................" of such other estimators.

Q.37 Let (t_n, t_n') be consistent estimators for the parameter θ such that :

$$t_n \sim N(\theta, \frac{\sigma^2}{n})$$

$$t_n' \sim N(\theta, \frac{\sigma'^2}{n})$$

then t_n is said to be more efficient them to if ...., for $\forall\ n$.

Q.38 Let e denotes the efficienty of an estimator t_n than e is defined as :

$$e = \frac{Sampling\ Variance\ of\ most\ efficient\ estimator}{Samplig\ Variance\ of\ the\ estimator\ t_n} < \dots\dots\dots\dots\dots\dots\dots$$

Q.39 Let e be the efficienty of two unbiased estimators (t_1, t_2) with equal sampling variance then we have the following relation :

Correlation (t_1, t_2) > (.... – ...)

Q.40 Let (t_1, t_2) be two most efficient estimators for the parameter θ then :

$$\int (t_1, t_2) = \dots\dots\dots$$

i.e. (t_1, t_2) are linearly related.

Q.41 In case of normal distribution N(μ, σ^2), the sample mean $\bar{x}_n$ is more efficient than the sample (......................) to estimate the (...................)μ.

Q.42 Uniformly minimum variance estimator (U.M.V.E.) of a parameter is "....................".

Q.43 An estimator $t_i(\overline{x})$ of parameter θ_i, is said to be U.M.V.E. of θ_i if :

$V\ (t_i(\overline{x})$..........V (hi (x))

for any other estimator hi(x) of $\theta_i\ \forall\ \theta i \in \Omega\ \theta$ is the parameter space.

Q.44 Let (t_1, t_2) be two consistent/unbiased estimators of a parameter θ and t_1 is more efficient than t_1 with efficiency e then :

Correlation (t_1, t_2) = ...

Q.45 Let (t_1, t_2) be two unbiased estimators of parameter θ with efficiencies (e_1, e_2) respectively with correlation $(t_1, t_2) = \rho$ then we have the relation :

$$\sqrt{e_1 e_2} - \sqrt{(.....)\ (.....)} \le \rho \le \sqrt{e_1\ e_2} + \sqrt{(.....)\ (.....)}$$

Q.46 Let t_n be the (M.V.U.E.) of parameter θ and t_n' be an other estimator of θ with efficiency e. then e is defined as :

$e(t_n'/t_n)$ = (..........) < 1.0

Q.47 In Cauchy distribution, let $t_n(x)$ denotes the (mle) of a parameter θ then $t_n(x)$ is most efficient estimator with

V(tn (x)) = ...

Q.48 In Cauchy distribution, we have

V (Sample median x) = ...

Q.49 In Cauchy distribution, the efficiency of sample median $\overline{x}$ is defined as :

$$e(x) = \frac{(2/n)}{(..........)} = \frac{8}{\pi^2} = 0.8 \text{ approximately}$$

Q.50 In normal distribution N (μ, σ^2), let the sample mean deviation

$$s^* = \frac{1}{n} \sum_{i}^{n} |\, x_i - \overline{x}\, | \text{ then we have}$$

$$V(s^*) = (..........)\ \frac{\sigma^2}{n} < \text{V (Sample mean } \overline{x}_n)$$

Q.51 An (U.M.V.U.E.) of a parameter θ may not exist but an (..........). does exist.

Q.52 The correlation between (U.M.V.U.E.) and any other unbiased estimator is always (..........).

Q.53 Let (t_1, t_2) be two (U.M.V.U.E.) of two parametric functions (g_1 (θ), g_2 (θ) respectively then an (U.M.U.V.E.) "t" of parametric function a_1 g_1

$(\theta) + a_2\, g_2\, (\theta)$ will be given as :

$$t = \sum_{i=1}^{n} \text{........, where } (a_1, a_2) \text{ are fixed constants.}$$

Q.54 An U.M.V.U.E. may not attain the ".........." lower bound.

Q.55 Let $t_i\ (i = 1, 2, ..., k)$ be k independent unbiased estimators of a parametric function $\phi(\theta)$ and $V(t_i) = \sigma_i^2 < \sigma^2\ \forall\ i$ then an unbiased estimator t of (θ) will be given as :

$t =$ ....

1.2.1 Cramer, Rao Inequality

Q.56 Ig $t_n\ (\underset{\sim}{x})$ is an unbiased estimator of a parametric function $\psi(\theta)$ and $V\ (t_n)$ then C.R. inequality states that :

$$V(t_n) \geq \frac{\left(\frac{2}{2\theta}\psi\ (\theta)\right)^2}{E\ (..........)^2} = \frac{\{..........\}^2}{I_\theta}$$

where 1_θ is the information on θ supplied by the random sample.

Q.57 One of the regularity conditions for C.R. inequality to hold good is that : "The range of integration is (..........) of the parameter θ so that

$\int \prod_{i}^{n} f(x_i, \theta)\, dx$ can be (..........) under integral sign."

Q.58 For C.R. inequality to hold good, the following criteria :

$E\ [....................]^2 > \theta\ \forall\ \theta \in \Omega_\theta$

is true.

Q.59 For C.R. inequality to hold good, the function :

$$\int t_n\ (\underset{\sim}{x}) \prod_{i}^{n} f\ (x_i, \theta)\, dx$$

is (..........) under (..........) sign.

Q.60 In case of Poisson distribution $f(x, \lambda) = \dfrac{e^{-\lambda}\ \lambda^x}{x1}$, x = 0, 1, 2, ..., a complete and sufficient statistic for parameter λ is given as :

$t_n =$ ....

Q.61 In case of Poisson distribution $f(x, \lambda)$, the M.V.U.E. of $e^{-\lambda}$ is given by a statistic T defined as :

$$T = (..........)^{\sum_i^n x_i} = (..........)^{Tn}$$

Q.62 Let $T = \left(\frac{n-1}{n}\right)^{\sum^n x_i}$ is the MV.U.E. of $e^{-\lambda}$ in case of Poisson distribution

$f(x, \lambda), x = 1, 2,$ ...., then $V(T) = e^{-2\lambda}$ (..........)

Q.63 In Q.(62), the C.R. lower for T is equal to (..........).

Q.64 If an unbiased estimator t_n of parameter θ has the C.R. lower bound as its sampling variance, then t_n must be a (..........) statistic for θ.

Q.65 C.R. lower bound attends equality sign if :

$$\frac{\partial}{\partial\theta}\log\prod_{i=1}^{n} f(x_i, \theta) = \kappa(\theta, n) \text{ (..........)}$$

holds good.

Q.66 Let $t_n(x)$ be an unbiased estimator of a parametric function ψ(θ). then under C.R. inequality, we have :

$$V(t_n(x)) = \frac{(............................)}{\kappa\ (\theta, n)}$$

Q.67 C.R. inequality gives lower bound to the (..........) of an unbiased estimator t_n (x) of parameter θ.

Q.68 Information I_θ on parameter θ is provided by the expression written as :

$$I_\theta = V\left(\frac{1}{L}\frac{\partial L}{\partial\theta}\right) = \text{nE [..........]}^2$$

Q.69 Let t_n (x) be an unbiased estimator of parametric function ψ(θ) then the following relation exists :

$$E(t_n - \theta)^2 \geq [...........]^2 + \frac{(.................)^2}{I_\theta}$$

Q.70 Let ψ(θ) = θ + b(θ) and E(tn) = ψ(θ) then we have :

$V(t_n) \geq [..................] / I_\theta$

Q.71 Let $E(\theta-\theta)^2$ denotes the mean squared deviation of an estimator θ of the parameter θ and $b(\theta)$ denotes the bias of θ then we have :

$$E\,(\theta-\theta)^2 \geq \frac{[\ldots\ldots\ldots\ldots]}{I_\theta} + (\ldots\ldots\ldots)^2$$

Q.72 If $\frac{\partial}{\partial\theta}\log L = \frac{1}{\lambda}\left(t_n - \psi(\theta)\right)$, then C.R. inequality states that 't_n' is (............) of ψ(θ). with :
$V\,(t_n) =$ ....

Q.73 Let $x \sim N(\mu, \sigma^2)$ then C.R. inequality states that the M.V.U.E. of μ is given by (......................) with minimum $V\left(\overline{x}_n\right) =$ .... for known σ^2.

Q.74 In Poisson distribution $f(x, \lambda)$, the MVUE of λ is given by (..........) with sampling variance $\frac{\lambda}{n}$ where n is the sample size.

Q.75 In case of Poisson distribution $f(x, \lambda)$, we have :

$$\frac{\partial}{\partial\lambda}\log L = \frac{n}{\lambda}\;(\ldots\ldots\ldots\ldots)$$

where L is the likelihood function.

Q.76 In case of binomial distribution B (x, n, p), we have :

$$\frac{\partial}{\partial p}\log \mathrm{L} = \frac{(\ldots\ldots\ldots)}{p\,(1-p)/n}$$

Q.77 In case of binomial distribution B (x, n, p), the M.V.U.E. of p is given by :

$$t_n = \ldots\ldots\ldots\ldots \text{ with } V(t) = \frac{(\ldots\ldots\ldots)}{n}.$$

Q.78 In case of Cauchy distribution $f(x, \theta) = \frac{1}{\pi}\,\frac{1}{1+(x-\theta)^2}$, the M.V.U.E. of θ ....

Q.79 Let $x \sim N(\mu, \sigma^2)$, then we have :

$$\frac{\partial}{\partial\sigma^2} \log L = \frac{n}{2\sigma^4}[.........- \sigma^2]$$

Q.80 In case of normal distribution $N(\mu, \sigma^2)$, the M.V.U.E. of σ^2 is given as :

$$\hat{s}^{*2} = ...$$

$$\text{with } V(\hat{s}^{*2}) = \frac{............}{n}$$

Q.81 Consider the p.d.f. of type III population given as $P(x, p, \theta) =$

$$\frac{1}{\overline{|p}\ \theta p} x^{p-1} e^{\frac{-x}{\theta}}, p > 0, 0 < x < \infty$$

then M.V.B.E. of θ for fixed p is given as :

$$t_n = ... \text{ with } V(t_n) = \frac{\theta^2}{(..........)}$$

1.2.2 Minimum Variance Bound Estimators

Q.82 A condition for the existance of M.V.B.E. for $\psi(\theta)$ is given as :

$$\frac{\partial}{\partial\theta}\log L = A(\theta)\ \{..............\}$$

where $A(\theta)$ is independent of sample values (x) and may be the function of parameter θ.

Q.83 If $\frac{\partial}{\partial\theta}$ log L can not be put in the form as :

$$\frac{\partial}{\partial\theta}\log L = A(\theta)\ \{t - \psi(\theta)\}$$

then there may still exist an estimator t* of $\psi(\theta)$ which has uniformaly smaller varience than that of any other estimator. Such estimator t* is known as ".................".

Q.84 The least attainable variance of any estimator say t* may be greater than that of (.................) of the parametric function $\psi(\theta)$.

Q.85 In binomial distribution B. (x, n, p), an unbiased estimator 't' of 'pq' is given as :

$$t = \frac{r}{n}.\ (..........) \text{ where } r = \sum_{i=1}^{n} x_i$$

Q.86 In case of $N(\mu, \sigma^2)$, the M.V.B.E. of σ^2 exists if μ is (..........).

Q.87 In case of $N(\mu, \sigma^2)$, the M.V.B.E. of σ (..........)

Q.88 Consider the logistic distribution :

$$f(x, \theta) = e\frac{-(x-\theta)}{1+e^{-(x-\theta)}}$$

the V(sample mean $\overline{x}_n$) is given as :

$V(\overline{x}_n)$ = ...

Q.89 In case of Q(88), the exact V(sample median x) is given as :

V(x) = ...

Q.90 In case of $N(\mu, \sigma^2)$ distribution, the sufficient statistic for μ is given as :

t_n = ...

Q.91 In case of normal distribution $N(\mu, \sigma^2)$, for a set of unknown constants (ai, i = 1, 2, ..., n), the statistic $t_n = \sum_{i=1}^{n} a_i x_i$ is an unbiased estimator for the parameter μ if :

$\sum_{i=1}^{n} a_i$ = ...

Q.92 The M.V.B.E. of a parametric function $\psi(\theta)$ exists only if a (..........) statistic for $\psi(\theta)$ exists.

Q.93 Let 't_n' be a sufficient statistic for the parameter θ, then one to one function of 't_n' will also be a (..........) statistic.

Q.94 If a M.V.B.E. for a parameter θ exists, then it is given by the method of (.............................).

Q.95 Consider the geometric distribution :

$$f(x, \theta) = \theta\,(1-\theta)^{x-1}, \quad 0 < \theta < 1$$

then an unbiased estimat of $\frac{1}{\theta}$ is given by a statistic t_n defined as :

t_n = ...

Q.96 Consider the distribution :

$$f(x, \theta) = \frac{1}{\Gamma p}(x-\theta)^{p-1}\, e^{-(x-\theta)}, \qquad 0 \leq x < \infty$$

$$p > 2$$

then M.V.B.E. for the parameter θ holds and is given by a sample statistic $t_n =$ ....

Q.97 For the distribution given in Q.(96), the M.V.B.E. for θ is not attainable because no single (..........) exists for θ.

Q.98 If the zero frequency for the Poisson distribution does not exist then the distribution is known as (..........).

Q.99 In case of truncated Poisson distribution $\frac{e^{\theta} . \theta^{x}}{x!}$, from a sample observations (x = 1, 2, 3, ...) when x is even then an unbiased estimate of $(1 - e^{-\theta})$ is equal to ....

Q.100 For Poisson distribution $f(x, \lambda)$ the λ can not be estimated by the method of (..........)

Q.101 The C.R. inequality provides a means of judging whether or not a given unbiased estimator is also (..................).

1.2.3. Blackwellization and M.V.U.E

Q.102 Rao-Blackwell theorem enables us to obtain a ".........." from any other unbiased estimator with the help of a sufficient statistic say 't_n' for parameter θ.

Q.103 Rao-Blackwell theorem enables us to obtain a M.V.U.E. from any other unbiased estimator with the help of a "..........". Say 't_n' for the parameter θ.

Q.104 To apply Rao-Blackwell theorem, the only condition to be satisfied is that the sufficient $t_n(x)$ should have a (..................).

Q.105 An application of "Rao-Blackwell Theorem" is that if we are given an unbiased estimator 'U' then we can improve upon 'U' by forming a new statistic $f(t_n(x))$ based upon 'U' and a ".....................". $t_n(x)$.

Q.106 The process of finding an improved estimator (from unbiasedness and minimum variance point of view) starting from an unbiased estimator is known as (...............) after D. Blackwell.

Q.107 In case of Rao-Blackwell theorem, the new estimator 'ϕ' $(t_n(x))$ will not only be a better estimator than 'U' but it will also be "Best estimator" if $t_n(x)$ has a ".......................".

Q.108 This is the fact that "If an estimator is not M.V.B. estimator" it does not mean that "It is also not a "....................".

Q.109 C.R. imequality method is too much restrictive due to the "...................."

to be satisfied.

Q.110 Let (x, y) be two r.v.'s such that $E(y) = \mu$ and $V(y) = \sigma_y^2$. Again let $E(y/x) = \phi(x)$, then Rao-Blackwell theorem ststes that :

$E(\phi(x)) =$ ...., $V(\phi(x))$ ....σy^2.

Q.111 The statistic $\phi(x)$ defind as in Q(110) is an (..........) with (.........) variance for parameter μ.

Q.112 In Q.(110) the random variable x has been treated as a (..........) for the parameter μ.

Q.113 If a M.V.B.E. exists for a parameter θ, then it is the function of (..........) statistic.

Q.114 Consider the binomial distribution b(x, n,p). Then according to Rao-Blackwell theorem, a M.V.U.E of parameter p is given by a statistic $t_n(x)$ defined as :

$t_n(x) =$ ....

Q.115 In case of Poisson distribution $P(x, \lambda)$, a sufficient statistic for parameter λ is given as :

$t_n(x) =$ ....

Q.116 In Q.(115). a sufficient statistic $t_n(x)$ for $e\text{–}\lambda$ is given as :

$t_n(x) =$ ....

Q.117 For Q.(115), Rao-Blackwell theorem states that a M.V.U.E. of $e^{-\lambda}$ is guven by a statistic $t_n(x)$ defined as :

$$t_n(x) = (..........)^r \text{ for } r = \sum_{i=1}^{n} x_i$$

Q.118 In Q.(115), a sufficient statistic $t_n(x)$ for λ^2 is given as :

$t_n(x) =$ (..........).

Q.119 Consider the distribution :

$$f(x, \theta) = \frac{1}{\theta} e^{\frac{-x}{\theta}}, \qquad 0 < x < \infty$$

According to C.R. inequality, an unbiased estimator for the parameter θ is given as :

$t_n(x) =$ ....

Q.120 In case of distribution considered in Q.(119), the sampling variance of the estimator $t_n(x) = \overline{x}_n$ is given as :

$V(t_n(x)) = V(\overline{x}_n) =$ ....

Q.121 If $t_n(x)$ is a M.V.U.E. and 'U' is another unbiased estimator of parameter θ then :

Correlation $(t_n(x), U - tn(x)) =$ ....

1.2.4 Sufficient Statistics and Completeness of Sufficient Statistic

Q.122 Let $f(x, \theta)$ be the p.d.f. of a r.v.x. If a statistic $t_n(x)$ gives all relevant information about the parameter θ then $t_n(x)$ is called (...............) for θ.

Q.123 In Q.(122), the conditional distribution of $(y_2, y_3, \ldots\ldots y_n)$ given $t_n(x)$, is independent of (..........) θ.

Q.124 Let $t_n(x)$ be a sufficient statistic with p.d.f. denoted by $G(tn(x), \theta)$. If range of x is independent of θ. then "Fisher Neyman Factorization Theorem" states that $t_n(x)$ is sufficient for θ if :

$$\prod_{i=1}^{n} f(x_i, \theta) = \ldots\ldots\ldots\ldots H(x)$$

where $H(x)$ is the function of x and is independent of θ.

Q.125 If $(t_1(x), t_2(x))$ are two sufficient statistics for parameter θ then there exists a (..........) between $t_1(x)$ & $t_2(x)$.

Q.126 A sufficient statistic $t_n(x)$ is said to be complete, if no function of it has (..........) unless it is zero itself almost every where.

Q.127 A statistic $t_n(x)$ is said to be sufficient and complete if :

$E[q(t_n(x))] \Leftrightarrow P[\ldots\ldots\ldots\ldots = 0] = 1$

1.2.5 Distributions Possessive Sufficient Statistic

Q.128 The statisticians Dormois, Pitman and Koopman have given the following form of exponential distribution as :

$f(x, \theta) = e^{A(\theta)B(x) + C(x) + D(\theta)}$

which is the condition for the existence of M.V.B.E. for some function $\phi(\theta)$ in case if the range of x is (..........) of θ.

Q.129 If range of x is independent of parameter θ then :

$$\frac{\partial}{\partial\theta}\log f(x, \theta) = p(\theta)\,k(x) + \frac{q(\theta)}{n}$$

is the sufficient condition for the existance of (..........) for the parameter θ for (p, q) being orbitrary constants.

Q.130 If the lower terminal of the range of r.v.x depends on θ then frequency function is of the form :

$f(x, \theta) = a(\theta) < x < b.$

then lowest $x_{(1)}$ of $x_{(1)} < x_{(2)} < \dots\dots\dots < x(n)$ is the (..........) statistic for θ.

Q.131 If the range of x is like a < x < b (θ) where b(θ) is monotone function in θ then highest order statistic $x_{(n)}$ of $x_{(1)} \leq x_{(2)} \leq \dots\dots\dots \leq x_{(n)}$ is the (..........) statistic for θ.

Q.132 If the range of r.v.x is like $a(\theta) \leq x \leq b\ (\theta)$ and $L(x/\theta) =$

$$\prod_{i=1}^{n} f(x_i / \theta) \cup \left(x_{(1)} - a_{(\theta)}\right) \cup \left(b(\theta) - x_{(n)}\right)$$ then the pair $(x_{(1)}, x_{(n)})$

is (..........) for θ iff $f(x, \theta) = \dfrac{g\ (\dots\dots\dots)}{h\ (\dots\dots\dots)}$

Q.133 Let $f(x, \theta) = g(x)/h(\theta)$. Given $f(x, \theta) = a(\theta) \leq x \leq$ b, where a(θ) is monotone function in θ then $x_{(1)}$ is single (..........) and (..........) statistic for θ.

Q.134 Let $f(x, \theta) = g(x)/h(\theta)$. Given $f(x, \theta)$ = a < x < b (θ) where b(θ) is a monotone function of θ then $x_{(n)}$ is single (..........) and (..........) statistic for θ.

Q.135 Given $f(x, \theta) = a(\theta) \leq x \leq$ b(θ) where $a(\theta)$ and b(θ) are monotone function in θ (either both increasing or both decreasing) function θ then $(x_{(1)}, x_{(n)})$ are jointly (..........) and (..........) statistic for θ.

Q.136 Given $f(x, \theta) = a(\theta) \leq x \leq$ b(θ) and if $a(\theta)$ is decreasing and b(θ) is increasing function in θ then the sufficient and complete statistic for θ is given as :

$t_n(x)$ = Max. {.........., ...}.

Q.137 In Q.(136), if $a(\theta)$ is increasing function and $b(\theta)$ is decreasing function in θ then a sufficient and complete statistic for θ is given by :

$t_n(x)$ = Min. {.........., ...}

1.2.6 Sufficient Statistic for Several Parameters

Q.138 Let us define each observation xi as a vector variate, $t(x)$ as s component vector of statistics and θ as k component vector of parameters. If we can write $L(x/\theta) = g(t(x)/\theta)\ h(\theta)$.

then we call the components of $t(x)$ as a set of jointly (..........) for parameter θ.

Q.139 In Q.(138), if we put $t_{(x)} = x$, we say that the observations x themselves always constitute a set of (..........) for θ with $ = n.

Q.140 The statistic $\hat{s}^2 = \frac{1}{n}\sum_{i=1}^{n}(x_i - \overline{x}_n)^2$ alone is not (..........). for σ^2 when μ is unknown.

Q.141 If a single sufficient statistic $t(x)$ exists for parameter θ, then, $\hat{\theta}$ is the (..........) of θ and is a function of $t(x)$ alone.

Q.142 The M.L.E. θ is unique if likelihood function has a (..........).

Q.143 All the optimum properties of a single sufficient statistic, are conformed upon (..........). which are one to one function of them.

Q.144 A sufficient statistic '$t_n(x)$' is said to be (..........) for parameter θ, if it is the function of every other (...........) statistic.

Q.145 A sufficient statistic t(x) is said to be (...........) if the data represented by x can not be condence beyond $t(x)$ without sacrificing its sufficiency.

Q.146 The statistic $t(x)$ is said to be "complete" if for any measureable function $\phi(t_{(x)})$, we have

$E_\theta\, \phi(t_{(x)}) =$ .... $\forall\ \theta \in \Omega_\theta$.

and it implies that $f(t(x)) =$ ...., almost every where $\forall\ \theta \in \Omega\ \theta$.

Q.147 In Q.(146), the concept of completeness utilises both concepts of (1). "estimation" and (2) ".........." of hypothesis."

Q.148 The family $\{P_\theta(t(x)/\theta \in \Omega\ \theta\}$ of distribution is said to be complete if for any measurable function $\phi(t(x))$, we have :

$$\int_\theta \phi(t(x))\ P_\theta(t(x))\ d_\theta = \text{.........}\ \forall\ \theta \in \Omega\ \theta$$

which implies that $\phi(t_{(x)}) =$ .... $\forall\ \theta \in \Omega\ \theta$

Q.149 A sample statistic $t(x)$ is said to be (..........) if there exists a measurable function $\phi(t_{(\tilde{x})})$ such that :

$|\phi(t_{(x)})| < M\ \forall$ some M.

and $E_\theta(\phi(tx)) =$ .... $\forall\ \theta \in \Omega\ \theta$.

which implies that $f\,\phi(t_{(x)}) =$ ...., $\forall\ \theta \in \Omega_\theta$.

almost every where.

Q.150 A family {P (t(x), θ), θ ∈ Ω θ} of distributions is said to be ".........." complete if there exists a measurable function ϕ (t(x)) of sufficient statistic $t(x)$ such that :

$\left|\phi\left(t(x)\right)\right| < M$ for some M

and $\int\limits_{\theta} \phi(t(x))\ P(t(x), \theta)\ d_\theta$ = ..., $\forall\ \theta \in \Omega\ \theta$.

which implies that $\phi\ (t(x)\ ) = 0\ \forall\ \theta \in \Omega\ \theta$.

Q.151 If $t(x)$ is a complete statistic then it is necessarily a (..........) complete but converse is not true.

Q.152 Let (t_1, t_2) be jointly sufficient statistics for two parameters (θ_1, θ_2) respectively and further let t_1 be sufficient for θ_1 when θ_2 is known. Then conditional distribution of t_2 for a given value of t_1, is (..........) of θ_1.

Q.153 A sufficient estimator t (x) for the parameter θ is "most efficient estimator" if (..........). exists.

Q.154 The method of "Minimum Variance" will provide a solution only if there exist a (..........) t (x) for parameter θ.

Q.155 Let $(t_1, t_2, ..., t_k)$ be the set of jointly sufficient statistics for the parameter θ, then any set of one to one fuction or transformation of $(t_1, t_2, ..., t_k)$ is also jointly (..........) for θ.

Q.156 Let $(x_i, i = 1, 2, ..., k)$ be i.i.d. r.v.'s from a continuous distribution $f(x, \theta)$. Then the set of ordered statistic $(x_{(1)}, x_{(2)}, ..., x_{(k)})$ is (..........) for θ.

Q.157 Let $f(x, \theta)$ be one parametric exponential family of distributions. Let $t(x_i)$ be sufficient statistic for θ, then $\sum_{i=1}^{n} t(x_i)$ is complete and (..........) for θ.

Q.158 Let $f(x, \theta_1, \theta_2, ..., \theta_k)$ be the regular case of k parameters exponential family of distributions. Then $\sum_{i=1}^{n} t_1(x_i), \sum_{i=1}^{n} t_2(x_i), ..., \sum_{i=1}^{n} t_k(x_i)$ is a set of complete and (...........) for a set $(\theta_1, \theta_2, ..., \theta_k)$ of k parameter respectively.

Q.159 Bounded completeness implies the (..........) but "Minimal sufficieny" does not implies the (..........) of the statistic.

Q.160 There can not be more than one boundedly complete (..........) for parameter θ.

Q.161 Let $x, x_2, ..., x_n$ be i.i.d. r.v's with p.d.f, $f(x, \theta)$. Let S = S(x) be a complete and sufficience statistics for θ and if $t = t$(S) is a function of S and is an unbiased estimator of $\phi(\theta)$ then t(S) is (..........) of $\phi(\theta)$.

Q.162 Let (x, y) be independently normally distributed as $N(\mu_x, \mu_y, \sigma_x^2, \sigma_y^2)$ respectively. Then $t = \{\bar{x}, s_x^2, \bar{y}, s_y^2\}$ is (..........) but not complete statistic.

Q.163 Let (x_1, x_2) be two r.v's distributed as $N(\theta, 1)$. Let a new variable y be defined as $y = (x_1 + x_2)$ then y is (..........) statistic for parameter θ.

Q.164 Let $x \sim N(\mu, \sigma^2)$ where σ^2 is known. Then sufficient statistic $t(x)$ for μ is defined as :

$t(x) =$ ....

Q.165 Let $x \sim N(\mu, \sigma^2)$ where μ is known. Then $s^2 = \frac{1}{n}\sum_i^n (x_i - \bar{x})^2$ is (..........) for σ^2.

Q.166 Let $x \sim N(\mu, \sigma^2)$ where μ is known. Then the statistic $\$^{*2} = \frac{1}{n}\sum_i^n (x_i - \mu)^2$ is (...........) for σ^2.

Q.167 Let $x \sim N(0, \theta)$. Then an unbiased statistic for θ is given as :

$t(x) =$ .... with $V(t(x)) = \frac{2\theta^2}{n}$

Q.168 Let $x \sim N(0, \theta)$ then a sufficient statistic t(x) for θ is defined as :

$t(x) =$ ....

Q.169 Consider the p.d.f. defined as

$f(x, \theta) = \theta.(1-\theta)^x$, $x = 0, 1, 2, \ldots\ldots, 0 < \theta < 1$.

Then a sufficient statistic $t(x)$ for parameter θ is defined as :

$t(x) =$ ....

Q.170 Consider the exponential distribution as :

$f(x, \theta) = e^{-(x-\theta)}, 0 < \theta < \infty$

Then the sufficient statistic $t(x)$ for parameter θ is given as :

$t(x) =$ (...................).

1.2.7 Ordered Statistics

Q.171 Let the random sample $x, x_2, \ldots\ldots$, xn be arranged in ascending order of magnitude as $y_1 < y_2 < y_3 < \ldots\ldots < y_n$.

Then $y_{(r)}$, $(r = 1, 2, \ldots\ldots, n)$ is known as r^{th} (..........).

Q.172 The order statistic $y(r)$ $(r = 1,2, \ldots\ldots, n)$ are the function of (..........), $(x_i, i = 1, 2, \ldots\ldots, n)$.

Q.173 The ordered statistic $y_{(r)}$ is particularly used in ".........." methods.

Q.174 Let a statistic M be defined as :

$$M = \begin{cases} x_{\left(\frac{n+1}{2}\right)} & \textit{if n is odd.} \\ \dfrac{x(n) + x(n+1)}{2} & \textit{if n is even} \end{cases}$$

Then M is known as the (..........) of sample values $x_1, x_2, \ldots\ldots, x_n$.

Q.175 Let R_m be defined as :

$R_m = x_{(n-m+1)} - x_{(m)}$

Then R_m is known as (..........) of the sample values $x_1, x_2, \ldots\ldots, x_n$.

Q.176 Let $R_m = x_{(n-m+1)} - x_{(m)}$. If we put m = 1 then Rm reduces to :

$R_1 = x_{(n-m+1)} - x_{(m)}$

and this R_1 is known as (..........) of the sample values.

Q.177 Let R* be defined as :

$$R^* = \frac{1}{2}\ [x(n - m + 1) + x(m)]$$

then R* is known as (..........).

Q.178 Even if $x_1, x_2, \ldots\ldots, x_n$ are independent r.v's then order statistics generated from them are (..........) r.v's.

Q.179 Let $(x_1, x_2, \ldots\ldots, x_n)$ be iid r.v.'s from a continuous distribution. Then the set of ordered statistics say $(y_1 < y_2 < \ldots\ldots < y_n)$ is both complete and (..........).

Q.180 Let x be a continuous or discreter r.v. and let $(x_1, < x_2 < \ldots\ldots < x_n)$ be a set of ordered statistics. with parameter μ. If μ does not exist then $\mu_{(r)}$ may exist for certain 'r' but not for (..........) of r where $\mu_{(r)} = E(X_{(r)})$.

Q.181 The population mean μ does not exist but $\mu_{(r)}$ exists for the distribution known as (..........). defined as :

$$f(x, \theta) = \frac{1}{\pi} \cdot \frac{1}{(1+x^2)}$$

Q.182 Let $y_1 < y_2 < \ldots\ldots < y_n$ be a set of ordered statistics, then the distribution given as :

$$f(y_1, y_n) = n(n-1)[F(y_n) - F(y_1)]^{n-2} f(y_1) f(y_n)$$

(is the joint pdf of (..........), for $a \le y_1 \le y_2 \le - y_n \le b$)

Q.183 If $y_1 < y_2 < ..... <$ yn is a set of ordered statistics then the p.d.f. of largest variable y_n is given as :

$$f(y_n) = n\,[..........]^{n-1} ...., \; a \le y_n < b$$

Q.184 Let $y_1 < y_2 < ..... < y_n$ be a set of ordered statistics. Then p.d.f. of any orbitrary variate y_k is given as :

$$f(y_k) = \frac{\overline{|(n+1)}}{\overline{|k}\;\overline{|(n-k+1)}}[..........]^{k-1}[..........]^{n-k} f(y_k)$$

Q.185 Consider the rectangular distribution :

$$f(x, \theta) = \begin{cases} \frac{1}{\theta} & \text{for } 0 < x < \theta \\ 0, & \text{otherwise} \end{cases}$$

Let $(x_1, < x_2 < ..... < x_n)$ be a set of ordered statistics. Then a sufficient statistic for θ is given by t written as :

$t =$ .... $(x_1, x_2, ..., x_n)$

Q.186 In Q.(185), a M.V.U.E for parameter q is given as :

$$\lambda = \left(\frac{n+1}{n}\right) t$$

where $t =$ .... $(x_1, x_2, ..., x_n)$.

Q.187 In case of binomial distribution $b(x, n, p)$, the complete and sufficient statistic for parameter p is given as :

$t(x) =$ ....

Q.188 In case of Poisson distribution $f(x, \lambda)$, a complete and sufficient statistic for parameter λ is given as :

$t(x) =$ ....

Q.189 Consider the negative binomial distribution :

$$f(x, p, r) = \binom{r+x-1}{x} p^r (1-p)^x, \qquad x = 0, 1, 2, ...$$

$$r > 0$$

$$0 \le p \le 1$$

If r is known then a complete and sufficient statistic for parameter p is given as

t(x) = ...

Q.190 Consider the negative binomial distribution written in Q.(189). Then for known r, am U.M.V.U.E. for thc paramctcr ‘p’ is given as :

$$t^* = \left(\frac{(..........)}{(..........) + \sum_{i}^{n} x_i} \right)$$

Q.191 Consider the following uniform discrete distribution $f(x, k) = \frac{1}{k}$, x = a, a + h, a + 2h, ..., a +(k–1) h.

Then population mean μ is defined as :

$\mu = a +$ ....

Q.192 For the distribution defined as in Q.(191), the population variance σ^2 is defined as :

$$\sigma^2 = \frac{1}{12} (..........)$$

Q.193 From the uniform distribution defined in Q.(191), let $x_1 < x_2 < .... < x_n$ be a set or ordered sample values. Then a complete and sufficient statistic for the parameter ‘k’ is given as :

$t =$ ....

Q.194 For the uniform distribution defined in Q.(191), an U.M.V.U.E. for the parameter k is given as :

$$t = \frac{\left(x_n^{n+1} - (x_{n-1})^{n+1}\right)}{(.....................)}$$

Q.195 Consider the distribution defined in Q.(191). Then an U.M.V.U.E. for $\left(\frac{k+1}{2}\right)$ is given as :

t* = (..........)

Q.196 Consider the hyper geometric distribution defined as :

$$f(x, \theta) = \frac{\binom{\theta}{x}\binom{N-\theta}{n-x}}{\binom{N}{n}}, \quad x = 0, 1, 2, 3, \ldots.. \quad \theta = 1, 2, 3, \ldots.., N$$

Then a complete and sufficient statistic for parameter θ is given as — $t(\underset{\sim}{x}) = (\ldots\ldots\ldots)$.

Q.197 For the distribution given in Q.(196), let $t_n(x)$ be a complete and sufficient statistic for parameter θ. Then an U.M.V.U.E. for θ is given as : $t = (\ldots\ldots\ldots.)\ t_n(x)$

Q.198 Consider the following power series distribution :

$$f(x, \nu, \lambda) = \frac{a(x).\lambda^x}{h(\nu, \lambda)}, \ x = \nu, \nu+1, \ldots\ldots\ldots 0$$

Consider the ordered sample $x_1 < x_2 < \ldots.. < x_n$.

Let us define $t_1 = x_1$, and $t_2 = \sum_i^n x_i$, then the complete and sufficient statistics jointly for the parameters (ν, μ) is given respectively as $t^* = (\ldots\ldots\ldots., \ldots\ldots\ldots.)$.

Q.199 Consider the normal distribution $N(x, \mu, \sigma^2)$. Then $f(x, \mu, \sigma^2)$ is complete w.r. to μ when σ^2 is (.....).

Q.200 In case of normal distribution $N(x, \mu, \sigma^2)$, let μ to be known. Then $f(x, \mu, \sigma^2)$ is not even (..........) w.r. to σ^2.

Q.201 Consider normal distribution $N(x, \mu, \sigma^2)$ where σ^2 is known. Then a complete and sufficient statistic for μ is given as : $t = \ldots\ldots\ldots$

Q.202 Let $(x_1, x_2, \ldots\ldots\ldots., x_n)$ be a random sample from normal distribution $N(x, \mu, \sigma^2)$. Then the jointly complete and sufficient statistics for (μ, σ^2) are given as $t\ (x) = (\ldots\ldots\ldots., \ldots\ldots\ldots.)$ respectively.

Q.203 In case of x~ $N(\mu, \sigma^2)$, when μ is known then complete and sufficient statistic for σ^2 is given by $t(x) = (\ldots\ldots\ldots.)$

Q.204 In case of normal distribution $N(\mu, \sigma^2)$ when μ is unknown then an

U.M.V.U.E. for population

s^2 = ...

Q.205 In Q.(204), if μ is known then an U.M.V.U.E. for σ^2 is given as :

s*2 = ...

Q.206 In Q.(204), if σ^2 is known then an U.M.V.U.E. for μ^2 is given as :

$$t = \frac{\left(\sum^{n} x_i\right)^2 \quad \ldots\ldots\ldots\ldots}{n^2}$$

Q.207 Consider the log normal distribution given as :

$$f(x, \mu, \sigma^2) = \frac{1}{\sqrt{2\pi}\sigma} e^{-\frac{1}{2\sigma^2}(\log x - \mu)^2} \qquad \begin{matrix} x > 0 \\ \sigma^2 > 0 \end{matrix}$$

Then $f(x, \mu, \sigma^2)$ is complete and sufficient w.r. to μ when σ^2 is ....

Q.208 If in Q.(207), σ^2 is known then a complete and sufficient statistic for parameter μ is given by :

$t(x)$ = ...

Q.209 For the log normal distribution given in Q.(207), a jointly complete and sufficient statistic for (μ, σ^2) is given respectively as :

$t(x)$ = {.........., ...}

Q.210 For the distribution given as in Q.(207), an U.M.V.U.E for parameter μ is given as :

$t(x)$ = ...

Q.211 For the distribution given in Q.(207), for the known value of μ, an U.M.V.U.E. of population variance σ^2 is given by the statistic t^* defined as :

t^* = ...

Q.212 Consider the distribution given in Q.(207). If μ is unknown then an U.M.V.U.E. for parameter σ2 is given by a statistic $\phi(x)$ defined as :

$$\phi(x) = \frac{n\sum^{n}(\log x_i)^2 - (\ldots\ldots\ldots\ldots)^2}{(\ldots\ldots\ldots\ldots)}$$

Q.213 Consider the power series distribution given as :

$$f(x, \theta, \alpha) = \frac{\theta . x^{\theta-1}}{\alpha^{\theta}}, \quad \begin{matrix} 0 < x < \alpha \\ \theta > 0 \end{matrix}$$

Then $f(x, \theta, \alpha)$ is complete w.r. to θ when α is ....

Q.214 In Q.(213), the p.d.f. $f(x, \theta, \alpha)$ is complete w.r. to α when θ is ....

Q.215 In Q.(213), when α is known then a complete and sufficient statistic for parameter θ is given as :

$t(x)$ = ...

Q.216 Let $x_1 < x_2 < ..... < xn$ be an ordered sample drawn from distribution given in Q.(213). Then for the parameter α, a complete and sufficient statistic is given as :

t(x) = ...

Q.217 Consider Cauchy distribution defined as :

$$f(x, \mu, k) = \frac{k}{\pi[k^2 + (x - \mu)^2]}, \quad \begin{matrix} -\infty < x < \infty \\ -\infty < \mu < \infty \\ k > 0 \end{matrix}$$

when k = 1 then $f(x, \mu, k)$ is complete w.r. to ....

Q.218 In Q.(217), when $\mu = 0$ then f(x, μ, k) is not even (..........). complete w.r. to K^2.

Q.219 Consider the Cauchy distribution as :

$$f(x, \mu) = \frac{1}{\pi[1+(x-\mu)^2]}, \quad \begin{matrix} -\infty<x<\infty \\ -\infty<\mu<\infty \end{matrix}$$

Then M.L.E. of parameter μ is obtainable using the method of (..........).

Q.220 In case of distribution given in Q.(219), the consistant estimator of population median 'μ' is given by the statistic t(x) defined as :

$t(x)$ = ...

Q.221 Let x_1, x_2 be a sample of size 2 drawn from Cauchy distribution. Then Pitman closure estimator for μ than x_1 is given by the statistic t(x) defined as :

$t(x)$ = ...

Q.222 The Cauchy distribution defined in Q.(219) is symmetrical with mode and median equal to ....

Q.223 Consider the Gamma distribution defined as :

$$f(x, \theta) = \frac{\theta^x \, x^{m-1} \, e^{-\theta x}}{\overline{|m}}, \quad \begin{matrix} x > 0 \\ \theta > 0 \\ m > 0 \end{matrix}$$

Then for m = 1, $f(x, \theta)$ is known as (..........................) distribution.

Q.224 In Q.(223), if m is known then a complete and sufficient statistic $t(x)$ for parameter θ is given as :

$t(x)$ = ...

Q.225 In Q.(223), if m is known then an U.M.V.U.E. for $\frac{1}{\theta}$ is given by a statistic $t(x)$ defined as :

$t(x)$ = ...

Q.226 Consider beta distribution defined as :

$$f(x, p, q) = \frac{\overline{|(p+q)} \, x^{p-1} \, (1-x)^{q-1}}{\overline{|(p)}\,\overline{|(q)}}, \quad \begin{matrix} 0 < x < 1 \\ p > 0 \\ q > 0 \end{matrix}$$

then distribution $f(x, p, q)$ is complete w.r. to p when q is (..........) and vice versa.

Q.227 In Q.(226), when q is known then a complete and sufficient statistic for the parameter 'p' is given by a statistic $t(x)$ defined as :

$f(x)$ = ...

Q.228 In Q.(226), if p is known then a complete and sufficient statistic for the parameter q is given by a statistic $t(x)$ given as :

$$t^*(x) = \sum_{i}^{n} \ldots\ldots\ldots\ldots$$

Q.229 In Q.(226) for q = 1, and n > 1, an U.M.V.U.E. for the parameter p is given by a statistic t(x) defined as :

$$t(x) = \frac{1}{\frac{1}{n-1}\sum^{n}(..........)}$$

Q.230 In Q.(226), for $p = 1$, and $n > 1$, the statistic $t^*(x)$ defined as:

$$t^*(x) = \frac{1}{\frac{1}{n-1}\sum_{i=1}^{n}\log(..........)}$$

is an U.M.V.U.E. for the parameter 'q'.

Q.231 Consider the following uniform distribution $f(x, \theta)$ as $f(x, \theta) = \frac{1}{\theta}, 0 < x < \theta, \theta > 0$

Then an U.M.V.U.E for θ^r is given by the statistic $t(x)$ given as :

$$t(x) = (..........)\ x_{(n)}^{r}$$

Q.232 Consider the uniform distribution defined as :

$$f(x, \theta_1, \theta_2) = \frac{1}{(\theta_2 - \theta_1)}, \quad \theta_1 < x < \theta_2$$

Then $f(x, \theta_1, \theta_2)$ is complete w.r. to θ_1 when θ_2 is (..........) and vice-versa.

Q.233 In Q.(232), for an ordered sample $(x_1 < x_2 < ..... < x_n)$ and θ_2 being known, an U.M.V.U.E. for θ_1 is given by a statistic $t(x)$ defined as :

$$t(x) = \frac{(n+1)\ x_1 - ....}{n}$$

Q.234 In Q.(232), for an ordered sample $x_1 < x_2 < ..... < x_n$, and θ_1 being known, an U.M.V.U.E. for θ_2 is given by a statistic $t(x)$ defined as :

$$t(x) = \frac{(..........)\ x_n - ....}{n}$$

Q.235 Consider the following p.d.f defined as :

$$f(x, \theta_1, \theta_2) = \frac{1}{\theta_2}, \quad \begin{array}{c} \theta_1 - \frac{\theta_2}{2} \le x \le \theta_1 + \frac{\theta_2}{2} \\ -\infty < \theta_1 < 0, \theta_2 > 0 \end{array}$$

Let $(x_1 < x_2 < ..... < x_n)$ be an ordered sample then an U.M.V.U.E. for parameter θ_1, the statistic $t(x)$ is given as :

$$t(x) = \frac{(..........)}{2}$$

Q.236 In Q.(235), an U.M.V.U.E. for parameter θ_2 is given by a statistic $t(x)$ defined as :

$$t(x) = \left(\frac{n+1}{n-1}\right) \ (..........)$$

Q.237 Consider the p.d.f. $f(x, \theta)$ defined as :

$$f(x, \theta) = 1, \qquad \left(\theta - \frac{1}{2}\right) \begin{array}{l} < x < (\theta + \frac{1}{2}) \\ -\infty < \theta < 0 \end{array}$$

consider a set of ordered sample $(x_1 < x_2 < ..... < x_n)$

Then a sufficient but not even boundedly complete statistic for parameter θ is given is given by $t(x)$ written as :

$t(x) = (..........)$

Q.238 In Q.(237), a M.V.U. linear estimator for parameter θ is given by a statistic $t(x)$ written as :

$$t(x) = \frac{(..........)}{2}$$

Q.239 Consider the following p.d.f. given as :

$f(x, \theta) = 1, \theta < x < \theta + 1, -\infty < \theta < \infty$

Consider an ordered sample $(x_1 < x_2 < .... < x_n)$ then a sufficient statistic t(x) for parameter θ is given as :

$t(x) = (..........)$

Q.240 In Q.(239), then a M.V.U. linear estimator for the parameter θ is given by a statistic t(x) written as :

$$t(x) = \frac{1}{2}(.......... + ...)$$

Q.241 Consider the p.d.f. given as :

$$f(x, \theta) = \frac{1}{2\theta}, \quad -\theta < x < \theta, \theta > 0$$

Then a complete and sufficient statistic $t(x)$ for parameter θ is given as :

t(x) = Max (.........., ...)

Q.242 In case of p.d.f. given in Q.(241), an U.M.V.U.E. for parameter θ is given by a statistic $t(x)$ written as :

$t(x)$ = (..........) max $(-x_1, x_n)$

Q.243 Consider the Pareto distribution defined as :

$$f(x, \alpha, \beta) = \frac{\alpha\beta^x}{x(\alpha+1)}, x > 1, \beta > 0, \alpha > 1.$$

Then $f(x, \alpha, \beta)$ is (..........). w.r. to α when β is (..........) and vice-versa.

Q.244 In case of Q.(243) a complete and sufficient statistic $t(x)$ for α when β is known, is given as :

$$t(x) = \sum_{i}^{n} \ldots\ldots\ldots\ldots$$

Q.245 Consider an ordered sample $x_1 < x_2, < \ldots\ldots < x_n$, from Pareto distribution given in Q(243). Then a complete and sufficient statistic $t(x)$ for parameter b when a is known, is given by a statitic t(x) defined as :

t(x) = ...$(x_1 < x_2 < \ldots\ldots < x_n)$

Q.246 Consider the p.d.f. $f(x, \theta)$ defined as :

$$f(x, \theta) = \frac{1}{\theta}, k\,\theta < \times < (k+1)\,\theta, k > 0$$

Then the sufficient statistic for parameter q is given by a statistic $t(x)$ defined as :

$t(x)$ = (.........., ...) jointly from an ordered sample $(x_1 < x_2 < \ldots\ldots < x_n)$ from the distribution under study.

Q.247 In Q.(246) no (..........) statistic for parameter θ exists.

Q.248 In Q.(246), let θ be the M.L.E. of parameter θ then θ is defined as :

θ = (.........)

Q.249 In Q.(246), due to symmetry, the (x_1, x_2) have the sample variance, say V, then we have :

$$V(\theta) = (\ldots\ldots\ldots.)$$

Q.250 Let a $r\ v\ x \sim N(0, \sigma^2)$ defined as :

$$f(x, 0, s^2) = \frac{1}{\sqrt{2\pi}\,\sigma} e^{-\frac{1}{2}\frac{x^2}{\sigma^2}}, \quad \begin{matrix} -\infty < x < \infty \\ \sigma^2 > o \end{matrix}$$

Then m.l.e. and sufficient statistic for σ is given by $t(x)$ as :

$t(x) = \sigma = \sqrt{\ldots\ldots\ldots\ldots}$

Q.251 Consider the p.d.f. of double exponential distribution as $f(x, \theta) = \frac{1}{2} e^{-|x-\theta|}, \ -\infty < \theta < \infty$

Then the (..........) statistic $t(x)$ for the parameter θ is given as :

$t(x) = (x_1 < x_2 < \ldots\ldots < x_n)$ considered together.

Q.252 In Q.(251), the m.l.e. of parameter θ is given by θ as :

$\theta = \ldots\ldots\ldots \{x_1, x_2, \ldots\ldots\ldots, x_n\}$

Q.253 Consider the following p.d.f. as :

$$f(x, \theta) = \frac{1}{\theta} e^{-\frac{x}{\theta}}, 0 < x < \infty, 0 < \theta < \infty$$

Then the sufficient statistic $t(x)$ for parameter θ is given as :

$t(x) = \ldots\ldots\ldots$

Q.254 In Q.(253), an unbiased estimator $t(x)$ for parameter q is given as :

$$t(x) = (\ldots\ldots\ldots\ldots) \frac{1}{x}$$

Q.255 Consider the following two parametric exponential p.d.f. given as :

$$f(x, \theta_1, \theta_2) = \frac{1}{\theta_2\, e^{\frac{(x-\theta_1)}{\theta_2}}}, \quad \begin{matrix} \theta_1 < x < \infty \\ -\infty < \theta_1 < \infty \\ 0 < \theta_2 < \infty \end{matrix}$$

Let $x_1 < x_2 < \ldots\ldots < x_n$ be an ordered sample from the distribution. Then the sufficient statistics for the parameters (θ_1, θ_2) are given jointly and respectively as :

t(x)= (.........., ...).

Q.256 Let 't' be the best estimator for λ(θ) then for any other estimator t^*, we have :

$P_\theta [\, 1\, t - \lambda(\theta)\,] < \ldots\ldots] \geq P_\theta [\, | \ldots\ldots\ldots | < \in] \ \forall\ \theta \in \Omega\ \theta$

Q.257 Let us consider the following p.d.f.

$f(x, \alpha) = e{-}(x - \alpha),\ \alpha \leq x < \infty$

Then a sufficient statistic t(x) for lower terminal of α is given as :

$t(x) = \text{..........}\ (x_1 < x_2 < \text{..........} < x_n)$

Q.258 Consider the following two parametric distribution function :

$$f(x, \alpha, \beta) = \frac{1}{(\beta - \alpha)},\ \alpha \leq x \leq \beta$$

Let $x_1 < x_2 < \text{.....} < x_n$ be an ordered sample drawn from the population (distribution). Then for the fiven β, the sufficient statistic $t(x)$ for α is given as :

$t(x) = \text{..........}(x_1, x_2, \text{..........}, x_n)$

Q.259 In Q.(258), for given α, a sufficient statistic $t(x)$ for parameter b is given as :

$t(x) = \text{..........}(x_1, x_2, \text{..........}, x_n)$.

Q.260 In Q.(258), the joint sufficient statistics for the parameters (α, β) respectively are given as :

$t(x) = (\text{..........}, \text{..........})$

Q.261 Let t be a complete statistic for parameter θ, then any other statistic say t* depending on random sample $x_1, x_2, \text{.....}, x_n$ through 't' is also (..........) for θ.

Q.262 Consider Bernoulli distribution with parameter p defined as :

$P(x = 1) = p \qquad 0 < p < 1$

$p\ (x = 0) = (1 - p) = q$

Then the sample statistic $t(x) = \sum^{n} x_i$ is (..........) for p.

Q.263 In standardized bivariate normal distribution, consider a paired sufficient statistic t(x) defined as :

$$t(x) = \left(\sum_i^n x_i^2 + \sum_i^n y_i^2,\ \sum_i^n x_i\ y_i \right)$$

Then t(x) is a (..........) for population correlation ρxy.

Q.264 Consider the following p.d.f. :

$$f(x, \theta) = 1, \qquad \left(\theta - \frac{1}{2}\right) \le x \le (\theta + \frac{1}{2})$$

Let $x_1 < x_2 < ..... < x_n$ be an order sample drawn from the distribution (population). Then joint sufficient statistic $t(x)$ for parameter θ is given as :

$t(x)$ = (.........., ...).

Q.265 For the distribution given in Q.(264), likelihood function (L.F.) is maximum at any value lying in the interval (.........., ...).

Q.266 For the distribution given in Q.(264), the mid point of the interval $\left(x_n - \frac{1}{2}, x_1 + \frac{1}{2}\right)$ is an (..........) of the parameter θ.

Q.267 For the distribution given in Q.(264), the sample mean $\bar{x}_n$ is given as an estimate of parameter θ ontained through a method known as (..........).

Q.268 Consider the power series distribution written as :

$$f(x, \theta) = \frac{a_x \ \theta^x}{f(x)^n}, \text{ x = c, c+1, c+2, ...}$$

consider a random sample drawn from the distribution. Then a complete sufficient statistic $t(x)$ for parameter θ is given as :

$t(x)$ = ...

Q.269 For the distribution given in Q.(268), the statistic 't' has distribution of the same form written as :

$$g(t, \theta) = \frac{b_t \ \theta^t}{(..........)^n}, \ t = h, h+1, h+2, ...$$

Q.270 Consider the binomial distribution as :

$$B(x, n, p) = {}^{n}C_x \ p^x \ q^{n-x}, \quad x = 0, 1, 2, ..., n$$

$$0 < p < 1, 0 < q < 1$$

$$p + q = 1.$$

When $p = q = \frac{1}{2}$, the value of n is least for $B(x, n, p)$ to be approximated by a distribution known as (..........).

1.3 Multiple Choice Questions and Answers

Q.1 The process of knowing all the relevant information from known sample to unknown population, is known as :

(a) Statistical inference (b) Point estimation
(c) Interval estimation (d) Level of significance

Q.2 A problem to guess completely about the unknown feature of a population from a random sample drawn from population, is known as the :

(a) Problem of point estimation
(b) Problem of interval estimation
(c) Problem of estimation
(d) Problem of discrete distribution

Q.3 An estimation of population parameter, say θ by a single value (an estimate) from a random sample drawn from population, is known as :

(a) Point estimation (b) Interval estimation
(c) Statistical decision (d) Deciding Level

Q.4 A process of getting an interval within which the population parameter θ, is expected to lie with certain confidence coefficient, is known as :

(a) Point estimation (b) Interval estimation
(c) Statistical decision (d) Linear estimation

Q.5 An estimator having the properties of Unbiasedness, consistency, sufficiency and efficiency, is known as :

(a) Sufficient estimator (b) Efficient estimator
(c) Unbiased estimator (d) Best estimator

Q.6 An estimator whose mathematicl expectation is exactly equal to the true value of the population parameter, say θ, is known as :

(a) A consistent estimator (b) Sufficient estimator
(c) Efficient estimator (d) Unbiased estimator

Q.7 An estimator whose mathematical expectation is not equal to the population parameter θ, is known as :

(a) Consistent estimator (b) Biased estimator
(c) Interval estimator (d) Sufficient estimator

Q.8 Let $t_n(x)$ be an estimator of the parameter θ, then $t_n(x)$ is said to be an unbiased estimator if :

(a) $E(t_n(x)) > \theta$ (b) $E(t_n(x)) < \theta$
(c) $E(tn(x) \neq \theta$ (d) $E(tn(x)) = \theta$

Q.9 An estimator $t_n(x)$ of a parameter θ, is said to be a biased estimator for θ if :

(a) $E\ (tx(x)\ ) > \theta$ (b) $E\ (t_n(x)\ ) < \theta$

(c) $E\ (t_n(x) \neq \theta$ (d) $E\ (t_n(x)\ ) > \theta$

Q.10 The amount of bias of an estimator $t_n(x)$ for parameter θ, can be worked out using the expression "Bias (θ)" equal to :

(a) $t_n(x) - \theta$ (b) $E\ (t_n(x)\ ) - \theta$

(c) $\theta - \mathrm{E}\ (\mathrm{tn}(\mathrm{x})\ )$ (d) $\theta - t_n(x)$

Q.11 An estimator $t_n(x)$ is said to be positively biased estimator for the parameter θ if :

(a) $E\ (t_n(x)\ ) < \theta$ (b) $E\ (t_n(x)\ ) = \theta$

(c) $E\ (t_n(x)\ ) > \theta$ (d) $E\ (t_n(x)\ ) \geq \theta$

Q.12 An estimator $t_n(x)$ is said to be negatively biased estimator for the parameter θ, if :

(a) $E\ (t_n(x)\ ) \geq \theta$ (b) $E\ (t_n(x)\ ) \leq \theta$

(c) $E\ (t_n(x)\ ) < \theta$ (d) $E\ (t_n(x)\ ) = \theta$

Q.13 An estimator $t_n(x)$ is said to be asymptotically unbiased for the parameter θ if :

(a) $\lim\limits_{n\to\infty} E\ (t_n(x)\ ) < \theta$ (b) $\lim\limits_{n\to\infty} E\ (t_n(x)\ ) \geq \theta$

(c) $\lim\limits_{n\to\infty} E\ (t_n(x)\ ) \geq \theta$ (d) $\lim\limits_{n\to\infty} E\ (t_n(x)\ ) = \theta$

Q.14 An estimator $s^{*2} = \frac{1}{n}\sum_{i=1}^{n}(x_i - \overline{x}_n)^2$ is asymptotically unbiased for the population variance σ_x^2 because :

(a) $\lim\limits_{n\to\infty} E(\$^{*2}) \geq \sigma_x^2$ (b) $\lim\limits_{n\to\infty} E(\$^{*2}) \leq \sigma_x^2$

(c) $\lim\limits_{n\to\infty} E(\$^{*2}) > \sigma_x^2$ (d) $\lim\limits_{n\to\infty} E(\$^{*2}) = \sigma_x^2$

Q.15 The sample mean $\overline{x}_n$ is an absolutely unbiased estimator for population mean μ_x because :

(a) $E(\overline{x}_n) = \underset{x}{\mu}$ (b) $E(\overline{x}_n) > \underset{x}{\mu}$

(c) $E(\overline{x}_n) < \underset{x}{\mu}$ (d) $E(\overline{x}_n) \neq \underset{x}{\mu}$

Q.16 Let $(x_1, x_2, ..., x_n)$ be a random sample from $N(\mu, 1)$ and $t_n(x) = \sum_{i=1}^{n} x_i^2$ is an estimator then we have :

(a) $E\,(t_n(x)\,) = \mu^2+1$ (b) $E\,(t_n(x)\,) = 1+\mu$

(c) $E\,(t_n(x)\,) = 1+\mu^2$ (d) Both (a) & (c)

Q.17 Let $(x_1, x_2, ..., x_n)$ be a random sample taken on Bernoulli trial such that :

$P\,(x = 1) = \theta$

$P\,(x = 0) = (1 - \theta)$

Then an unbiased estimator for θ^2 is given as :

(a) $t^* = \dfrac{t\,(t-1)}{n\,(n-1)}$ (b) $t^* = \dfrac{t^2\,(t-1)}{n\,(n-1)}$

(c) $t^* = \dfrac{t\,(t^2-1)}{n\,(n-1)}$ (d) $t^* = \dfrac{t}{n}$

Q.18 The difference between the estimated value and true value of the parameter is described by the property known as :

(a) Unbiasedness (b) Sufficiency

(c) Efficiency (d) Consistency

Q.19 An estimator $t_n(x)$ of the parameter θ is said to be consistent for θ if :

(a) $\lim_{n\to\infty} P[|t_n(x)-\theta| \leq \in] > (1-\delta)$

(b) $\lim_{n\to\infty} P[|t_n(x)-\theta| \leq \in] < (1-\delta)$

(c) $\lim_{n\to\infty} p[|t_n(x)-\theta| \geq \in] > (1-\delta)$

(d) $\lim_{n\to\infty} P[|t_n(x)-\theta| \geq \in] \geq (1-\delta)$

Where $(\in, \delta)$ are infinitely small positive quantities.

Q.20 An estimator $t_n(x)$ is said to be consistent for the parameter θ if it :

(a) Diverges to θ is probability
(b) Converges to θ is probability
(c) Diverges to θ with probability equal to 0.5
(d) Diverges to θ with probability 1.0

Q.21 An estimator $t_n(x)$ is said to be consistent for the parameter θ if :

(a) $\lim_{n\to\infty} P[|t_n(x) - \theta| < \in] = 1.0$

(b) $\lim_{n\to\infty} P[\,|\,t_n(x) - \theta\,| < \in\,] \to 1.0$

(c) $\lim_{n\to\infty} P[\,|\,t_n(x) - \theta\,| < \in\,] \geq \frac{1}{2}$

(d) $\lim_{n\to\infty} P[\,|\,t_n(x) - \theta\,| < \in\,] \geq 1.0$

Where $\in$ is a infinitely small positive value.

Q.22 A consistent estimator (..........) an unbiased for parameter θ.

(a) Is always (b) May or may not be
(c) Is never (d) None of the three

Q.23 A consistent estimator $t_n(x)$ in a large sample with finite mean value, may tend to be :

(a) An unbiased estimator (b) A consistent estimator
(c) An efficient estimator (d) A sufficient estimator

Q.24 If $t_n(x)$ is an unbiased estimator of parameter θ with $V\left(t_n(x)\right) = \sigma_n^2$ then $t_n(x)$ will be consistent for θ if :

(a) $\theta_n \to \theta$ and $\sigma_n^2 \to 1$ for $n \to \infty$

(b) $\theta_n \to \theta$ and $\sigma_n^2 \to 0$ for $n \to \infty$

(c) $\theta_n \to \theta$ and $\sigma_n^2 \to \theta$ for $n \to \infty$

(d) $\theta_n = \theta$ and $\sigma_n^2 = 0$ for $n = \infty$

Q.25 In normal distribution $N(\mu, \sigma^2)$, the sample median x is :

(a) An unbiased estimator for μ.
(b) An consistent estimator for μ.
(c) A sufficient estimator for μ.
(d) An efficient estimator for μ.

Q.26 In the distribution $N(\mu, \sigma^2)$, the sample variance

$$s^{*2} = \frac{1}{n} \sum^{n} (x_i - \bar{x}_n)^2 \text{ is :}$$

(a) An unbiased estimate of σ^2.

(b) A consistent estimate of σ^2.

(c) Not a consistent estimate of σ^2

(d) A consistent estimate of σ.

Q.27 In Cauchy distribution : $f(x, \theta) = \frac{1}{\pi} \frac{1}{1+(x-\theta)^2}$, $-\infty < x < \infty$, the sample mean $\bar{x}_n$ is :

(a) Not a consistent estimate of θ.

(b) Not an unbiased estimate of θ.

(c) Not a sufficient estimate of θ.

(d) All the three.

Q.28 In Cauchy distribution :

$$f(x, \theta) = \frac{1}{\pi} \frac{1}{1 + (x - \theta)^2}, -\infty < x < \infty$$

the sample median x is :

(a) An unbiased estimate of θ.

(b) A consistent estimate of θ.

(c) Not a consistent estimate of θ.

(d) None of the above.

Q.29 In Cauchy distribution :

$$f(x, \theta) = \frac{1}{\pi} \frac{1}{1 + (x - \theta)^2}, -\infty < x < \infty$$

the sample mean $\bar{x}_n$:

(a) Does not exist

(b) Always exists

(c) Is an unbiased estimate of θ.

(d) Is a consistent estimate of θ.

Q.30 In case of Poisson distribution :

$$f(x, \lambda) = \frac{e^{-\lambda} \lambda^x}{x!}, \text{x} = 0.\ 1, 2, \text{..........}$$

the consistent estimator for $\frac{1}{\lambda}$ is given by :

(a) Sample mean $\overline{x}_n$ (b) $\frac{1}{\text{Sample mean } \overline{x}_n}$

(c) Sample median x (d) None of the three.

Q.31 Let t be a consistent estimator for any parameter θ then a consistent estimator for e^{θ} will be given by :

(a) e^{t^2} (b) e^{-t}

(c) e^t (d) None of the three

Q.32 A function of sample value say $t(x)$ to estimate the parameter θ is known as :

(a) A statistic (b) An estimator

(c) An estimate (d) Predicted value

Q.33 A sample statistic whose sampling distribution concentrates as closely as possible to the true value θ_0, may be considered as :

(a) A biased estimator (b) An unbiased estimator

(c) An estimator (d) None

Q.34 In a class of consistent estimators for the parameter θ, if there exists an estimator t^* whose sampling variance is less than that of any such estimator in the class, then t^* is known as :

(a) Most efficient estimator (b) An unbiased estimator

(c) A sufficient estimator (d) A consistent estimator

Q.35 A most estimator t^* for parameter θ provides a criteria to measure the

(a) Unbiasedness of other estimator

(b) Efficiency of other estimator

(c) Sufficiency of other estimator

(d) Consistency of other estimator

Q.36 Let (t_n and t_n') be consistent estimators for parameter θ such that :

$$t_n \sim N(\theta, \frac{\sigma^2}{n})$$

$$t_n' \sim N(\theta, \frac{\sigma'^2}{n})$$

then t_n is said to be more efficient estimator then t_n' if :

(a) $\sigma^2 < \sigma'^2$ (b) $\sigma^2 > \sigma'^2$

(c) $\sigma^2 \geq \sigma'^2$ (d) $\sigma^2 \not< \sigma'^2$

Q.37 Let '*e*' denotes the efficiency of an estimator t_n, then *e* is defined as :

(a) $e = \dfrac{\text{V (Most efficient estimator)}}{\text{V (Other estimator } t_n)}$

(b) $e = \dfrac{\text{V (Estimator } t_n)}{\text{V (Most efficient estimator)}}$

(c) $e = \dfrac{\text{V(Estimator } t_n)}{\text{V (Consistent estimator)}}$

(d) $e = \dfrac{\text{V(Consistent estimator)}}{\text{V (Estimator } t_n)}$

Q.38 Let e denotes the efficiency two unbiased estimators (t_1, t_2) with equal sampling variance, then we have the relation :

(a) $\rho(t_1, t_2) \leq (2e - 1)$ (b) $\rho(t_1, t_2) \geq (2e - 1)$

(c) $\rho(t_1, t_2) \leq (1 - e^2)$ (d) $\rho(t_1, t_2) \geq (1 - e^2)$

Q.39 Let (t_1, t_2) be most efficient estimators of parameter θ then :

(a) $\rho(t_1, t_2) = \dfrac{-1}{2}$ (b) $\rho(t_1, t_2) = \dfrac{1}{2}$

(c) $\rho(t_1, t_2) = 1.00$ (d) $\rho(t_1, t_2) \leq 1.0$

Q.40 In case of $N(\mu, \sigma^2)$, the sample mean $\bar{x}_n$ and sample median x are the estimators of parameter μ. Then we have :

(a) $V(\bar{x}_n) > V(x)$ (b) $V(\bar{x}_n) < V(x)$

(c) $V(\bar{x}_n) = V(x)$ (d) $V(\bar{x}_n) \geq V(x)$

Q.41 Let $t_i(x)$ be an U.M.V.E. of parameter θ_i and $h_i(x)$ be an other estimator of θ_i. Then $\forall\ \theta_i \in \Omega\ \theta$, we have :

(a) $V(t_i(x)) > V(h_i(\theta_i))$ (b) $V(t_i(x)) < V(h_i(x))$

(c) $V\left(t_i\,(x)\right) \geq V\left(h_i\,(\theta_i)\right)$ (d) $V\,(t_i\,(x)) \ngeq V\,(h_i\,(x))$

Q.42 Let (t_1, t_2) be two consistent/unbiased estimators of a parameter θ and t_1 is more efficient than t_2 with efficiency $(t_2/t_1) = e$ then we have :

(a) Correlation $(t_1, t_2) \geq e$ (b) Correlation $(t_1, t_2) = \sqrt{e}$

(c) Correlation $(t_1, t_2) \leq e$ (d) Correlation $(t_1, t_2) < \sqrt{e}$

Q.43 Let (t_1, t_2) be two unbiased estimators of a parameter $\Psi(\theta)$ with efficiencies (e_1, e_2) respectively with $\rho\,(t_1, t_2) = \rho$ then the following relation is true.

(a) $\sqrt{e_1\,e_2} - \sqrt{(1-e_1)\,(1-e_2)} \leq \rho \leq \sqrt{e_1\,e_2} + \sqrt{(1-e_1)\,(1-e_2)}$

(b) $\sqrt{e_1\,e_2} + \sqrt{(1-e_1)\,(1-e_2)} \leq \rho \leq \sqrt{e_1\,e_2} - \sqrt{(1-e_1)\,(1-e_2)}$

(c) $\sqrt{e_1\,e_2} - \sqrt{(1+e_1)\,(1+e_2)} \leq \rho \leq \sqrt{e_1\,e_2} + \sqrt{(1+e_1)\,(1+e_2)}$

(d) $\sqrt{e_1\,e_2} - \sqrt{(1-e_1)\,/\,(1-e_2)} \leq \pi \leq \sqrt{e_1\,e_2} + \sqrt{(1-e_1)\,/\,(1-e_2)}$

Q.44 Let t be a M.V.U.E. of parameter θ and t^* be another estimator for θ with efficiency 'e' then e is defined as :

(a) $e\,(t^*/t) = \dfrac{V(t^*)}{V(t)}$ (b) $e\,(t^*/t) = \dfrac{V(t)}{V(t^*)}$

(c) $e\,(t^*/t) = \sqrt{\dfrac{V(t^*)}{V(t)}}$ (d) $e\,(t^*/t) = \sqrt{\dfrac{V(t)}{V(t^*)}}$

Q.45 In Cauchy distribution, let $t_n(x)$ denotes the (M.L.E.) of parameter θ then $t_n(x)$ is most efficient estimator with :

(a) $V\left(t_n\,(x)\right) = \dfrac{1}{n}$ (b) $V\left(t_n\,(x)\right) = \dfrac{2}{n}$

(c) $V\left(t_n\,(x)\right) = \dfrac{1}{\sqrt{n}}$ (d) $V\left(t_n\,(x)\right) = \sqrt{\dfrac{2}{n}}$

Q.46 In Cauchy distribution, the variance (sample median x) is equal to :

(a) $\dfrac{\pi^2}{4n}$ (b) $\dfrac{4n}{\pi^2}$

(c) $\dfrac{\pi^2}{n}$ (d) $\dfrac{\pi}{n}$

Q.47 In Cauchy distribution, the efficiency of sample median is given as :

(a) $e(x) = \frac{8}{\pi^2}$ (b) $e(x) = \frac{4}{\pi^2}$

(c) $e(x) = \frac{16}{2\pi^2}$ (d) Both (a) & (c)

Q.48 Consider normal distribution $N(x, \mu, \sigma^2)$ and let sample mean deviation be defined as :

$$D = \frac{1}{n}\sum_{i}^{n} |x_i - \bar{x}_n|$$

Then we have :

(a) $V(D) = V(\bar{x}_n)$ (b) $V(D) < V(\bar{x}_n)$

(c) $V(D) > V(\bar{x}_n)$ (d) $V(D) \leq V(\bar{x}_n)$

Q.49 In case of normal distribution $N(x, \mu, \sigma^2)$, the variance of sample mean deviation D is given as :

(a) $V(D) = \frac{\pi}{\pi - 2} \frac{\sigma^2}{n}$ (b) $V(D) = \frac{\pi - 2}{\pi} \cdot \frac{\sigma^2}{n}$

(c) $V(D) = \frac{\pi^2}{\pi - 2} \cdot \frac{\sigma^2}{n}$ (d) $V(D) = \frac{\pi - 1}{\pi} \cdot \frac{\sigma^2}{n}$

Q.50 The; correlation between an U.M.V.U.E. and any other unbiased estimator of parameter is always :

(a) Negative (b) Positive

(c) Equal to 1.0 (d) Equal to 0.5

Q.51 Let $t_i (i = 1, 2, ..., k)$ be k independent unbiased estimators of a parametric function $\Psi(\theta)$ and $V(t_i) = \sigma_i^2 < \sigma^2 \ \forall_i = 1, 2, ..., k$. Then an unbiased estimator t of $\psi(\theta)$ will be given as to as :

(a) $t = \frac{1}{K}\sum^{K} t_i$ (b) $t = \sum_{i-1}^{K} t_i$

(c) $t = \sum_{i-1}^{k} \left(\frac{t_i}{k}\right)$ (d) Both (a) & (c)

1.3.1 Cramer, Rao Inequality

Q.52 If $t_n(x)$ is an unbiased estimator of a parametric function $\psi(\theta)$, then C.R. inequality states that :

(a) $$V(t_n(x)) \leq \frac{\left(\frac{\partial}{\partial\theta}\psi(\theta)\right)^2}{E\left(\frac{\partial}{\partial\theta}\log L\right)^2}$$

(b) $$V(t_n(x)) \geq \frac{\left(\frac{\partial}{\partial\theta}\psi(\theta)\right)^2}{E\left(\frac{\partial}{\partial\theta}\log L\right)^2}$$

(c) $V(t_n(x)) \geq \frac{[\psi'(\theta)]^2}{I_{(\theta)}}$

(d) Both (b) & (c)

Where I_θ is the information supplied on θ by the sample.

Q.53 One of the regularity conditions for C.R. inequality to hold good is that :

(a) $\frac{\partial}{\partial\theta} f(x, \theta) = 0 \quad \forall\ \theta \in \Omega\ \theta$

(b) $\frac{\partial}{\partial\theta} f(x, \theta)$ exists $\forall\ \theta \in \Omega$

(c) $\frac{\partial}{\partial\theta} f(x, \theta)$ does not exist

(d) $\frac{\partial}{\partial\theta} f(x, \theta) \to 1, \quad \forall\ \theta \in \Omega\ \theta$

Q.54 One of the regularity conditions holds good is that the range of integration

$\int \prod_i^n f(x_i, \theta)\, dx$ is :

(a) Dependent of parameter θ

(b) Independent of parameter θ.

(c) May or may not be dependent θ.

(d) Both (a) & *(c)

Q.55 For C.R. inequality to hold good, we have the following result to be true.

(a) $E\left[\frac{\partial}{\partial\theta}\log f(x,\theta)\right]^2 < 0 \quad \forall\, \theta \in \Omega\, \theta$

(b) $E\left[\frac{\partial}{\partial\theta}\log f(x,\theta)\right]^2 > 0 \quad \forall\, \theta \in \Omega\, \theta$

(c) $E\left[\frac{\partial}{\partial\theta}\log f(x,\theta)\right]^2 = 0 \quad \forall\, \theta \in \Omega\, \theta$

(d) None of the above.

Q.56 For C.R. inequality to hold good, the function $\int t_n(x)\prod_{i}^{n} f(x_i,\theta)\, dx$ is :

(a) Differentiable under integral sign.

(b) Not differentiable under integral sign

(c) May or may not be differentiable under integral sign.

(d) None of the above.

Q.57 In case of Poisson distribution $f(x, \lambda)$, a complete and sufficient statistic for parameter is given by $t_n(x)$ defined as :

(a) $t_n(x) = \sum_{i=1}^{n} x_i$ (b) $t_n(x) = \sum_{i=1}^{n} x_i^2$

(c) $t_n(x) = \sum_{i=1}^{n} \frac{1}{x_i}$ (d) $t_n(x) = \sum_{i=1}^{n} \frac{1}{x^2}$

Q.58 In case of Poisson distributio $f(x, \theta)$, the M.V.U.E. of $e^{-\lambda}$ is given by a statistic $t_n(x)$ defined as :

(a) $t_n(x) = \left(\frac{n-1}{n}\right)^{\sum_{i}^{n} x_i}$ (b) $t_n(x) = \left(\frac{n}{n-1}\right)^{\sum^{n} x_i}$

(c) $t_n(x) = \left(\frac{n-1}{n}\right)^{1/\sum_i^n x_i}$ (d) $t_n(x) = \left(\frac{n}{n-1}\right)^{\sum^n x_i}$

Q.59 Let $t_n(x) = \left(\frac{n-1}{n}\right)^{\sum^n x_i}$ is the M.V.U.E of $e^{-\lambda}$ in case of Poisson distribution $f(x, \lambda)$ the variance $(t_n(x))$ is equal to :

(a) $e^{-2\lambda}\left(e^{\lambda/n} - 1\right)$ (b) $e^{2\lambda}\left(e^{\lambda/n} - 1\right)$

(c) $e^{-2\lambda}\left(e^{\lambda/n} + 1\right)$ (d) $e^{2\lambda}\left(e^{\lambda/n} + 1\right)$

Q.60 In case of Q(58), the C.R. lower bound for $t_n(x) = \left(\frac{n-1}{n}\right)^{\sum^n x_i}$ is equal to :

(a) $e^{-2\lambda}\left(\frac{n}{\lambda}\right)$ (b) $e^{2\lambda}\left(\frac{n}{\lambda}\right)$

(c) $e^{-\lambda}\left(\frac{n}{\lambda}\right)$ (d) $e^{-2\lambda}\left(\frac{\lambda}{n}\right)$

Q.61 If an unbiased estimator $t_n(x)$ for parameter θ has C.R. lower bound as its "sampling variance", then t_n must be a :

(a) Sufficient statistic for θ (b) Not sufficient statistic for θ

(c) Unbiased statistic for θ. (d) Both (a) & (c)

Q.62 The C.R. lower bound attains equality if :

(a) $\frac{\partial}{\partial\theta} \log \prod_i^n f(x_i, \theta) = k(\theta, n)\left(t_n - \psi(\theta)\right)$

(b) $\frac{\partial}{\partial\theta} \log \prod_i^n f(x, \theta) > k.(\theta, n)\left(t_n - \psi(\theta)\right)$

(c) $\frac{\partial}{\partial\theta} \log f(x, \theta) = k.(\theta, n)\,(t_n - \Psi(\theta))$

(d) Both (a) & (c)

where t_n is an unbiased estimator for $\psi(\theta)$.

Q.63 If there exists an estimator t_n for parametric function $\psi(\theta)$ such that the condition "d" exists in Q(62), then it is :

(a) Sufficient statistic for $\psi(\theta)$

(b) An U.M.V.U.E. for $\psi(\theta)$

(c) Both (a) & (b)

(d) None of the above.

Q.64 In Q.(62), the sampling variance of the estimator t_n is equal to :

(a) $\dfrac{\psi'(\theta)}{k(\theta,n)}$ (b) $\dfrac{k.(\theta, n)}{\Psi'(\theta)}$

(c) $\dfrac{\frac{\partial}{\partial\theta}\psi(\theta)}{k(\theta, n)}$ (d) Both (a) & (c)

Q.65 Cramer, Rao inequality gives lower bound to the sampling variance of :

(a) An unbiased estimator of θ.

(b) Sufficient statistic for θ.

(c) U.M.V.U.E. for θ.

(d) None of the above.

Q.66 The information I_θ on parameter θ is provided by the expression :

(a) $I_\theta = V\left(\frac{1}{L}\frac{\partial L}{\partial\theta}\right)$ (b) $I_\theta = n\,E\left(\frac{\partial}{\partial\theta}\log f\right)^2$

(c) $I_\theta = \frac{n}{2}\,E\left(\frac{\partial}{\partial\theta}\log f\right)^2$ (d) Both (a) & (b)

Q.67 Let t_n be an unbiased estimator of parametric function $\psi(\theta)$. then following retation is true.

(a) $E(t_n - \theta)^2 > [\theta - \psi(\theta)]^2 + \dfrac{(\psi'(\theta))^2}{I\theta}$

(b) $E(t_n - \theta)^2 \geq [\theta - \psi(\theta)] + \dfrac{\psi'(\theta)}{I_\theta}$

(c) $E(t_n - \theta)^2 \leq [\theta - \psi(\theta)]^2 + \dfrac{(\psi'(\theta))^2}{I_\theta}$

(d) $E(t_n - \theta)^2 \geq [\theta - \psi(\theta)]^2 + \dfrac{\psi'(\theta)}{I_\theta^2}$

Q.68 Let $\psi(\theta) = \theta + b(\theta)$ and $E(t_n) = \psi(\theta)$, then following relation is true.

(a) $V(t_n) \geq \dfrac{1+b'(\theta)}{I_\theta}$ (b) $V(t_n) \geq \dfrac{[1+b'(\theta)]^2}{I_\theta}$

(c) $V(t_n) \leq \dfrac{[1+b'(\theta)]}{I_\theta}$ (d) None.

Q.69 Let E $[\hat{\theta} - \theta]^2$ denotes the mean squared deviation of an estimator θ of parameter θ and $b(\theta)$ denotes the bias of θ then following relation is true :

(a) $E(\theta - \theta)^2 \geq \dfrac{1+b'(\theta)}{I_\theta} + (b(\theta))^2$

(b) $E(\theta - \theta)^2 \geq \dfrac{(1+b'(\theta))^2}{I_\theta} + (b(\theta))^2$

(c) $E(\theta - \theta)^2 \leq \dfrac{1+b'(\theta)}{I_\theta} + (b(\theta))^2$

(d) $E(\theta - \theta)^2 \leq \dfrac{(1+b'(\theta))^2}{I_\theta} + (b(\theta))^2$

Q.70 If $\dfrac{\partial}{\partial\theta} \log L = \dfrac{1}{\lambda}(t_n - \psi(\theta))$, then C.R. inequality states that t_n is:

(a) M.V.U.E. of $\psi(\theta)$ (b) M.V.U.E. of (θ)

(c) Not M.V.U.E. of $\psi(\theta)$ (d) M.L..E. of $\psi(\theta)$

Q.71 If $\dfrac{\partial}{\partial\theta} \log L = \dfrac{1}{\lambda}(t_n - \psi(\theta))$, then C.R. inequality states that :

(a) $V(t_n) \geq \lambda$ (b) $V(t_n) = \lambda$

(c) $V(t_n) \geq \lambda^2$ (d) $V(t_n) = \dfrac{\lambda}{n}$

Q.72 Let $x \sim N(\mu, \sigma^2)$, then C.R. inequality states that M.V.U.E. of μ is given by the statistic t_n given as :

(a) $t_n = \sum_{i}^{n} x_i$ (b) $t_n = \dfrac{1}{n}\sum_{i}^{n} x_i$

(c) t_n = Sample mean $\overline{x}_n$ (d) Both (b) & (c)

Q.73 Let $x \sim N(\mu, \sigma^2)$ where σ^2 is known. Then C.R. inequality states that the minimum variance of M.V.U.E. say t_n is equal to.

(a) $\dfrac{\sigma^2}{n}$ (b) $\left(\dfrac{\sigma}{\sqrt{n}}\right)^2$

(c) Both (a) & (b) (d) None of the above.

Q.74 Consider Poisson distribution $P(x, \lambda)$. Then M.V.U.E. of parameter λ is given by a statistic t_n defined as :

(a) t_n = sample median x (b) $t = x_n$

(c) $t_n = \dfrac{1}{n}\sum_{i=1}^{n} x_i$ (d) Both (b) & (c)

Q.75 In Poisson distribution, the sampling variance of M.V.U.E. of λ is given by $V(\hat{\lambda})$ as :

(a) $V(\lambda) = \dfrac{\lambda^2}{n}$ (b) $V(\lambda) = \dfrac{\lambda}{n}$

(c) $V(\lambda) = \left(\sqrt{\dfrac{\lambda}{n}}\right)^2$ (d) Both (b) & (c)

Q.75 In case of Poisson distribution $P(x, \lambda)$, the following result is true :

(a) $\dfrac{\partial}{\partial\lambda}\log L = \dfrac{n}{\lambda}(\overline{x}_n - \lambda)$ (b) $\dfrac{\partial}{\partial\lambda}\log L = \dfrac{\lambda}{n}(\overline{x}_n - \lambda)$

(c) $\frac{\partial}{\partial\lambda}\log L = \frac{n^2}{\lambda}(\bar{x}_n - \lambda)$ (d) $\frac{\partial}{\partial\lambda}\log L = \frac{n}{\sqrt{\lambda}}(\bar{x}_n - \lambda)$

where L denotes the likelihood function of sample values.

Q.76 Let x be a r.v.. following b (x, n, p) and L be the likelihood function of sample values then following result is true.

(a) $\frac{\partial}{\partial p}\log L = \frac{\left(\frac{r}{n} - p\right)}{pq/n}$ (b) $\frac{\partial}{\partial p}\log L = \frac{\left(\frac{r}{n} - p\right)}{p(1-p)/n}$

(c) $\frac{\partial}{\partial p}\log L = \frac{\left(\frac{r}{n} - q\right)}{pq/n}$ (d) Both (a) & (b)

Q.77 Let $x \sim b(x, n, p)$. Then M.V.U.E. of p is given by a statistic t_n defined as :

(a) $t_n = \frac{n}{r}$ (b) $t_n = \frac{r}{n}$

(c) $t_n = \frac{\sum^{n} x_i}{n}$ (d) Both (b) & (c)

Q.78 In case of b (x, n, p), the sampling variance of M.V.U.E. say t_n of parameter 'p' is given as :

(a) $V(t_n) = \frac{n}{pq}$ (b) $\frac{pq}{n}$

(c) Both (b) & (d) (d) $V(t_n) = \frac{p(1-p)}{n}$

Q.79 Let $X \sim N(\mu, \sigma^2)$ and $s^{*2} = \frac{1}{n}\sum^{n}(x_i - \mu)^2$ and μ is known. Then the following result is true.

(a) $\frac{\partial}{\partial\sigma^2}\log L = \frac{n}{2\sigma^4}\left(s^{*2} - \sigma^2\right)$

(b) $\frac{\partial}{\partial \sigma^2} \log L = \frac{n}{2\sigma^2}\left(s^{*2} - \mu^2\right)$

(c) $\frac{\partial}{\partial \sigma^2} \log L = \frac{n}{2\sigma^2}\left(s^{*2} - \sigma^2\right)$

(d) None of the above.

Q.80 Let x~ $N(\mu, \sigma^2)$. Where μ is known. Let $s^{*2} = \frac{1}{n}\sum_{i}^{n}(x_i - \mu)^2$. Then minimum variance (s^{*2}) is equal to :

(a) Min $V\left(s^{*2}\right) = \frac{2\sigma^4}{n}$ (b) Min $V\left(s^{*2}\right) = \frac{2\sigma^2}{n}$

(c) Min $V\left(s^{*2}\right) = \left(\frac{\sqrt{2}}{\sqrt{n}}\sigma\right)^2$ (d) Both (b) & (c)

Q.81 Let x~ $N(\mu, \sigma^2)$ where m is known. Then M.V.U.E. of population variance σ^2 is given by a statistic t_n defined as :

(a) $t_n = \frac{1}{n-1}\sum_{i}^{n}(x_i - \mu)^2$ (b) $t_n = \frac{1}{n}\sum_{i}^{n}(x_i - \mu)^2$

(c) $t_n = \frac{1}{n}\sum_{i}^{n}(x_i - \overline{x}_n)^2$ (d) None of the above.

Q.82 Consider the p.d.f. of type III population given as :

$$P(x, p, \theta) = \frac{1}{\overline{|p}\,\theta p},\ x^{p-1} e^{-\frac{x}{\theta}}\ p > 0, 0 < x < \infty$$

Then M.V.B.E. of parameter θ for given p is given by a statistic t_n defined as :

(a) $t_n = \frac{\overline{x}_n}{p}$ (b) $t_n = \frac{p}{\overline{x}_n}$

(c) $t_n = \left(\sqrt{\frac{x_n}{p}}\right)^2$ (d) Both (a) & (c)

Q.83 In Q.(82), the sampling variance of $t_n = \frac{\overline{x}_n}{p}$ is equal to :

(a) $\frac{\theta}{np}$ (b) $\frac{\theta^2}{np}$

(c) $\left(\frac{\theta}{\sqrt{np}}\right)^2$ (d) Both (b & (c)

1.3.2 Minimum Variance Bound Estimators

Q.84 A condition for the existance of M.V.B.E. for $\phi(\theta)$ is written as :

(a) $\frac{\partial}{\partial\theta} \log L = A(\theta)\left(t_n - \phi(\theta)\right)$

(b) $\frac{\partial}{\partial\theta} \log L = A(\theta)\left(t_n + \phi(\theta)\right)$

(c) $\frac{\partial}{\partial\theta} \log L = A(\theta)\left(t_n - \phi(\theta)\right)^2$

(d) None.

Where A(θ) may be the function of θ but is independent of sample values (x).

Q.85 The least attainable variance of an estimator say, t_n may be greater than that of :

(a) M.V.B.E. of parameter $\phi(\theta)$

(b) M.V.U.E. of parameter $\phi(\theta)$.

(c) Sufficient estimator of parameter $\phi(\theta)$.

(d) Both (a) & (b)

Q.86 If the regularity condition for the existence of M.V.B.E. do not exist for parameter $\phi(\theta)$ then least attainable variance of any estimator may be less than that of :

(a) M.V.B.E. (b) M.V.U.E.

(c) Both (a) & (b) (d) None.

Q.87 In binomial distribution $b(x, n, \text{p})$ an unbiased estimator of pq is given by statistic t_n defined as :

(a) $t_n = \dfrac{r}{n}$ (b) $t_n = \dfrac{r}{n}\dfrac{(n-r)}{(n-1)}$

(c) $t_n = \dfrac{n}{r}\dfrac{(n-r)}{(n-1)}$ (d) $t_n = \dfrac{r}{n}\dfrac{(n-1)}{(n-r)}$

Q.88 In case of normal distribution $N(x, \mu, \sigma^2)$, the M.V.B.E. of σ^2 exists when :

(a) μ is unknown

(b) μ is known

(c) Sufficient statistic of μ is known

(d) None

Q.89 Consider the logistic curve defined as :

$$f(x, \theta) = e\,\frac{-(x-\theta)}{1+e^{-(x-\theta)}}$$

Then exact sampling variance of sample mean $\bar{x}_n$ is equal to :

(a) $\dfrac{\pi}{3n}$ (b) $\dfrac{2\pi}{3n}$

(c) $\dfrac{\pi}{2n}$ (d) $\dfrac{3}{2}\dfrac{\pi}{n}$

Q.90 In Q.(89), the exact sampling variance of the sample median x is given as :

(a) $V(x) = \dfrac{n}{4}$ (b) $V(x) = \dfrac{4}{n}$

(c) $V(x) = \dfrac{n}{2}$ (d) $V(x) = \dfrac{2n}{3}$

Q.91 Consider normal distribution $N(x, \mu, \sigma^2)$. Where σ^2 is known. then M.V.B.E. of μ^2 is given by ϕ defined as :

(a) $\phi = \left(\bar{x}^2 + \dfrac{\sigma^2}{n}\right)$ (b) $\phi = \left(\bar{x}^2 - \dfrac{\sigma^2}{n}\right)$

(c) $\phi = \left(\dfrac{n\,\overline{x}^2 - \sigma^2}{n}\right)$ (d) Both (b) & (c)

Q.92 In Q.(91), the sufficient statistic t_n for μ^2 is given as :

(a) $t_n = sample\ median\ x$ (b) $t_n = sample\ median\ x_n$

(c) $t_n = \dfrac{\overline{x}_n}{2}$ (d) $t_n = \dfrac{x}{2}$

Q.93 Consider normal distribution $N(\mu, \sigma^2)$ and a statistic t is defined as :

$t = \sum_{i-1}^{n} a_i\ x_i$, where ai's are unknown constants.

Then t will be an unbiased estimator for μ if.

(a) $\sum_{i-1}^{n} a_i = 1.0$ (b) $\sum_{i-1}^{n} a_i^2 = 1.0$

(c) $\sum_{i-1}^{n} a_i \neq 1.0$ (d) $\sum_{i-1}^{n} a_i^2 \leq 1.0$

Q.94 A M.V.B.E. for parametric function $\psi(\theta)$ exists only if a :

(a) Sufficient statistics exists for $\psi(\theta)$

(b) Consistent estimator exists for $\psi(\theta)$

(c) Unbiased estimator exists for $\psi(\theta)$

(d) Biased estimator exists for $\psi(\theta)$

Q.95 Let t be sufficient statistic for parameter θ. Then one to one function of t, will be :

(a) An unbiased estimator (b) A sufficient statistic

(c) Not a sufficient statistic (d) Efficient estimator

Q.96 If a M.V.B.E. exists for parameter θ, then it is given by the :

(a) Method of moment

(b) Method of M.L.E.

(c) Method of Scoring

(d) Method of factorisation

Q.97 Consider the distribution defined as :

$$f(x, \theta) = \frac{1}{\Gamma(p)} (x - \theta)^{p-1} e^{-(x - \theta)}, \quad 0 \leq x < \infty, p > 2$$

Then MVBE of parameter θ will by given t as

p > 2

(a) $t = \frac{p-1}{2}$ (b) $t = \frac{p-2}{2}$

(c) $t = \frac{p-1}{3}$ (d) $t = \frac{(p-1)^2}{2}$

Q.98 Consider the geometric distribution :

$f(x, \theta) = \theta.(1 - \theta)^{x-1}, \ 0 < \theta < 1$

Then an unbiased estimator t for $\frac{1}{\theta}$ will be given as :

(a) t = sample mean $\overline{x}_n$ (b) t = sample median $\tilde{x}$

(c) t = sample mode (d) none

Q.99 If no zero frequency exists for Poisson distribution, then the distribution is know as :

(a) Non truncated Poisson distribution

(b) Truncated Poisson distribution

(c) Disproportionate Poisson distribution

(d) None.

Q.100 In case of truncated Poisson distribution $\frac{e^{-\lambda} \lambda^x}{x!}$, consider a sample ($x$ = 1, 2, 3, ...), Then only an unbiased estimate of $(1 - e^{-\lambda})$ tanes the value zero if :

(a) x is odd (b) x is even

(c) x is equal to $\frac{n}{2}$ (d) None

Q.101 In case of Q.(100), when x is even then unbiased estimate of $(1 - e^{-\lambda})$ is equal to :

(a) 2.0 (b) 1.0

(c) 3.0 (d) 4.0

Q.102 In case of Poisson distribution $P(x, \lambda)$, the parameter λ can not be estimated through :

(a) Method of m.l.e. (b) Method of moment

(c) C.R. inequality (d) All the three

Q.103 The C.R. inequality provides a means for judging whether or not a given unbiased estimator is also a :

(a) M.V.U.E. (b) Consistent estimator

(c) Sufficient estimator (d) m.l.e.

1.3.3. Blackwellization and M.V.U.E.

Q.104 Rao-Blackwell theorem enables us to obtain a M.V.U.E for parameter θ, from any other unbiased estimator by using a :

(a) Sufficient estimator for θ (b) Unbiased estimatro for θ

(c) Consistent estimator for θ (d) Biased estimator for θ

Q.105 To apply Rao-Blackwell theorem, the only condition to be satisfied is that the "sufficient statistic $t_n(x)$ should have a :

(a) Continuous sampling distribution

(b) Discrete sampling distribution

(c) Complete family of distribution

(d) Monotonicity property

Q.106 The implication of "Rao-Blackwell theorem" is that if we are given an unbiased estimator 'U' then we can improve upon 'U' by considering a new statistic $\phi(t)$ based upon 'U' and a :

(a) Sufficient statistic t (b) Efficient statistic t

(c) Consistent statistic t (d) M.L.E. t

Q.107 The process of obtaining an improved estimator (from unbiasedness) and (minimum variance point of view) starting from an unbiased estimator, is known as :

(a) Unbiasedness property (b) Consistency property

(c) Factorisation theorem (d) Blackwellisation

Q.108 In case pf Rao-Blackwell theorem, the new estimator ϕ(t) will not only be a better estimator than 'U' but will also be "Best estimator" if 't' has :

(a) A complete family of distribution

(b) Continuous sampling distribution

(c) Discrete sampling distribution

(d) Unbiasedness property

Q.109 C.R. inequality method is too much restrictive due to :
(a) Unbiasedness property (b) Consistency property
(c) Regularity conditions to be satisfied
(d) None

Q.110 If an estimator t_n is not M.V.B. estimator, it does not mean that it is also not a :
(a) M.V.U. estimator (b) Consistent estimator
(c) Biased estimator (d) M.L. estimator

Q.111 Let (x, y) be two random variables such that $E(y) = \mu$, and $V(y) = \sigma_y^2$. Let $E(y/x) = \phi(x)$, then Rao-Blackwell theorem states that :
(a) $E(\phi(x)) = \mu$ and $V(\phi(x)) \leq \sigma_y^2$
(b) $E(\phi(x)) = \mu$ and $V(\phi(x)) \geq \sigma_y^2$
(c) $E(\phi(x)) = \mu$ and $V(\phi(x)) > \sigma_y^2$
(d) $E(\phi(x)) \neq \mu$ and $V(\phi(x)) \geq \sigma_y^2$

Q.112 The statistic $\phi(x) = E(y/x)$ defined in Q.(111) is :
(a) Not an unbiased estimator for μ
(b) An unbiased estimator for μ
(c) A consistent estimator for μ
(d) A sufficient statistic for μ

Q.113 In Q.(111), the variable ϕ(x) for parameter μ has been treated as :
(a) An unbiased estimator for μ (b) A sufficient statistic for μ
(c) As a M.V.U.E. for μ (d) As a M.L.E. for μ

Q.114 If a M.V.B.E. exists for parameter θ, then it is the function of :
(a) A biased estimator (b) Sufficient Statistic
(c) M.L.E. (d) None

Q.115 Let $x \sim b(n, p)$. Then according to Rao-Blackwell theorem, a M.V.U.E. of parameter p is given as :
(a) $t = \frac{n}{r}$ (b) $t = \frac{r}{n}$

(c) $t = \frac{r-1}{n-1}$ (d) $t = \frac{r}{n}\frac{(r-1)}{(n-1)}$

Q.116 Consider Poisson distributiom $P(x, \lambda)$. Then a statistic $t_n = \sum_{i}^{n} x_i$ for parameter λ is a :

(a) Unbiased estimator (b) Sufficient statistic

(c) Consistent estimator (d) Maximum likelihood estimator

Q.117 For Poisson $P(x, \lambda)$, the statistic $t_n = \sum^{n} x_i$ for $e^{-\lambda}$ is a :

(a) Sufficient statistic (b) Consistent estimator

(c) Biased estimator (d) Efficient estimator

Q.118 In case of Poisson distribution $P(x, \lambda)$, the Rao-Blackwell theorem states that M.V.U.E. of $e^{-\lambda}$ is given by a statistic t_n defined as :

(a) $t_n = \left(\frac{n-1}{n}\right)^r$ (b) $t_n = \left(\frac{n}{n-1}\right)^r$

(c) $t_n = \left(\frac{n-2}{n-1}\right)^r$ (d) $t_n = \left(\frac{n}{n-2}\right)^r$

Q.119 In case of Poisson distribution, the sufficient statistic for λ^2 is given as :

(a) $t_n = \left(\frac{\overline{x}^2 - \overline{x}}{n}\right)$ (b) $t_n = \left(\frac{n}{\overline{x}^2 - \overline{x}}\right)$

(c) $t_n = \left(\frac{n}{\overline{x}^2 - \overline{x}}\right)^{-1}$ (d) Both (a) & (c)

Q.120 Consider the distribution $f(x, \theta) = \frac{1}{\theta}.e^{\frac{-x}{\theta}}, 0 < x < \infty$, then C.R. inequality states that an unbiased estimator of θ is given as :

(a) $t_n = \frac{1}{n}\sum_{i}^{n} x_i$ (b) $t_n = \bar{x}_n^2$

(c) $t_n = \bar{x}_n$ (d) Both (a) & (c)

Q.121 In case of p.d.f. as $f(x, \theta) = \frac{1}{\theta}.e^{\frac{-x}{\theta}}$, $0 < x < \infty$, the sampling variance of the sample mean $\bar{x}_n$ is given as :

(a) $V(\bar{x}_n) = \frac{\theta}{n}$ (b) $V(\bar{x}_n) = \frac{\theta^2}{n}$

(c) $V(\bar{x}_n) = \left(\frac{\sqrt{n}}{\theta}\right)^{-2}$ (d) Both (b) & (c)

Q.122 Let 't' be a M.V.U.E. and 'U' be an other unbiased estimator of a parameter θ, we have :

(a) Correlation $(t, U - t) \neq 0$ (b) Correlation $(t, U - t) = 0$

(c) Correlation $(U - t, t) = 0$ (d) Both (b) & (c)

1.3.4 Sufficient Statistics and Completeness of Sufficient Statistics

Q.123 Let $f(x, \theta)$ be the p.d.f. with parameter θ and a statistic $t_n(x)$ gives all relevant information about parameter θ, then $t_n(x)$ is known as :

(a) An unbiased estimator for θ. (b) A sufficient statistic for θ

(c) A consistent statistic for θ (d) A efficient estimator for θ

Q.124 Let $f(x, \theta)$ be the p.d.f. with parameter θ, and $t_n(x)$ is a sufficient statistic for θ. Let $y_2(x)$, $y_3(x)$, ... $y_n(x)$ are other $(n - 1)$ statistics then conditional distribution of $y_2(x), y_3(x), \ldots, y_n(x)$ is :

(a) Dependent of parameter θ (b) Independent of parameter θ

(c) Not dependent of parameter θ (d) Both (b) & (c)

Q.125 Let $t(x)$ be a sufficient statistic with p.d.f. as $g(t, \theta)$. If the range of x is independent of θ, then "Fisher-Neyman Factorization Theorem" states that $t(x)$ is sufficient for θ iff :

(a) $\prod_{i-1}^{n} f(x_i, \theta) = g(t(x), \theta)\, h(x)$

(b) $\prod_{i-1}^{n} f(x_i, \theta) = h(x)\, g(t(x), \theta)$

(c) $\prod_{i-1}^{n} f(x_i, \theta) > g(t(x)) \,.\, h(x)$

(d) Both (a) & (b)

Where $h(x)$ is a function of (x) and is independent of θ.

Q.126 Let (t_1, t_2) be two sufficient statistics for parameter θ. Then there exists :

(a) A function relationship between (t_1, t_2)
(b) No function relation between (t_1, t_2)
(c) Correlation (t_1, t_2)
(d) Both (a) & (c).

Q.127 If no function of a sufficient statistic t(x) has zero expectation unless it is zero itself atmost every where, then such statistic is said to be :

(a) Unbiased (b) Efficient
(c) Complete (d) Consistent

Q.128 Let t(x) be a statistic such that :

$$E(q(t(x)) \Leftrightarrow P[q(t(x)=0] = 1$$

Then the statistic $t(x)$ is :

(a) Sufficient (b) Complete
(c) Consistent (d) Both (a) & (b)

1.3.5 Distribution Possessing Sufficient Statistics

Q.129 Let us consider the following precise form of exponential distribution :

$f(x, \theta) = e^{A(\theta)\, B(x) + D(\theta) + C(x)}$

Then it is the condition for the existence of M.V.B. estimator for some parametric function $\phi(\theta)$ in case of range of variate x is :

(a) Independent of θ (b) Dependent of θ
(c) Not dependent of θ (d) Both (a) & (c)

Q.130 If the range of variate x is independent of θ then :

$$\frac{\partial}{\partial\theta} \log f(x, \theta) = p(\theta)\, k(x) + \frac{q(\theta)}{n}$$

is the sufficient condition for the existence of :

(a) Sufficient statistic for parameter θ
(b) Not sufficient statistic for θ

(c) Consistent statistic for θ

(d) Efficient statistic for θ.

Q.131 Let the lower terminal of the range of r.v.x. depends on parameter θ and p.d.f. is of the form :

$f(x, \theta),\ a(\theta) \leq x \leq b$

and $f(x, \theta)$ can be factorised as

$f(x, \theta) = g(x)/h(\theta)$. ...(A)

Then sufficient statistic for θ is the :

(a) $x_{(1)}$ (b) $x_{(n)}$

(c) $x_{(n)} - x_{(1)}$ (d) $x_{(1)} + x(n)$

Q.132 Consider the following p.d.f.

$f(x, \theta),\ a < x < b(\theta)$

Where $b(\theta)$ is the monotone function in θ and $f(x, \theta)$ can be factorised as :

$f(x, \theta) = g(x) / l(\theta)$ ...(B)

Then sufficient statistic for parameter θ is the :

(a) $x_{(1)}$ (b) $x_{(n)} + x_{(1)}$

(c) $x_{(n)} - x_{(1)}$ (d) $x_{(n)}$

Q.133 Consider the following p.d.f.

$f(x, \theta),\ a(\theta) < x < b(\theta)$

$$L(x/\theta) = \prod_{i-1}^{n} f(x_i/\theta) \,.\, \cup (x_{(1)} - \theta)\ \mathrm{U}\ (\mathrm{b}(\theta) - x_{(n)})$$

and $f(x, \theta) = g(x)/k(\theta)$

Then pair of sufficient statistic for θ is given as :

(a) Extream observations $(x_{(1)}, x_{(n)})$

(b) Extream observations $(x_{(n)}, x_{(1)})$

(c) $\frac{1}{2}(x_{(1)}, x_{(n)})$

(d) Both (a) & (b)

Q.134 Consider a p.d.f. $f(x, \theta),\ a(\theta) < x < b$ where $a(\theta)$ is a monotone function in θ and $f(x, \theta)$ is factorised as :

$f(x, \theta) = g(x)/h(\theta)$.

Then lowest order statistic $x_{(1)}$ is :

(a) Sufficient and efficient statistic for θ
(b) Sufficient and complete statistic for θ
(c) Unbiased and sufficient for θ
(d) Consistent and efficient for θ

Q.135 Consider a p.d.f. $f(x, \theta) = a < x < b\ (\theta)$ where b(θ) is a monotone function in θ and and $f(x, \theta)$ is factorised as $f(x, \theta) = g\ (x) / h\ (\theta)$.

Then order statistic x_n of $x_1 \leq x_2 \leq .... \leq x_n$ is :

(a) Sufficient and unbiased for θ (b) Complete and unbiased for θ
(c) Compete and consistent for θ (d) Complete and sufficient for θ

Q.136 Consider a p.d.f. $f(x, \theta)$, $a(\theta) \leq x \leq b(\theta)$ where $a(\theta)$ and $b(\theta)$ are monotone functions in θ (both increasing or both decreasing). Then pair $(x_{(1)}, x_{(n)})$ are jointly :

(a) Sufficient & complete for θ
(b) Sufficient & efficient for θ
(c) Complete & consistent for θ
(d) Sufficient & consistent for θ

Q.137 Let $f(x, \theta)$ be a p.d.f. of r.v.x. and $a(\theta) \leq x \leq b(\theta)$. If $a(\theta)$ is decreasing and $b(\theta)$ is an increasing function in θ then a sufficient and complete statistic for θ is given as :

(a) Max. $\{ a^{-1}\ (x_{(1)}), b^{-1}\ (x_{(n)}) \}$
(b) Max. $\{ a\ (x_{(1)}), b\ (x_{(n)}) \}$
(c) Min. $\{ a(x_{(1)}), b(x_{(n)}) \}$
(d) Min. $\{ a^{-1}\ (x_{(1)}), b^{-1}\ (x_{(n)}) \}$

Q.138 If in Q.(137), $a(\theta)$ is an increasing and $b(\theta)$ is a decreasing function in θ then a sufficient and complete statistics for θ is given as :

(a) Max. $\{ a\ (x_{(1)}), b\ (x_{(n)}) \}$
(b) Max. $\{ a^{-1}\ (x_{(1)}), b^{-1}\ (x_{(n)}) \}$
(c) Min. $\{ a(x_{(1)}), b(x_{(n)}) \}$
(d) Min. $\{ a^{-1}\ (x_{(1)}), b^{-1}\ (x_{(n)}) \}$

1.3.6 Sufficient Statistic for Several Parameters

Q.139 Let us define each observation x_i as a vector avariate, t as s component vector of statistics and θ as k component vector of parameters.

Suppose we can write :

$L\ (x/\theta) = g\ (t/\theta).\ h(x).$

Then the components of 't' will be a set of jointly

(a) Consistent estimator for θ. (b) Sufficient ststistics for θ

(c) Efficient estimation for θ (d) B.L.UE's for θ

Q.140 In Q.(139), we may have s as s ≤ *k* or s ≥ If = 1, then we may call '*t*' as :

(a) Single sufficient statistic for θ

(b) Joint sufficient statistics for θ

(c) Joint unbiased estimator for θ

(d) Joint consistent estimator for θ

Q.141 In Q.(139), if we have t = x, we say that the sample $x = (x_1, x_2, \ldots\ldots\ldots, x_n)$ itselves always constitute a set of :

(a) Sufficient statistics for θ with $ = *n*.

(b) Sufficient statistics for θ with $ = 1

(c) Consistent estimators for θ with $ = *n*.

(d) Efficient estimators for θ with $ = *n*

Q.142 Let x~ $N(\mu, \sigma^2)$ where μ is known. Then the statistic $\$ = \frac{1}{n}\sum^{n}(x_i - \bar{x})^2$ is :

(a) Alone a sufficient statistic for σ^2.

(b) Alone not a sufficient statistic for σ^2.

(c) Alone a consistent estimator for σ^2

(d) Alone an efficient estimator for σ^2

Q.143 If a single sufficient statistic 't' exists for θ, then MLE say θ of θ is :

(a) A function of *t* alone

(b) A function of (t & θ)

(c) A function of t^2 alone

(d) Not a function of *t* alone.

Q.144 If likelihood function $L = \sum_{i}^{n} f(x_i, \theta)$ has unique maxima then :

(a) M.L.E. θ is not unique for θ.

(b) M.L.E. θ is unique for θ.

(c) M.L.E. θ is unbiased for θ.

(d) M.L.E. θ is biased for θ.

Q.145 All the optimum properties of a single sufficient statistic are conferred upon :

(a) Efficient estimators (b) M.L. Estimators

(c) Unbiased estimators (d) Biased estimators

Q.146 If a sufficient statistic 't' for parameter θ is the function of every other sufficient statistic, then 't' is said to be :

(a) Not a sufficient statistic for θ

(b) Minimal sufficient statistic for θ

(c) M.L. estimator for θ

(d) Consistent estimator for θ.

Q.147 If the data represented by a sample x = (x1, x2, ..., xn) can not be condenced beyond a statistict(x). without sacrificing sufficiency then the statistic t(x) is said to be :

(a) A sufficient statistic

(b) A consistent estimator

(c) A minimal sufficient statistic

(d) A.M.L. estimator

Q.148 A statistic $t(x)$ is said to be complete for any measureable function $\psi(t)$ if:

$E_\theta (\psi(t)) = 0 \ \forall \ \theta \in \Omega \ \theta$

implies that :

(a) $\psi(t) = 0$, almost every where $\forall$ θ.

(b) $\psi(t) \neq 0$, almost every where $\forall$ θ.

(c) $\psi(t) \geq 1.0$, almost every where $\forall$ θ.

(d) $\psi(t)$ is an unbiased estimator.

Q.149 In Q.(148), the concept of completeness utilises both the concepts of :

(a) Point estimation & confidence interval

(b) Point estimation & testing of hypothesis

(c) Point estimation & unbiasedness

(d) Consistency and sufficiency

Q.150 The family $\{ P_\theta (t) / \theta \in \Omega \ \theta \}$ of distributions is said to be complete if for an measureable function $\phi(t)$,

$$\int_{\theta} \phi(t).\ P_{\theta}\big(t(x)\big)d\theta = 0 \ \forall\ \theta \in \Omega\ \theta$$

for which it implies that $\phi(t(x)) = 0\ \forall\ \theta \in \Omega\ \theta$ is said to be :

(a) Incomplete family of distribution (b) Complete family of distribution

(c) Monotone family of distribution (d) Efficient family of distribution

Q.151 A sample statistic $t(x)$ is said to be boundedly complete if there exists a measurable function $\phi(t)$ such that :

$|\ \phi(t)\ | < M$ for some M

and $E_{\theta}\ (\phi\ (t) = 0\ \ \forall\ \theta \in \Omega\ \theta$

which implies that :

(a) $\phi(t) \neq 0,\ \forall\ \theta \in \Omega\ \theta$ almost every where

(b) $\phi(t) = 0,\ \forall\ \theta \in \Omega\ \theta$ almost every where

(c) $\phi(t) = 1,\ \forall\ \theta \in \Omega\ \theta$ almost every where

(d) $\phi(t) \geq 1,\ \forall\ \theta \in \Omega\ \theta$ almost every where

Q.152 A family $\{P(t, \theta), \theta \in \Omega\ \theta\}$ of distribution is said to be boundedly complete, if there exists a measureable function $\phi(t)$ of any statistic $t(x)$ such that :

$|\ \phi(t)\ | < M$ for some M.

and

$$\int_{\theta} \phi(t).\ P\big(t, \theta)\big)d\theta = 0\ \forall\ \theta \in \Omega\ \theta$$

and it implies that $\phi(t) = 0\ \forall\ \theta \in \Omega\ \theta$ then statistic $t(x)$ is said to be :

(a) A sufficient statistic

(b) An incomplete sufficient statistic

(c) A consistent statistic

(d) An efficient statistic

Q.153 If $t(x)$ is a complete statistic for parameter θ, then it is necessarily :

(a) A bounded complete statsitic

(b) Not a bounded complete statistic

(c) A maximum likelihood estimator

(d) An efficient statsitic.

Q.154 Let (t_1, t_2) be jointly sufficient statistics for parameters (θ_1, θ_2) respectively and further let t1 is sufficient for θ_1 when θ_2 is known. Then conditional distribution $f(t_2/t_1)$ is

(a) Dependent of θ_1 (b) Independent of θ_1

(c) Not dependent of θ_1 (d) Both (b) & (c)

Q.155 A sufficient statistic t(x) for parameter θ is most efficient estimator of most efficient estimator :

(a) Does not exist (b) Exists

(c) Is obtainable (d) Both (b) & (c)

Q.156 The method of minimum variance will provide a solution for θ only if a :

(a) Sufficient statistic $t(x)$ exists for θ.

(b) Sufficient statistic $t(x)$ does not exist for θ.

(c) Consistent estimator exists for θ.

(d) Efficient estimator exists for θ.

Q.157 If $(t_1. t_2. \ldots\ldots\ldots., t_r)$ is the set of jointly sufficient statistics for parameter θ, then any set of one to one function or transformation of $(t_1, t_2, \ldots\ldots\ldots., t_r)$ is

(a) Not jointly sufficient statistic for θ.

(b) Jointly sufficient statistic for θ.

(c) Jointly consistent statistic for θ.

(d) Jointly efficient statistic for θ.

Q.158 Let $(X_1, X_2, \ldots\ldots\ldots., X_n)$ be i.i.d. r. v's from a continuous distribution $f(x, \theta)$. Then the set of ordered sample $(x_{(1)} < x_{(2)} < \ldots\ldots\ldots < x_n)$ is

(a) Not a sufficient statistic for θ.

(b) A sufficient statistic for θ.

(c) A consistent statistic for θ.

(d) An efficient statistic for θ.

Q.159 The original sample $(x_1, x, \ldots.., x_n)$ drawn from a population $f(x, \theta)$ is always :

(a) A sufficient statistic for θ.

(b) Not a sufficient statistic for θ.

(c) Not an unbiased statistic for θ.

(d) Both (a) & (c)

Q.160 Let $f(x, \theta)$ belongs to one parametric exponential family of distribution.

Let $t(x_i)$ be a sufficient statistic for θ. Then the statistic $t(x) = \sum_i^n t(x_i)$

is

(a) Complete and minimal sufficient statistic for θ.

(b) Not a complete and minimal sufficient statistic for θ.

(c) Complete sufficient statistic for θ.

(d) Both (a) & (c).

Q.161 Let $f(x, \theta_1, \theta_2, ... \theta_k)$ be the regular case of k parametric exponential family of distribution. Then the complete and minimal sufficient statistics for $(\theta_1, \theta_2, ... \theta k)$ respectively are given as :

(a) $\sum_i^n t_1(x_i), \sum_i^n t_2(x_i), ..., \sum_i^n t_k(x_i)$

(b) $\sum^n t_1(x_i^2), \sum_i^n t_2(x_i^2), ..., \sum^n t_k(x_i^2)$

(c) $\sum_i^n t_1(x_i^3), \sum^n t_2(x_i^3), ..., \sum_i^n t_k(x_i^3)$

(d) $\sum_i^n t_1(e^{x_i}), \sum^n t_2(e^{x_i}), ..., \sum_i^n t_k(e^{x_i})$

Q.162 Bounded completeness implies the :

(a) Efficiency (b) Consistency

(c) Sufficiency (d) Minimal sufficiency

Q.163 Minimal sufficiency of a statistic $t(x)$ for a parameter θ.

(a) Implies the bounded completeness of $t(x)$

(b) Does not emply the bounded completeness of t(x)

(c) Implies consistency of $t(x)$

(d) Implies efficiency of $t(x)$

Q.164 Let $x_1, x_2, ...,x_n$ be i i d. r.v's with p.d.f. f(x, θ). If $S = S(x)$ be a complete and sufficient statistic for θ and let $t = t(s)$ and is an unbiased estimator of $\psi(\theta)$ then $t(s)$ is

(a) U.M.V.U.E. of $\psi(\theta)$

(b) A consistent estimator of $\psi(\theta)$

(c) An unbiased estimator of $\psi(\theta)$.

(d) Both (a) & (c).

Q.165 Let $(x_i, y_i, i = 1, 2, \ldots\ldots, n)$ be independently normaly distributed with same mean μ and variances (σ_1^2, σ_2^2) respectively. Then the minimal and sufficient statistic for $(\mu, \sigma_1^2, \mu, \sigma_2^2)$ respectively are given as :

(a) $(\bar{x}, \bar{y}, s_x^2, s_y^2)$

(b) $(\bar{x}, \bar{y}, s_y^2, s_x^2)$

(c) $(\bar{x}, s_x^2, \bar{y}, s_x^2)$

(d) $(\bar{x}, s_y^2, \bar{y}, s_x^2)$

Q.166 Let (x_1, x_2) be two r,v's distributed as $N(\theta, 1)$. Let a new variable y be defined as $y = (x_1 + x_2)$. Then y is

(a) Unbiased estimator for θ.

(b) M.L.E. for θ.

(c) Efficient statistic for θ.

(d) Sufficienct statistic for θ.

Q.167 Let $x \sim N(\mu, \sigma^2)$. Then sample $\bar{x}$ is

(a) Sufficient for μ if σ^2 is known.

(b) Sufficient for μ for σ^2 is unknown.

(c) Unbiased for μ.

(d) Both (a) & (c)

Q.168 Let $x \sim N(\mu, \sigma^2)$. Then sample variance $s^2 = \frac{1}{n}\sum^{n}(x_i - \bar{x})^2$ is

(a) Sufficient statistic for σ^2.

(b) Not sufficient statistic for σ^2.

(c) $s^{*2} = \frac{1}{n}\sum^{n}(x_i - \mu)^2$ is sufficient for σ^2.

(d) Both (b) & (c).

Q.169 Let $x \sim N(0, \theta)$. Then an unbiased statistic for parameter θ is given by $t(x)$ defined as :

(a) $t(x) = \dfrac{\sum^{n} x_i}{n}$

(b) $t(x) = \dfrac{\sum^{n} x_i^2}{n}$

(c) $t(x) = \dfrac{1}{n-1} \sum^{n} x_i^2$

(d) None.

Q.170 Let x~ N(0, θ) and $t(x) = \sum^{n} x_i^2$ is an unbiased estimator for θ. Then sampling variance of $t(x)$ is given as :

(a) $V(t(x)) = \dfrac{2\theta^2}{n}$

(b) $V(t(x)) = \dfrac{\theta^2}{n}$

(c) $V(t(x)) = \dfrac{3\theta^2}{n}$

(d) $V(t(x)) = \dfrac{3}{2\,n}\theta^2$

Q.171 Let x~ N(0, θ). Then a statistic $t(x) = \sum^{n} x_i^2$ is

(a) Unbiased estimator for θ.
(b) Sufficient estimator for θ.
(c) Efficient estimator for θ.
(d) M.V.U.E. for θ.

Q.172 Consider the p.d.f. $f(x, \theta)$ defined as $f(x, \theta) = \theta\,(1-\theta)^x$, $x = 0, 1, 2, \ldots\ldots$, $0 < \theta < 1$.

Then the statistic $t(x) = \sum_{i-1}^{n} x_i$ is

(a) An unbiased estimator for θ.

(b) Efficient estimator for θ.

(c) Sufficient statistic for θ.

(d) None of the above.

Q.173 Consider a random sample $(x_1, x_2, \ldots\ldots, x_n)$ be drawn from an exponential distribution:

$f(x, \theta) = e^{-(x-\theta)}$, $0 < \theta < \infty$

Then a sufficient statistic for parameter θ is given by :

(a) Min. $(x_1, x_2, \ldots\ldots, x_n)$

(b) Max. $(x_1, x_2, \ldots\ldots, xn)$

(c) $\sum_{i-1}^{n} x_i$

(d) $\sum^{n} x_i^2$

1.3.7 Ordered Statistics

Q.174 Let $x_{(1)} < x_{(2)} < \ldots\ldots < x_{(n)}$ be an ordered sample. Then ordered statistic $t(x) = x(r)$ is used in :

(a) Parametric method

(b) Non. parametric method

(c) Finding sufficient statistic

(d) Both (b) & (c)

Q.175 Let a statistic M is defined as :

$$M = \begin{cases} x_{\frac{x+1}{2}} \; if \; n \; is \; odd \\ \dfrac{x_{(n)} + x_{(n+1)}}{2} \; if \; n \; is \; even \end{cases}$$

Then M is known as :

(a) Median of the sample values

(b) Mode of the sample values

(c) G.M. of the sample values

(d) A.M. of the sample values

Q.176 Let R_m be defined as :

$R_m = x(n - m + 1) - x_{(m)}$

Where $x_{(1)} < x_{(2)} < ..... < x_{(n)}$ is an ordered sample. Then R_m is known as :

(a) Range (b) Semi quartile range

(c) Quasi range (d) None of above.

Q.177 Let R_m^* be defined as :

$$R_m^* = \frac{1}{2}[x_{(n-m+1)} + x_{(m)}]$$

Then R_m^* is known as the :

(a) Average value (b) m^{th} mid range

(c) Quasi range (d) Measure of central tendency

Q.178 If R_1^* is defined as :

$$R_1^* = \frac{1}{2}(x_{(n)} + x_{(1)})$$

(a) Mid range (b) Range

(c) Quasi range (d) None

Q.179 If $x_1, x_2, ..., x_n$ are independent r.v's, then order statistics generated from them are :

(a) Independent r.v's (b) Dependent r.v.'s

(c) Not independent r.v. (d) Both (b) a& (c)

Q.180 Let $(x_1, x_2, ..., x_n)$ be i.i.d. r.v's from a continuous distribution. Then the set of ordered statistics say $(y_1 < y_2 < ..... < y_n)$ is both :

(a) Complet and unbiased statistic

(b) Complete and efficient statistic

(c) Complete and consistent statistic

(d) Complete and sufficient statistic

Q.181 Let X be a continuous or discrete r.v. and let $(x_1 < x_2 < ..... < x_n)$ be a set of ordered statistics such that :

$E(x_{(v)}) = \mu_{(r)}$

Then $\mu_{(r)}$ exists provided μ :

(a) Does not exist

(b) Exists

(c) Is not measurable

(d) Is mode value

Q.182 In Q.(181), if μ does not exist then $\mu_{(r)}$:

(a) May exist for all r.

(b) May exist for certain r.

(c) May not exit for all r.

(d) None.

Q.183 Let $y_1 < y_2 < ..... < y_n$ be a set of ordered statistic.

Then joint p.d.f. of (y_n, y_1) is given as :

(a) $f(y_1, y_n) = n(n-1)\ \{F(y_n) - F(y_1)\}^{n-2} f(y_1)\, f(y_n)$

(b) $f(y_1, y_n) = (n-1)n\ \{F(y_n) - F(y_1)\}^{n-2} f(y_n)\, f(y_1)$

(c) $f(y_1, y_n) = n(n-1)\ \{F(y_1) - F(y_n)\}^{n-2} f(y_n)\, f(y_1)$

(d) Both (a) & (b).

Where $f(y_1), f(y_2)$ $F(y_1)$ & $F(y_n)$ have usual meanings in the theory of distribution function.

Q.184 In Q.(183), the p.d.f. of largest ststatic y_n is given as :

(a) $f(y_n) = n\ [F\ (y_n)\ ]^{n-1} f(y_n)$

(b) $f(y_n) = n\, f(y_n)\ [\ F\ (y_n)\ ]$n–1

(c) $f(y_n) = n\, f(y_1)\ [F\ (y_n)]^{n-1}$

(d) Both (a) & (b)

Q.185 Consider the rectangular distribution defined as :

$$f(x, \theta) = \begin{cases} \dfrac{1}{\theta}. & for\ 0 < x < \theta \\ 0, & otherwise \end{cases}$$

Let $x_1 < x_2 < ..... < x_n$ be a set of ordered statistic. Then a statistic $t(x) = x_n$ is :

(a) An unbiased statistic for θ.

(b) An efficient statistic for θ.

(c) A sufficient statistic for θ.

(d) A consistent estimator for θ.

Q.186 In Q.(185), consider a statistic $t(x)$ defined as $t(x) = \left(\dfrac{n+1}{n}\right) y_n$

Then t(x) is

(a) An U.M.V.U.E. for θ.

(b) An unbiased estimator for θ.

(c) An efficient estimator for θ.

(d) All the above.

Q.187 Consider binomial distribution :

$$f(x, n, p) = {}^{n}C_{x}\, p^{x} q^{n-x}, \quad \begin{array}{l} x = 0, 1, 2, \ldots\ldots n \\ 0 < p < 1, 0 < q < 1 \\ p + q = 1 \end{array}$$

Define a statistic t(x) as :

$$t(x) = \sum_{i=1}^{n} x_i \text{, then } t(x) \text{ is a :}$$

(a) Complete statistic for θ.

(b) Sufficient statistic for θ.

(c) Complete & sufficient statistic for θ.

(d) All the above.

Q.188 Consider Poisson distribution :

$$f(x, \lambda) = \frac{e^{-\lambda} \lambda^{x}}{x!}, \text{ x} = 0, 1, 2, \ldots\ldots$$

$$\lambda > 0.$$

Define a statistic $t(x) = \sum^{n} x_i$. Then $t(x)$ is a :

(a) Sufficient statistic for λ.

(b) Complete statistic for λ.

(c) Complete and sufficient statistic for λ.

(d) All the above.

Q.189 Consider negative binomial distribution :

$$f(x, p, r) = \binom{r + x - 1}{x} p^{r} (1-p)^{x}, \quad \text{x} = 0, 1, 2,$$

$$\text{r} > 0$$

$$0 \le \text{p} \le 1$$

Let r be known and define a statistic t as $t = \sum_{i}^{n} x_i$. Then t is

(a) Sufficient statistic for p.
(b) Complete statistic for p.
(c) Complete sufficient statistic for p.
(d) All the three.

Q.190 In Q(189), if a statistic t* is defined as :

$t^* = \dfrac{n-1}{(n-1+t)}$, then t* is :

(a) An unbiased estimator for p.
(b) An efficient estimator for p.
(c) An U.M.V.U.E. for p.
(d) All the above.

Q.191 Consider the uniform discrete distribution defined as :

$f(x, k) = \dfrac{1}{k}, x = a, a + h, a + 2h, \ldots\ldots, a + (k-1)\, h$

Then population mean μ is defined as :

(a) $\mu = a + \dfrac{h\,(k-1)}{2}$ (b) $\mu = \dfrac{h\,(k-1)}{2} + a$

(c) $\mu = a + \dfrac{h}{2} \cdot (k-1).$ (d) All the above

Q.192 In Q(191), the population variance σ^2 is defined as :

(a) $\sigma^2 = \dfrac{h^2 (k^2 - 1)}{12}$ (b) $\sigma^2 = \dfrac{(k^2-1)\, h^2}{12}$

(c) $\sigma^2 = \dfrac{h^2}{12}\,(k+1)\,(k-1)$ (d) All the above.

Q.193 Let $x1 < x2 < \ldots\ldots < xn$ be a ordered sample drawn from the population defined in Q(191), and define a statistic $t(x) = x_n$. Then the statistic $t(x)$ is

(a) A complete statistic for k.
(b) A sufficient statistic for k.
(c) A complete and sufficient statistic for k.
(d) All the above.

Q.194 In case of Q.(191), we define a statistic t_1 as :

$$t_1 = \frac{(x_n^{n+1} - x_{n-1}^{n+1})}{(x_n^n - x_{n-1}^n)}$$

Then t_1 is

(a) Minimum variance unbiased estimater for k.

(b) An unbiased estimator for k.

(c) An efficient estimator for k.

(d) All the above i.e. (U.M.V.U.E.)

Q.195 In Q.(191), we define an other statistic t_2 as $t_2 = (x_n + 1)/2$.

Then for (k + 1)/2, t_2 is

(a) An U.M.V.U.E.

(b) Mininum variance estimator.

(c) An unbiased estimator.

(d) All the above.

Q.196 Consider the hyper geometric distribution defined as :

$$f(x, \theta) = \frac{\binom{\theta}{x}\binom{N-\theta}{n-x}}{\binom{N}{n}}, \quad x = 0, 1, 2, 3, \ldots$$

$$\theta = 1, 2, 3, \ldots\ldots N$$

Then distribution $f(x, \theta)$ is

(a) Sufficient w.r. to θ.

(b) Complete w.r. to θ.

(c) Continuous w.r. to θ.

(d) Truncated w.r. to θ.

Q.197 In Q.(196), the sample $x = (x_1, x_2, \ldots\ldots, x_n)$ is

(a) Complete for θ.

(b) Sufficient for θ.

(c) Complete and sufficient for θ.

(d) All the above.

Q.198 In Q.(196), let $t(x)$. be a complete and sufficient statistic for parameter θ. and we define an other statistic $t^*(x)$ as :

$$t^*(x) = \frac{N}{n}\, t(x).$$

Then t*(x) is

(a) An U.M.V.U.E. for θ.

(b) An unbiased estimator for θ.

(c) Minimum variance estimator for θ.

(d) All the above.

Q.199 Consider the following power series distribution $f(x, \upsilon, \lambda) = \dfrac{a(x) \,.\, \lambda^x}{h\,(\upsilon, \lambda)}$,

$x = \upsilon, \upsilon + 1, \ldots\ldots\ldots\ldots\ldots\ldots$

Consider an ordered sample $x_1 < x_2 < \ldots\ldots\ldots\ldots < x_n$ drawn from $f(x, \upsilon, \lambda)$.
Let us define two statistics as :

$$t_1 = x_1, t_2 = \sum_{i=1}^{n} x_i$$

Then (t_1, t_2) are jointly :

(a) Complete statistics for (υ, λ) respectively.

(b) Sufficient statistics for (υ, λ) respectively.

(c) Complete sufficient for (υ, λ) respectively.

(d) All the above.

Q.200 Consider the normal distribution : (x, μ, σ^2) where σ^2 is known. Then $f(x, \mu, \sigma^2)$ is

(a) Complete w.r. to μ.

(b) Sufficient w.r. to μ.

(c) Complete & sufficient w.r. to μ.

(d) Discrete w.r. to μ.

Q.201 In Q.(200), let m is known. Then $f(x, \mu, \sigma^2)$ is

(a) Complete w.r. to σ^2.

(b) Boundedly complete w.r. to σ^2.

(c) Not even boundedly complete w.r. to σ^2.

(d) Sufficient w.r. to σ^2.

Q.202 Let $(x_1, x_2, \ldots\ldots, x_n)$ be a random sample drawn from normal distribution and N(μ, σ²) and let σ^2 is known. Then a statistic $t(x) = \sum^{n} x_i$ is

(a) Complete for μ.
(b) Sufficient for μ.
(c) Complete sufficient for μ.
(d) All the three.

Q.203 In Q.(202), let (μ, σ2) are unknown. Then the statistics $\left(\sum^{n} x_i, \sum^{n} x_i^2\right)$

are jointly :
(a) Complete statistics for (μ, σ²) respectively.
(b) Sufficient for (μ, σ²) respectively.
(c) Complete and sufficient for (μ, σ²) respectively.
(d) All the above.

Q.204 In Q. (202), let μ is known. Then the statistic $t = \sum^{n} (x_i - \mu)^2$ is

(a) Complete for σ².
(b) Sufficient for σ².
(c) Complete & sufficient for σ².
(d) All the above.

Q.205 In Q.(202), let μ is unknown. Then the statistic $\hat{s}^2 = \frac{1}{(n-1)} \sum^{n} (x_i - \overline{x})^2$

is
(a) An unbiased estimator for σ².
(b) Minimum variance estimator for σ².
(c) An U.M.V.U.E. for σ².
(d) All the above.

Q.206 In Q.(202), let μ is known. Then the statistic $s^{*2} = \frac{1}{n} \sum^{n} (x_i - \mu)^2$ is

(a) An unbiased estimator for σ^2.
(b) An efficient estimator for σ^2.
(c) An U.M.V.U.E. for σ^2.
(d) All the above.

Q.206 Let in Q.(202), σ^2 be known. Then the statistic $t(x) = \dfrac{\left(\sum^{n} x_i\right)^2 - n\,\sigma^2}{n}$ is

(a) An unbiased estimator for μ^2.
(b) An efficient estimator for μ^2.
(c) An U.M.V.U.E. for μ^2.
(d) All the above.

Q.207 Consider the log normal distribution given as :

$$f(x, \mu, \sigma^2) = \frac{1}{\sqrt{2\pi}\,\sigma}\, e^{-\frac{1}{2}\frac{(\log x - \mu)^2}{\sigma^2}}, \quad x > 0,\ \sigma^2 > 0$$

Let σ^2 be known, then $f(x, \mu, \sigma^2)$ is
(a) Complete w.r. to μ.
(b) Sufficient w.r. to μ.
(c) Complete & sufficient w.r. to μ.
(d) All the above.

Q.208 Let in Q.(207), the population variance σ^2 be known. Then the statistic

$t(x) = \sum_{i}^{n} \log x_i$ is

(a) Complete for parameter μ.
(b) Sufficient for μ.
(c) Complete & sufficient for μ.
(d) All the above.

Q.209 If in Q.(207), a statistic $t(x)$ is defined as $t(x) = \left\{\sum_{i}^{n} \log x_i, \sum_{i}^{n} (\log x_i)^2\right\}$

Then the statistic $t(x)$ is jointly :

(a) Complete for (μ, σ^2) respectively.

(b) Complete and sufficient for (μ, σ^2) respectively.

(c) Complete and sufficient for (σ^2, μ) respectively.

(d) Both (a) & (b)

Q.210 In Q.(207), if μ is known and a statistic $t(x)$ defined as $t(x) = \sum_{i}^{n}(\log x_i - \mu)^2$ is

(a) Sufficient for σ^2.

(b) Complete for σ^2.

(c) Complete and sufficient for σ^2.

(d) None.

Q.211 In Q.(207). a statistic t is defomed as :

$t = \frac{1}{n}\sum_{i}^{n}\log x_i$. Then statistic t is

(a) U.M.V.U.E. for μ.

(b) Unbiased estimator for μ.

(c) Minimum variance estimator for μ.

(d) All the above.

Q.212 In Q.(207), let a statistic t^* be defined as :

$$t^* = \frac{1}{n}\sum_{i}^{n}(\log x_i - \mu)^2 \text{ wher } \mu \text{ is known.}$$

Then the statistic t* is

(a) U.M.V.U.E. for σ^2.

(b) Efficient estimator for σ^2.

(c) Unbiased estimator for σ^2.

(d) All the above.

Q.213 In Q.(207), let μ be unknown and a statistic t(x) is defined as :

$$t' = \frac{n\sum_{i}^{n}(\log x_i)^2 - (\sum^{n}\log x_i)^2}{n(n-1)}$$

Then the statistic t' is

(a) An unbiased estimator for σ^2.

(b) An effieicnt estimator for σ^2.

(c) U.M.V.U.E. for σ^2.

(d) All the above.

Q.214 Consider the power series distribution defined as $f(x, \theta, \alpha) =$

$$\frac{\theta \, . \, x^{\theta-1}}{\alpha^{\theta}}, \quad \begin{matrix} 0 < x < \infty \\ \theta > 0 \end{matrix}$$

Then in case of α being known, the $f(x, \theta, \alpha)$ is

(a) Complete w.r. to θ.

(b) Sufficient w.r. to θ.

(c) Efficient w.r. to θ.

(d) All the above.

Q.215 If in Q.(214), θ is known then $f(x, \theta, \alpha)$ is

(a) Sufficient w.r. to α.

(b) Complete w.r. to α.

(c) Complete and sufficient w.r. to α.

(d) All the above.

Q.216 In Q.(214), if α is known and a statistic t(x) is defined as :

$$t(x) = \sum_{i}^{n} \log x_i \text{ . Then } t(\text{x}) \text{ is}$$

(a) Complete statistic for θ.

(b) Sufficient statistic for θ.

(c) Complete sufficient statistic for θ.

(d) All the above.

Q.217 Let an ordered sample $(x_1 < x_2 < ..... < x_n)$ be drawn from a power series distribution defined in Q.(214). If the parameter θ is known then the statistic $t(x) = x_n$ is

(a) Complete for α.

(b) Sufficient for α.

(c) Complete and sufficient for α.

(d) All the above.

Q.218 Consider Cauchy distribution defined as :

$$f(x, \mu, k) = \frac{k}{\pi [k^2 + (x - \mu)^2]}, \quad -\infty < x < \infty$$

$$-\infty < \mu < \infty$$

$$k > 0$$

when k = 1 then $f(x, \mu, k)$ is

(a) Sufficient w.r. to μ.

(b) Complete w.r. to μ.

(c) Complete and efficient w.r. to μ.

(d) All the above.

Q.219 In Q.(218), let μ = O, then the statistic $t(x)$ defined as $t(x) = \sum_{i}^{n} x_i^2$ is

(a) Sufficient for k^2.

(b) Complete for k^2.

(c) Consistent for k^2.

(d) None.

Q.220 In Q.(218), for $k = 1$, the mle of μ is obtained by using the :

(a) Method of scoring.

(b) Method of moment

(c) Method of minimum χ^2.

(d) None.

Q.221 In Q.(218), the sample median $\tilde{x}$ is the :

(a) Consistent estimator for population median μ.

(b) In consistent for population median μ.

(c) Unbiased estimator for population median μ.

(d) Efficient estimator for population median μ.

Q.222 Let x_1, x_2 be a sample of size 2 drawn from Cauchy distribution defined in Q.(218). Then Pitman Closure estimator for μ than x_1 is given by $t(x)$ as :

(a) $t(x) = \frac{(x_1 + x_2)}{3}$ (b) $t(x) = \frac{(x_1 + x_2)}{2}$

(c) $t(x) = \frac{(x_1^2 + x_2^2)}{2}$ (d) $t(x) = \frac{(x_1^2 + x_2^2)}{3}$

Q.223 Consider the Cauchy distribution $f(x, \mu)$ defined as :

$$f(x, \mu) = \frac{1}{\pi\,[1 + (x - \mu)^2\,]}, \quad \begin{matrix} -\infty < x < \infty \\ -\infty < \mu < \infty \end{matrix}$$

Then $f(x, \mu)$ is symmetrical with mode and median given as :

(a) Mode & Median > μ.

(b) Mode & Median < μ.

(c) Mode & Median = μ.

(d) Mode & Median $\geq \mu$.

Q.224 Consider Gamma distribution defined as :

$$f(x, \theta) = \frac{\theta^x \; x^{m-1} \; e^{-\theta x}}{\overline{|m}}, \quad \begin{matrix} x > 0 \\ \theta > 0 \\ m > 0 \end{matrix}$$

Let m be known and a statistic $t(x)$ be defined as :

$$t(x) = \sum_i^n x_i \text{ is :}$$

(a) Complete for θ.

(b) Sufficient for θ.

(c) Complete & sufficient for θ.

(d) All the above.

Q.225 Let in Q.(224), m be known and a statistic $t(x)$ is defined as :

$$t(x) = \sum_i^n x_i \,/\, mn \text{, then } t(x) \text{ is :}$$

(a) U.M.V.U.E. for $\frac{1}{\theta}$.

(b) An efficient estimator for $\frac{1}{\theta}$.

(c) An unbiased estimator for $\frac{1}{\theta}$.

(d) All the above.

Q.226 Consider beta distribution defined as :

$$f(x, \text{p}, q) = \frac{\overline{|(p+q)}}{\overline{|p}\,\overline{|q}} \qquad \begin{matrix} 0 < x < 1 \\ p > 0 \\ q > 0 \end{matrix}$$

If p is known then $f(x$, p, q) is

(a) Complete w.r. to q.

(b) Complete w.r. to p.

(c) Incomplete w.r. to q.

(d) Sufficient w.r. to p.

Q.227 In Q.(226), let q be known and a statistic t(x) be defined as :

$$t(x) = \sum_{i=1}^{n} \log x_i$$

then $t(x)$ is

(a) Complete statistic for p.

(b) Sufficient statistic for p.

(c) Complete and sufficient statistic for p.

(d) All the above.

Q.228 Let in Q.(226), p be known and a statistic $t^*(x)$ be defined as :

$$\text{t}^*(\text{x}) = \sum_{i}^{n} \log(1 - x_i)$$

Then $t^*(x)$ is

(a) Sufficient statistic for q.

(b) Complete statistic for q.

(c) Complete & sufficient statistic for q.

(d) All the above.

Q.229 Let in Q.(226), $q = 1$, and $n > 1$ and a statistic t_1 is defined as :

$$t_1 = \frac{1}{\left(\frac{1}{n-1} \sum^{n} \log x_i\right)}$$

Then t_1 is

(a) An unbiased estimator for p.

(b) Minimum variance estimator for p.

(c) U.M.V.U.E. for p.

(d) All the above.

Q.230 Let in Q.(226), p = 1, and n > 1. and a statistic t_2 is defined as :

$$t_2 = \frac{1}{\frac{1}{n-1}\sum_{i}^{n} \log(1-x_i)}$$

Then the statistic t_2 is

(a) U.M.V.U.E. for q.

(b) An unbiased estimator for q.

(c) A minimum variance estimator for q.

(d) All the above.

Q.231 Consider a continuous uniform distribution defined as $f(x, \theta) = \frac{1}{\theta}, 0 < x < \theta, \theta > 0$

Let $x_1 < x_2 < \ldots.. < x_n$ be an ordered sample drawn from $f(x, \theta)$. And a statistic t_r is defined as :

$$t_r = \left(\frac{n+1}{n}\right) x_n^r$$

Then t_r is

(a) U.M.V.U.E. for θ^r.

(b) An unbiased estimator for θ^r.

(c) Minimum variance estimator for θ^r.

(d) All the above

Q.232 Consider an uniform distribution defined as :

$$f(x, \theta_1, \theta_2) = \frac{1}{(\theta_2 - \theta_1)}, \theta_1 < x < \theta_2$$

When θ_2 is known then $f(x, \theta_1, \theta_2)$ is

(a) Complete w.r. to θ_1.

(b) Incomplete w.r. to θ_1.

(c) Complete w.r. to θ_2.

(d) None.

Q.233 Let $x_1 < x_2 < ..... < x_n$ be an ordered sample drawn from the distribution $f(x, \theta_1, \theta_2)$ defined in Q.(232) and a statistic t_1 is defined as :

$$t_1 = \frac{(n-1)\, x_1 - \theta_2}{n} \text{ and } \theta_2 \text{ is known.}$$

Then the statistic t_1 is

(a) An U.M.V.U.E. for θ_1.

(b) An unbiased estimator for θ_1.

(c) An M.V.E. for θ.

(d) All the above.

Q.234 In Q(232), let θ_1 be known and a statistic t_2 is defined as $t_2 = \frac{(n+1)\, x_n - \theta_1}{n}$ Then statistic t_2 is

(a) An unbiased estimator for θ_2.

(b) An efficient estimator for θ_2.

(c) An UMVUE for θ_2.

(d) All the above.

Q.235 Consider the p.d.f. $f(x, \theta_1, \theta_2)$ defined as :

$$f(x, \theta_1, \theta_2) = \frac{1}{\theta_2},\ \theta_1 - \frac{\theta_2}{2} \le x \le \theta_1 + \frac{\theta_2}{2}$$

$$-\infty < \theta_1 < 0,\ \theta_2 > 0$$

Let $x_1 < x_2 < ..... < x_n$ be an ordered sample drawn from $f(x, \theta_1, \theta_2)$ and a statistic $t(x)$ is defined as $t(x) = \frac{(x_1 + x_n)}{2}$

Then the statistic $t(x)$ is

(a) U.M.V.U.E. for θ_1.

(b) An unbiased estimator for θ_1.

(c) Minimum variance estimator for θ_1.

(d) All the above.

Q.236 Let in Q.(235), a statistic t*(x) be defined as :

$$t^*(x) = \left(\frac{n+1}{n-1}\right)(x_1 + x_n) \text{ where } \theta_1 \text{ is known. Then } t^*(\underset{\sim}{x}) \text{ is}$$

(a) M.V.U.E. for θ_2.

(b) An unbiased estimator for θ_2.

(c) An efficient estimator for θ_2.

(d) All the above.

Q.237 Consider the p.d.f., $f(x, \theta)$ defined as :

$$f(x, \theta) = 1, \qquad (\theta - \frac{1}{2}) < x < (\theta + \frac{1}{2}),$$
$$-\infty < \theta < \infty$$

Let $(x_1 < x_2 ..... < x_n)$ be an ordered sample from $f(x, \theta)$, then the statistic $t(x) = (x_1, x_n)$ is

(a) Sufficient statistic for θ.

(b) Complete statistic for θ.

(c) Bounded complete statistic for θ.

(d) None

Q.238 For the p.d.f. defined in Q.(237), define a statistic $t(x)$ as :

$$t(x) = \frac{(x_1 + x_n)}{2}$$

Then the statistic t(x) is

(a) M.V.U. estimator for θ.

(b) An unbiased estimator for θ.

(c) Minimum variance estimator for θ.

(d) All the above.

Q.239 Consider the p.d.f. defined as

$f(x, \theta) = 1, \theta < x < \theta + 1, -\infty < \theta < \infty$

and define a statistic $t(\underset{\sim}{x}) = (x_1, x_n)$. Then t $(\underset{\sim}{x})$ is

(a) Jointly sufficient for θ.

(b) Minimum variance estimator for θ.

(c) Unbiased estimator for θ.

(d) Efficient estimator for θ.

Q.240 For the p.d.f. given in Q.(239), define a statistic t* as

$$t^* = \frac{(x_1 + x_{n-1})}{2}$$

then t* is

(a) M.V.U. estimator for θ.

(b) Minimum variance estimator for θ.

(c) An unbiased estimator for θ.

(d) All the above.

Q.241 Consider the p.d.f. defined as :

$$f(x, \theta) = \frac{1}{2\theta}, -\theta < x < \theta, \theta > 0$$

and define a statistic $t(x) = \max(-x_1, x_n)$, then $t(x)$ is

(a) Complete for θ.

(b) Sufficient for θ.

(c) Complete & sufficient statistic for θ.

(d) All the above.

Q.242 In case of p.d.f. given in Q.(241), consider an ordered sample ($x_1 < x_2 <$ ...... $< x_n$) and define a statistic $t^*(x) = \left(\frac{n+1}{n}\right) \max(-x_1, x_n)$.

Then $t^*(x)$ is for parameter θ :

(a) U.M.V.U.E.

(b) Minimum variance estimator

(c) Unbiased estimator

(d) All the above.

Q.243 Consider Pareto distribution given as :

$$f(x, \alpha, \beta) = \frac{\alpha\beta^x}{x(\alpha + 1)}, x > 1, \beta > 0, \alpha > 1$$

Let β be known then $f(x, \alpha, \beta)$ is :

(a) Complete w.r. to α.

(b) Sufficient w.r. to α.

(c) Complete w.r. to β.

(d) None.

Q.244 Let in Q.(243) for $f(x, \alpha, \beta)$, a statistic t_1 is defined as :

$t_1 = \sum_{i=1}^{n} \log x_i$ and β is known then t_1 is

(a) Complete for α.

(b) Complete and sufficient for α.

(c) Sufficient for α.

(d) All the above.

Q.245 Let $x_1 < x_2 < ..... < x_n$ be an ordered sample drawn from $f(x, \alpha, \beta)$ defined in Q.(243) and a statistic $t_2 = x_n$ is defined. Then for known α, the statistic t_2 is

(a) Complete for β.

(b) Complete & sufficient for β.

(c) Sufficient for β.

(d) All the above.

Q.246 Consider the p.d.f., $f(x, \theta)$ defined as $f(x, \theta) = \frac{1}{\theta}$, $k\,\theta \le \times \le (k+1)\,\theta$, $k > 0$

Let $x_1 < x_2 < ..... < x_n$ be an ordered sample from $f(x, \theta)$, and a statistic $t = \{x_1, x_n)$ is defined then t is the pair of joint :

(a) Sufficient statistic for θ.

(b) Efficient statistic for θ.

(c) Consistent statistic for θ.

(d) None

Q.247 In Q.(246), Let θ be the mle of θ then θ is equal to

(a) $\frac{x_n}{n+1}$ (b) $\frac{x_n}{k+1}$

(c) $\frac{x_n}{k^2}$ (d) $\frac{x_n}{k^2+1}$

Q.248 In Q.(247), due to symmetry, the (x_1, x_n) have the same variance, say v then $v\,(\theta)$ is equal to

(a) $\frac{v}{k^2}$ (b) $\frac{v}{(k+1)^2}$

(c) $\frac{v}{(k+1)}$ (d) $\frac{v}{(k^2+1)}$

Q.249 Consider the p.d.f. of normal distribution $N(0, \sigma^2)$ as $f(x, 0, \sigma^2) =$

$$\frac{1}{\sqrt{2\pi}\,\sigma}\,e\,\frac{1}{2}\,\frac{x^2}{\sigma^2}$$

Define a statistic $t(x) = \sqrt{\left(\frac{1}{n}\sum^{n} x_i^2\right)}$

Then the statistic $t(x)$ is for parameter σ as :

(a) M.L.E. for σ (b) Sufficient for σ

(c) M.L.E. & Sufficient for σ (d) All the three

Q.250 Consider the following p.d.f. of double exponential distribution defined as :

$$f(x, \theta) = \frac{1}{2} e^{-|x-\theta|}, \quad -\infty < \theta < \infty$$

Let $(x_1 < x_2 < \ldots\ldots < x_n)$ be an ordered sample from $f(x, \theta)$. Then the sample is for θ as :

(a) Sufficient statistic (b) Efficient statistic

(c) M.L.E. (d) Consistent estimator

Q.251 In Q.(250), the mel θ of θ is given as :

(a) θ = Mode $(x_1, x_2, \ldots., \ldots, x_n)$

(b) θ = Median $(x_1, x_2, \ldots., \ldots, x_n)$

(c) θ = A.M. $(x_1, x_2, \ldots., \ldots, x_n)$

(d) θ = G.M. $(x_1, x_2, \ldots., \ldots, x_n)$

Q.252 Consider the p.d.f. $f(x, \theta)$ defined as :

$$f(x, \theta) = \frac{1}{\theta} e^{-\frac{x}{\theta}}, \quad 0 < x < \infty, 0 < \theta < \infty$$

Define a statistic $t(x) = \sum_{i}^{n} x_i$ or $\bar{x}$.

Then $t(x)$ is for parameter θ as :

(a) Efficient estimator

(b) Sufficient statistic

(c) Unbiased estimator

(d) None.

Q.253 For $f(x, \theta)$ defined in Q.(251), consider a statistic. $t(x)$ defined as :

$t(x) = \left(\frac{n-1}{n}\right)\frac{1}{\bar{x}}$ then t(x) is for parameter θ as :

(a) Unbiased estimator (b) Sufficient estimator

(c) M.L.E. (d) Consistent estimator

Q.254 Consider the following two parametric exponential p.d.f. defined as :

$$f(x, \theta_1, \theta_2) = \frac{1}{\theta_2} e - \frac{(x-\theta_1)}{\theta_2} \qquad \theta_1 < x < \infty$$

$$-\infty < \theta_1 < \infty$$

$$0 < \theta_2 < \infty$$

Let $\underset{\sim}{x} = (x_1, < x_2 < ..... < x_n)$ be an orderd sample drawn from $f(x, \theta_1, \theta_2)$ then jointly sufficient statistics for (θ_1, θ_2) respectively are given as :

(a) $t(x) = (x_1, \sum^{n} x_i^2)$ (b) $t = (x_1, \sum^{n} x_i)$

(c) $t(x) = (x_1, \frac{\sum x_i^2}{n})$ (d) $t = (\sum^{n} x_i, x_1)$

Q.255 Let us consider the following p.d.f. $f(x, \alpha)$ as "

$f(x, \alpha) = e-(x-\alpha), \alpha < x < \infty$

Let $(x_1 < x_2 < ..... < x_n)$ be an ordered sample from the distribution, then the sufficient statistic for α is given as :

(a) $t(x) = x_1$ (b) $t(x) = x_n$

(c) $t(x) = \frac{x_1 + x_n}{2}$ (d) $t(x)$ = None.

Q.256 Consider the following two parametric distribution defined as :

$$f(x, \alpha, \beta) = \frac{1}{(\beta-\alpha)}, \qquad \alpha < x < \beta$$

Let $(x_1 < x_2 < ..... < x_n)$ be an ordered sample then for given β, the sufficient statistic for α is given as :

(a) $t(x) = x_n$ (b) $t(x) = x_1$

(c) $t(x) = \sum_{x}^{n} x_i$ (d) $t(x) = \sum^{n} x_i^2$

Q.257 In Q.(265), let α be known then a sufficient statistic $t(x)$ for β is given as :

(a) $t(x) = x_n$ (b) $t(x) = x_1$

(c) $t(x) = x_n^2$ (d) $t(x) = x_1^2$

Q.258 In Q.(265), let (α, β) be unknown then the joint sufficient statistic for (α, β) respectively are given by $t(x)$ as :

(a) $t(x) = (x_n, x_1)$ (b) $t(x) = (x_1, x_n)$

(c) $t(x) = (x_n^2, x_1^2)$ (d) $t(x) = (x_1^2, x_n^2)$

Q.259 Let $t(x)$ be a complete statistic for θ then any other statistic say t* depending on random sample $x_1, x_2, \ldots, x_n$ through $t(x)$ is

(a) Sufficient for θ (b) Efficient for θ

(c) Consistent for θ (d) Complete for θ

Q.260 Consider the Bernouille distribution with parameter p defined as :

$P(x = 1) = p \quad 0 < p < 1$

$P(x = 0) = (1 - p)$

Then sufficient statistic for p is given as :

(a) $t = \sum^{n} x_i$ (b) $t = \frac{1}{\sum^{n} x_i}$

(c) $t = \sum^{n} x_i^2$ (d) $t = \frac{1}{\sum^{n} x_i^2}$

Q.261 Consider a standardized bivariate normal distribution.

Then pair of sufficient statistics for population correlation coefficient ρ is given as :

(a) $t(x) = (\sum^{n} x_i^2 + \sum^{n} y_i^2, \sum^{n} x_i y_i)$

(b) $t(x) = (\sum^n x_i\, y_i, \sum^n x_i^2 + \sum^n y_i^2)$

(c) Both (a) & (b)

(d) None of the above.

Q.262 Consider the following p.d.f. defined as :

$f(x, \theta) = 1,\ (\theta - \frac{1}{2}) \leq \text{x} \leq (\theta + \frac{1}{2})$

Consider an ordered sample $(x_1, < x_2 < \ldots\ldots < x_n)$.

Then the joint statistic $\text{t(x)} = (x_1, x_n)$ is

(a) Efficient for θ.

(b) Sufficient for θ.

(c) Consistent for θ.

(d) Unbiased for θ.

Q.263 In Q.(261), the likelihood function L(x, θ) will be maximum at any value in the interval given as :

(a) $(x_n - \frac{1}{2},\ x_1 + \frac{1}{2})$

(b) $(x_1 + \frac{1}{2}, x_n - \frac{1}{2})$

(c) $(x_n + 1, x_1 + 1)$

(d) $(x_1 - 1, x_n - 1)$

Q.264 In Q.(251), the mid value of the interval $(x_n - \frac{1}{2}, x_1 + \frac{1}{2})$ will be :

(a) A consistant estimator for θ.

(b) An unbiased estimator for θ.

(c) A sufficient estimator for θ.

(d) A complete and sufficient estimator for θ.

Q.265 In Q.(251), the parameter θ is estimated by the method of moment by a sample value known as :

(a) Sample median

(b) Sample mode

(c) Sample mean

(d) Sample G.M.

Q.266 Consider a power series distribution defined as :

$$f(x, \theta) = \frac{a_x \ \theta^x}{(f(x))^n}, x = c, c+1, c+2, ...$$

Then the sample statistic $t = \sum^{n} x_i$ for parameter θ is

(a) Complete (b) Sufficient

(c) Complete and sufficient (d) All the three.

Q.267 Consider binomial distribution as :

$$B(x, n, p) = {}^{n}c_{x}\, p^x q^{n-x}, \qquad x = 0, 1, 2, ... n$$

$$0 < p, < 1, 0 < q < 1$$

$$p + q = 1$$

When $p = q = \frac{1}{2}$, then B (x, n, p) can be approximated by normal distribution for the value of n being :

(a) Maximum (b) Minimum or least

(c) n = 50 (d) n = 100

1.4 Key Answers

Write True/False

Q.	A.	Q.	A.	Q.	A.	Q.	A.
1	True	31	True	61	True	91	True
2	True	32	True	62	True	92	True
3	True	33	True	63	True	93	True
4	True	34	True	64	True	94	True
5	True	35	True	65	True	95	True
6	True	36	True	66	True	96	True
7	True	37	True	67	True	97	True
8	True	38	True	68	True	98	True
9	True	39	True	69	True	99	True
10	True	40	True	70	True	100	True
11	True	41	True	71	True	101	True
12	True	42	True	72	True	102	True
13	True	43	True	73	True	103	True
14	True	44	True	74	True	104	True
15	True	45	True	75	True	105	True
16	True	46	True	76	True	106	True
17	True	47	True	77	True	107	True
18	True	48	True	78	True	108	True
19	True	49	True	79	True	109	True
20	True	50	True	80	True	110	True
21	True	51	True	81	True	111	True
22	True	52	True	82	True	112	True
23	True	53	True	83	True	113	True
24	True	54	True	84	True	114	True
25	True	55	True	85	True	115	True
26	True	56	True	86	True	116	True
27	True	57	True	87	True	117	True
28	True	58	True	88	True	118	True
29	True	59	True	89	True	119	True
30	True	60	True	90	True	120	True

Q.	A.	Q.	A.	Q.	A.	Q.	A.
121	True	153	True	185	True	217	True
122	True	154	True	186	True	218	True
123	True	155	True	187	True	219	True
124	True	156	True	188	True	220	True
125	True	157	True	189	True	221	True
126	True	158	True	190	True	222	True
127	True	159	True	191	True	223	True
128	True	160	True	192	True	224	True
129	True	161	True	193	True	225	True
130	True	162	True	194	True	226	True
131	True	163	True	195	True	227	True
132	True	164	True	196	True	228	True
133	True	165	True	197	True	229	True
134	True	166	True	198	True	230	True
135	True	167	True	199	True	231	True
136	True	168	True	200	True	232	True
137	True	169	True	201	True	233	True
138	True	170	True	202	True	234	True
139	True	171	True	203	True	235	True
140	True	172	True	204	True	236	True
141	True	173	True	205	True	237	True
142	True	174	True	206	True	238	True
143	True	175	True	207	True	239	True
144	True	176	True	208	True	240	True
145	True	177	True	209	True	241	True
146	True	178	True	210	True	242	True
147	True	179	True	211	True	243	True
148	True	180	True	212	True	244	True
149	True	181	True	213	True	245	True
150	True	182	True	214	True	246	True
151	True	183	True	215	True	247	True
152	True	184	True	216	True	248	True

Q.	A.	Q.	A.	Q.	A.	Q.	A.
249	True	265	True	281	True	297	True
250	True	266	True	282	True	298	True
251	True	267	True	283	True	299	True
252	True	268	True	284	True	300	True
253	True	269	True	285	True	301	True
254	True	270	True	286	True	302	True
255	True	271	True	287	True	303	True
256	True	272	True	288	True	304	True
257	True	273	True	289	True	305	True
258	True	274	True	290	True	306	True
259	True	275	True	291	True	307	True
260	True	276	True	292	True	308	True
261	True	277	True	293	True	309	True
262	True	278	True	294	True	310	True
263	True	279	True	295	True	311	True
264	True	280	True	296	True	312	True

Fill in the Blanks

1	Statistical inference
2	Problem of estimation
3	Point estimation
4	Interval estimation
5	Best estimator
6	Unbiased estimator
7	Biased estimator
8	$t_n(x)$, θ
9	$t_n(x)$, θ
10	$E(t_n(x)) - \theta$
11	$> \theta$
12	$< \theta$
13	$E(tn(x))$.
14	$\left(\frac{n-1}{n}\right)$
15	$\overline{x}_n$, μ_x
16	$(m^2 + 1)$
17	$(t - \Delta)$
18	Consistency
19	$\mid t_n - \theta\ )$
20	Probability
21	$\mid t_n - \theta \mid$
22	Unbiased
23	Consistent
24	Unbiased
25	$(\theta, 0, \infty)$
26	Consistent
27	Consistent
28	Consistent
29	Consistent
30	(Does not)
31	Consistent
32	e^t
33	Estimator
34	Estimator
35	Efficient
36	Efficiency
37	¥^{2} < ¥'^{2}
38	t_n
39	(2e – 1)

40 1.0

41 Median, Population mean.

42 Unique

43 <

44 $\sqrt{e}$

45 $(1-e_1)(1\ e_2)$, $(1-e_1)(1-e_2)$

46 $v(t_n')/v(t_n')$

47 $\frac{2}{n}$

48 $\frac{\pi^2}{4n}$

49 $\frac{\pi^2}{4n}$

50 $\frac{(\pi-2)}{\pi}$

51 Unbiased

52 Positive

53 $\sum_{i=1}^{n} a_i\, t_i$

54 Cramer Rao

55 $\frac{1}{k}\sum_{i=1}^{k} t_i$

56 $\frac{\partial}{\partial\theta}\log L,\ \psi'(\theta)$

57 Independent, Differentiated

58 $\frac{\partial}{\partial\theta}\log f(x,\theta)$

59 Differentiable, Integral

60 $\sum_{i}^{n} x_i$

61 $\left(\frac{n-1}{n}\right), \left(\frac{n-1}{n}\right)$

62 $(e^{\lambda/n}-1)$

63 $e^{-2\lambda}\left(\frac{n}{\lambda}\right)$

64 Sufficient

65 $(t_n-\psi\ (\theta)$

66 $\psi'\ (\theta)$

67 Sampling variance

68 $\left[\frac{\partial}{\partial\theta}\log f\right]$

69 $[\theta-\psi(q)],\ (\psi'(\theta))$

70 $[1+b'\ (\theta)]$

71 $[1+b'\ (\theta)],\ (b\ (\theta))$

72 M.V.U.E., λ

73 $\bar{x}_n, \frac{\sigma^2}{n}$

74 $\bar{x}_n$

75 $(\bar{x}_n-\lambda)$

76 $(\bar{x}_n, p)$

77 $\frac{r}{n}$, pq

78 Does not exist

79 $\frac{1}{n}\sum^{n}(x_i-\mu)^2$

80 $\frac{1}{n}\sum_{i}^{n}(x_i-\mu)^2$, $2\ \sigma^4$

81 $\frac{\bar{x}_n}{b}$, np

82 $t-\lambda(\theta)$

83 Minimum variance estimator

84 M.V.B.E.

85 $\left(\frac{n-r}{n-1}\right)$

86 Known

87 Does not exist

88 $\frac{\pi}{3n}$

89 $\frac{4}{n}$

90 $\bar{x}_n$

91 1.0

92 Sufficient

93 Sufficient

94 M.L.E.

95 $\frac{1}{n}\sum^{n} x_i$

96 $\frac{(p-2)}{2}$

97 Sufficient

98 Truncated

99 2.0
100 Moment
101 M.V.U.E.
102 M.V.U.E.
103 Sufficient statistic
104 Complete family of distribution
105 Sufficient
106 Blackwellisation
107 Complete family of distribution
108 M.V.U.E.
109 Regularity conditions
110 μ, <
111 Unbiased, minimum
112 Sufficient
113 Sufficient
114 $\frac{r}{n}$
115 $\sum_{i}^{n} x_i$
116 $\sum^{n} x_i$
117 $\left(\frac{n-1}{n}\right)$
118 $\left(\overline{x}^2 - \frac{\overline{x}}{n}\right)$
119 $\frac{1}{n}\sum^{n} x_i$
120 $q2\frac{\theta^2}{n}$
121 Zero
122 Sufficient statistc
123 Parameter
124 $G(t_n, -\theta)$
125 Functional relationship
126 Zero expectation
127 Q (t(x)) = 0
128 Independent
129 Sufficient statistic
130 Sufficient
131 Sufficient
132 Pair of sufficient statistics, (x), (θ).
133 Sufficient, complete
134 Sufficient, complete
135 Sufficient, complete
136 $\{a^{-1}(x_1,)$, $b^{-1}(x_n)$
137 $\{a^{-1}(x_1), b^{-1}(x_n)\}$
138 Sufficient statistics
139 Sufficient statistics
140 Sufficient statistics
141 M.L.E.
142 Uniqie maxima.
143 M.L. estimator
144 Minimal sufficient
145 Minimal sufficient
146 Zero
147 Point, Testing
148 Zero.
149 Boundary complete
150 Zero
151 Boundedly
152 Independent
153 Most efficient estimator
154 Sufficient statistic
155 Sufficient
156 Sufficient
157 Minimal sufficient
158 Minimal sufficient statistics
159 Minimal sufficient, Bounded completeness
160 Sufficient
161 U.M.V.U.E.
162 Minimal sufficent
163 Sufficient
164 $\frac{1}{n}\sum^{n} x_i$
165 Not sufficient
166 Sufficient

167 $\frac{\sum_{i}^{n} x_2}{n}$

168 $\sum_{i}^{n} x_2$

169 $\sum_{i}^{n} x_i$

170 Min. $(x_1, x_2$ ----- $,x_n)$

171 Order Statistic

172 Random sample

173 Nonparametric

174 Median

175 QuasimRange

176 Range

177 mth mid range

178 Dependent

179 Sufficient

180 All values of r

181 Canchy distribution

182 $(y_n \cdot y_n)$

183 $[F(yn)]. f(y_n)$

184 (f(yn)] (1–F(yn)]

185 Maxm

186 Max.

187 $\sum_{i=1}^{n} x_i$

188 $\sum_{i}^{n} x_i$

189 $\sum_{i=1}^{n} x_i$

190 $(n–1, n– 1)$

191 $\frac{h\,(k-1)}{2}$

192 $h^2\,(k^2 - 1)$

193 x_n

194 $\left(x_n^n - x_{n-1}^n\right)$

195 $\frac{(x_{n\_}q*}{2}$

196 Sample $(x_1, x_2, \ldots..,$ $xn)$

197 $\frac{N}{n}$

198 $(xi\,, \sum_{i}^{n} x_i\,)$

199 Known

200 Boundedly complete

201 $\sum_{i}^{n} x_i$

202 $(\sum^{n} x_i, \sum_{i}^{n} x_i^2)$

203 $(\sum^{n} x_i, \sum_{i}^{n} x_i^2)$

204 $(\sum^{n} x_i, \sum_{i}^{n} x_i^2)$

205 $\frac{1}{n}\sum^{n}(x_i - \mu)^2$

206 $n\,\sigma^2$

207 Known

208 $\sum^{n} \log x_i$

209

$(\sum^{n} \log x_i, \sum^{n} (\log x_i)^2$

210 $\frac{1}{n}\sum^{n} \log x_i$

211

$\frac{1}{n}\sum^{n}(\log x_i - \mu)^2$

212 $\left(\sum^{n}(\log x_i)^2\right),$

$n\,(n-1)$

213 Known

214 Known

215 $\sum^{n} \log x_i$

216 x_n

217 μ

218 Boundedly complete
219 Scoring
220 Sample median
221 $\frac{(x_1 + x_2)}{2}$
222 μ
223 Exponential
224 $\sum_i^n x_i$
225 $\frac{\sum_i^n x_i}{mn}$
226 Known
227 $\sum^n \log x_i$
228 $log\ (1 - x_i)$
229 $log\ x_i$
230 $(1 - x_i)$
231 $\left(\frac{n+1}{n}\right)$
232 Known
233 θ_2
234 θ_1
235 $\frac{(x_i + x_n)}{2}$
236 $(x_1 + x_n)$
237 (x_1, x_2)
238 $\frac{1}{2}(x_1 + x_2)$
239 (x_1, xn)
240 (x_1, x_{n-1})
241 Max. $(-x_1, xn)$
242 $\left(\frac{n+1)}{n}\right)$
243 Complete, known.
244 $\sum_i^n \log x_i$
245 x_n
246 (x_1, x_n)
247 Single sufficient
248 $\frac{x_n}{k+1}$
249 $\frac{V}{(k+1)^2}$
250 $\frac{\sum^n x_i^2}{n}$
251 Sufficient
252 Median
253 $\sum^n x_i$
254 $\left(\frac{n-1}{n}\right)$
255 $(x_1, \sum_i^n x_i)$
256 $(|t^* - l(\theta)| < C)$
257 Min.
258 Min.
259 Max.
260 (x_1, x_n)
261 Complete
262 Sufficient statistic
263 Sufficient Statistic
264 (x_1, x_n)
265 $(x_n - \frac{1}{2}, x_1 + \frac{1}{2})$
266 Unbiased estimator
267 Method of moment
268 $\sum_i^n x_i$
269 $f(\theta)$
270 Normal distribution

Multiple Choice Questions and Answers

Q.	A.	33	(c)	66	(d)	99	(b)
1	(a)	34	(a)	67	(a)	100	(a)
2	(c)	35	(b)	68	(a)	101	(a)
3	(c)	36	(a)	69	(a)	102	(b)
4	(b)	37	(a)	70	(a)	103	(a)
5	(d)	38	(b)	71	(a)	104	(a)
6	(d)	39	(c)	72	(d)	105	(c)
7	(b)	40	(b)	73	(c)	106	(a)
8	(d)	41	(b)	74	(d)	107	(d)
9	(c)	42	(b)	75	(d)	108	(a)
10	(b)	43	(a)	76	(d)	109	(a)
11	(c)	44	(b)	77	(d)	110	(a)
12	(c)	45	(b)	78	(c)	111	(a)
13	(d)	46	(a)	79	(a)	112	(b)
14	(d)	47	(d)	80	(d)	113	(b)
15	(a)	48	(b)	81	(b)	114	(b)
16	(d)	49	(b)	82	(d)	115	(b)
17	(a)	50	(b)	83	(d)	116	(b)
18	(d)	51	(d)	84	(a)	117	(a)
19	(a)	52	(d)	85	(d)	118	(a)
20	(b)	53	(b)	86	(c)	119	(d)
21	(b)	54	(b)	87	(b)	120	(d)
22	(b)	55	(b)	88	(b)	121	(d)
23	(a)	56	(a)	89	(a)	122	(d)
24	(b)	57	(a)	90	(b)	123	(b)
25	(b)	58	(a)	91	(d)	124	(d)
26	(b)	59	(a)	92	(a)	125	(d)
27	(d)	60	(a)	93	(a)	126	(d)
28	(b)	61	(d)	94	(a)	127	(c)
29	(a)	62	(d)	95	(b)	128	(d)
30	(b)	63	(c)	96	(b)	129	( )
31	(c)	64	(d)	97	(b)	130	(d)
32	(b)	65	(a)	98	(a)	131	(a)

132 (d)
133 (d)
134 (b)
135 (d)
136 (a)
137 (a)
138 (d)
139 (b)
140 (a)
141 (a)
142 (b)
143 (a)
144 (b)
145 (b)
146 (b)
147 (c)
148 (a)
149 (b)
150 (b)
151 (b)
152 (a)
153 (a)
154 (d)
155 (d)
156 (a)
157 (b)
158 (b)
159 (d)
160 (d)
161 (a)
162 (d)
163 (b)
164 (d)
165 (c)
166 (d)
167 (d)
168 (c)
169 (b)
170 (a)
171 (b)
172 (c)
173 (a)
174 (d)
175 (s)
176 (c)
177 (b)
178 (a)
179 (d)
180 (d)
181 (b)
182 (b)
183 (d)
184 (d)
185 (c)
186 (d)
187 (d)
188 (d)
189 (d)
190 (d)
191 (d)
192 (d)
193 (d)
194 (d)
195 (d)
196 (b)
197 (d)
198 (d)
199 (d)
200 (a)
201 (c)
202 (d)
203 (d)
204 (d)
205 (d)
206 (d)
207 (7)
208 (d)
209 (d)
210 (a)
211 (d)
212 (d)
213 (d)
214 (a)
215 (b)
216 (d)
217 (d)
218 (b)
219 (b)
220 (a)
221 (a)
222 (b)
223 (c)
224 (d)
225 (d)
226 (a)
227 (d)
228 (d)
229 (d)
230 (d)
231 (d)
232 (a)
233 (d)
234 (d)
235 (d)
236 (d)
237 (a)
238 (a)
239 (a)
240 (d)
241 (d)
242 (d)
243 (a)
244 (a)
245 (d)
246 (a)
247 (b)
248 (b)
249 (d)
250 (a)
251 (b)
252 (b)
253 (a)
254 (b)
255 (a)
256 (b)
257 (a)
258 (b)
259 (d)
260 (a)
261 (c)
262 (b)
263 (a)
264 (b)
265 (c)
266 (d)
267 (b)

❑❑❑

CHAPTER - 2

Methods of Estimation

2.1 Write True/False

Q.1 From theoretical point of view, the most general method of estimation is the "Maximum Likelihood Estimation Method."

Q.2 The method of "maximum Likelihood Estimation" was initially formulated by the statistician named as "C.F. Gauss."

Q.3 The method of maximum likelihood estimation as a "General Method of Estimation" was first time introduced by "Prof. R.A. Fishor."

Q.4 Consider a random sample $(x_1, x_2, ..., x_n)$ from a distribution having p.d.f. as $f(x, \theta)$, $\theta \in \Omega\ \theta$. Then joint p.d.f. denoted by L (x, θ) is written as :

$$L\ (x, \theta) = \prod_{i=1}^{n} f\ (x_i, \theta),\ \theta \in \Omega\ \theta.$$

Q.5 The joint p.d.f say $L(x, \theta)$ may be regarded as the function of θ alone.

Q.6 The joint p.d.f. $L(x, \theta)$ is known as "Likelihood function" L of random sample x = $(x_1, x_2, ..., x_n)$.

Q.7 Let U(x) be a non-trivial function of random sample x, Then if $L\{U(x)\}/ x_1, x_2, ..., x_n\}$ is atleast as great as L $(x_1, x_2, ..., x_n\ \theta)\ \forall\ \theta \in \Omega\ \theta$, we call U($x$) as "maximum likelihood estimator" for the parameter θ.

Q.8 The maximum likelihood estimator (m.l.e.) is usually denoted by θ and is expressed as :

$$\theta = \text{U}\ (x_1, x_2, ... , x_n)$$

Q.9 The m.l.e. θ of the parameter θ is the solution of :

$$\frac{\partial}{\partial\theta}\log L = 0, \text{ and } \frac{\partial^2}{\partial\theta^2}\log_e L < 0$$

Q.10 The equation $\frac{\partial}{\partial\theta}\log_e L\,(x,\,\theta) = 0$, is usually known as "Likelihood Equation."

Q.11 In many instances there will be an unique m.l.e. θ of parameter θ.

Q.12 In general, m.l.e. θ of parameter θ may or may not be unique.

Q.13 One of the assumptions for existance of M.L.E. is that the derivatives $\frac{\partial}{\partial\theta}\log^L(x, \theta)$ and $\frac{\partial^2}{\partial\theta^2}\log^L(x, \theta)$ exist and are 'continuous functions' of θ in range R including true value of θ for almost all x.

Q.14 For the existance of M.L.E., the following inequality hold :

$$\left|\frac{\partial}{\partial\theta}\log L\,(x,\,\theta)\right| < F_1\,(x)$$

$$\text{and} \left|\frac{\partial^2}{\partial\theta^2}\log L\,(x,\,\theta)\right| < F_2\,(x)$$

where $F_1(x)$ & $F_2(x)$ are integrable functions of x over $(-\infty, \infty)$.

Q.15 For the existance of M.L.E., we assume that 3rd order derivative $\frac{\partial 3}{\partial\theta^3}$ $\log L\,(x,\,\theta)$ exists such that :

$$\left|\frac{\partial 3}{\partial\theta^3}\log L\,(x,\,\theta)\right| < M(x)$$

where $M(x) < k$ for k being a positive quantity.

Q.16 For every θ in R, we have :

$$E\left|\frac{\partial 2}{\partial\theta^2}\log L\,(x,\,\theta)\right| = \int_{-\infty}^{\infty}(\frac{\partial 2}{\partial\theta^2}\log L\,(x,\,\theta)\,dx = I_{(\theta)}$$

where $I_{(\theta)}$ is finite and non zero.

Q.17 In case of existance of M.L.E., we assume that the range of integration is independent of parameter θ.

Q.18 In case of M.L.E. if the range of integration depends upon parameter θ then $f(x, \theta)$ vanishes at extreme depending on θ.

Q.19 The assumption made in Q.18 is made to make the differentiation under integral sign valid.

Q.20 Cramer Rao theorem on consistency states that the likelihood equation

$\frac{\partial}{\partial\theta} \log L(x, \theta) = 0$ has a solution which converges in probability to the true value θ_0.

Q.21 M.L.E.'s are consistent estimator to the parameter θ.

Q.22 Hazoor Bazar's theorem states that "Any consistent solution of likelihood equation provides a maximum to likelihood function with probability p $\rightarrow$ 1 as sample size $n \rightarrow \infty$.

Q.23 A consistent solution of likelihood equation is asymptotically normally distributed about true value θ_0.

Q.24 If θ is the m.l.e. of parameter θ then :

$$\theta \sim \text{asymptotically } N(\theta_0, \frac{1}{I_{(\theta_0)}})$$

where θ_0 is the true value of θ.

Q.25 If t(x) is a sufficient statistic (estimator) of parameter θ then M.L.E. is the function of t(x). i.e. $\theta = \eta(t)$.

Q.26 If $\frac{\partial}{\partial\theta} \log L(x) = \psi(t, \theta)$ then t is regarded as sufficient statistic for the parameter θ.

Q.27 If it is not possible to write $\frac{\partial}{\partial\theta} \log L(x, \theta) = \psi(t, \theta)$ then there does not exist a sufficient statistic $t(x)$ for the parameter θ.

Q.28 If θ is the M.L.E. of parameter θ then the invariant property states that $g(\theta)$ is the M.L.E. of $g(\theta)$.

Q.28* When M.L.E. θ exits. then for large samples, θ is most efficient under the assumption that the sampling distribution of θ tends to follow normal distribution.

Q.29 In case of multivariate distribution, let θ be a vector as $\theta = (\theta_1, \theta_2, \ldots\ldots, \theta m)$. Then the necessary condition for the likelihood function to have an absolute maxima is that :

$$\frac{\partial}{\partial \theta_i} \log L\,(x/\theta_1, \theta_2, \ldots\ldots, \theta_m) = 0 \text{ for } i = 1, 2, \ldots\ldots, m$$

Q.30. In Q.29, a sufficient condition is that the matrix $\frac{\partial^2}{\partial \theta^2} \log L\,(x, \theta)$ is negative definite matrix.

Q.31 In Q.29, the E $\frac{(\partial^2 \log L\,(x, \theta))}{\partial \theta_i \, \partial \theta_j}$ is called the dispersion matrix.

Q.32 If the method of M.L.E. leads to a complicated equations then iterative method is adopted to get the solution.

Q.33 The iterative method adopted in case of Q.32, is known as "Method of Scoring."

Q.34 An estimator obtained through the method of scoring is often consistent but less efficient.

Q.35 In case of standard Cauchy distribution written as :

$$f(x, \theta) = \frac{1}{\pi} \frac{1}{1+(x-\theta)^2}, -\infty < x < \infty$$

the method of scoring is adopted to obtain the m.l.e.

Q.36 In Cauchy distribution $f(x, \theta) = \frac{1}{\pi} \frac{1}{1+(x-\theta)^2}$, the sample median is the consistent estimator of parameter θ.

Q.37 In Q.36, the sample median is the M.L.E. of the parameter θ.

Q.38 Let 't' be the sample median in Cauchy distribution $f(x, \theta) = \frac{1}{\pi} \frac{1}{1+(x-\theta)^2}, -\infty < x < \infty$ and θ is the M.L. estimator of parameter θ

then we have $\theta = t + \frac{4}{n} \sum_{i}^{n} \frac{(x_i - \theta)}{[1+(x_i-\theta)^2]}$

Q.39 In Q.38, the sampling variance of 't' is written as $V(t) = \frac{1}{I_{(\theta)}} = \frac{2}{n}$

Q.40 Under some regularity conditions and if range of x is independent of θ then likelihood equation

$$\frac{d}{d\theta}\log L(x, \theta) = 0$$

has an unique solution with probability one and that solution θ will be M.V.B.E.

Q.41 If the regularity conditions for the efficiency of M.L.E. do not exist, even then the M.L.E. may have "minimum variance property" conferred upon it by the "sufficient statistic" of which, it is always a function.

Q.42 If the M.L.E., θ is the root of $\frac{\partial}{\partial\theta}\log L(x, \theta) = 0$ for which $\hat{\theta} = \bar{x}$ (sample mean), then the most general form of the distribution "differentiable in θ" is expressed as :

$f(x, \theta) = e^{\{A(\theta) + A'(\theta)(x - \theta) + B(x)\}}$

Q.43 In Q.42, the sample mean $\bar{x}$ is sufficient statistic for θ.

Q.44 In Q.42, the θ has minimum variance bound (M.V.B.) equal to $= \frac{1}{n\, A''(\theta)}$

Q.45 Let x be a r.v. following $b(x, n, p)$ distribution. Then a consistent estimator 't' for p (1 – p) will be given as :

$$t = \frac{r}{n}\left(1 - \frac{r}{n}\right) \text{ where } r = \sum_{i}^{n} x_i$$

Q.46 In Q.45, the consistent estimator of $\frac{1}{p}$ is given as $t = \frac{n}{r}$.

Q.47 Let T and U be consistent estimator for λ(θ). Then the statistic T* defined as :

T* = aT + b U

with (a + b = 1) is also consistent estimator for λ(θ).

Q.48 If we add to M.L.E., a function of observations which tends in probability to zero then this modified estimator will also have the asymptotic property.

Q.49 Barkson has shown through sampling experiment that in some particular situation arised in "Bio-Essay", the minimum x^2 estimator has smaller M.S.E. while M.L.E. some times has infinite M.S.E.

Q.50 The most general form of the distribution for which the sample geometric mean :

$$\text{G.M.}(x) = \left(\prod_{x}^{n} x_i\right)^{\frac{1}{n}}$$

is the M.L.E. of θ is of the form written as :

$$f(x, \theta) = \left(\frac{x}{\theta}\right)^{\theta A'(\theta)} \quad e(\text{A}(\theta) + c(x))$$

Q.51 The most general form of the distribution for which the statistic :

$$t(x) = \frac{n}{\left(\sum_{i}^{n} \frac{1}{x_i}\right)}$$

is the M.L.E. of θ, has the p.d.f. of the form written as :

$$f(x, \theta) = e^{[\frac{1}{\theta}\{\theta \text{ A}'(\theta) - \text{A}(\theta)\} - \text{A}'(\theta) + c(x)]}$$

Q.52 A sufficient (though not necessary) condition that any of the stationary values (M.L.E.) say θ will be a "local maxima" is that :

$$\text{L}''(\text{x}, \theta) < 0$$

Q.53 If we find all the local maxima of L.F. in the manner written in Q.52, and if there are more than one, then one must prefer the "Largest" of them.

Q.54 If a M.V.B. estimator 't' exists for $\psi(\theta)$ and likelihood function has a solution θ then :

$$t = \psi(\theta)$$

and solution θ is unique.

Q.55 Any sequence of estimators say (θ_n) obtained through L.F. is consistent.

Q.56 The unbiasedness of an estimator has not the "Invariance property."

Q.57 If θ is an unbiased estimator for θ then in general, we have :

$$\text{E}(\psi(\theta)) \neq \psi(\text{E}(\theta))$$

Q.58 In case of normal distribution $N(\mu, \sigma^2)$, when σ^2 is known then M.L.E. of population mean μ is given by the sample statistic $\overline{x}_n$ written as :

$$\mu = \bar{x}_n = \frac{1}{n}\sum_{i}^{n} x_i$$

Q.59 In Q.58, if population mean μ is known the M.L.E. of population variance σ^2 is given by the sample statistic s*2 defined as :

$$\sigma^2 = s^{*2} = \frac{1}{n}\sum_{i}^{n}(x_i - \mu)^2$$

Q.60 In Q.58, if μ and σ^2 (both) are unknown, their simultaneous M.L.E.'s are given by the sample estimator written as :

$$\mu = \bar{x}_n = \frac{1}{n}\sum_{i}^{n} x_i$$

$$\sigma^2 = \frac{1}{n}\sum_{i=1}^{n}(x_i - \bar{x}_n)^2$$

Q.61 In case of Poisson distribution $f(x, \lambda)$ the M.L.E. of λ is given by the sample estimator.

$$\lambda = \bar{x}_n = \frac{1}{n}\sum_{i=1}^{n} x_i$$

Q.62 In case of binomial distribution $b(x, n, p)$, the M.L.E. of p is given by the sample estimator as :

$$\hat{p} = \frac{r}{n} = \frac{1}{n}\sum_{i}^{n} x_i, \quad x_i = 0, 1, 2, \ldots\ldots, n$$

Q.63 In Q.62, a consistent estimator for 'p^2' is given by the sample estimator defined as :

$$p^2 = \left(\frac{r}{n}\right)^2 \text{ where r} = \sum_{x}^{n} x_i$$

Q.64 Consider the following p.d..f :

$$f(x, 0, \theta) = \frac{1}{\sqrt{2\pi}\,\theta}\, e^{-\frac{1}{2}\left(\frac{x}{\theta}\right)^2}, \quad -\infty < x < \infty$$

Then M.L.E. of θ is given by sample estimator defined as :

$$\theta = \sqrt{\frac{1}{n}\sum_{i}^{n} x_i^2}$$

Q.65 Consider the power series distribution defined as :

$$p\,(x = r) = \frac{a_r\,\theta^r}{f\,(\theta)},\ \text{r} = 0,\ 1,\ 2,\ \ldots\ldots\ldots$$

Then M.L.E. of the parameter θ is given as :

$$\theta = \sqrt{\overline{x}} = \sqrt{\frac{\theta f'\,(\theta)}{f\,(\theta)}}$$

Q.66 Consider the two normal population defined as $N(\mu, \sigma^2)$ and $N\,(\lambda\,\mu, \sigma^2)$ respectively. Then M.L.E. of μ is given as :

$$\mu = \left(\frac{n_1\overline{x}_1 + n_2\,\lambda\,\overline{x}_2}{n_1 + \lambda\,n_2}\right)$$

Q.67 In Q.66, the sampling variance of the estimator μ is given as :

$$\text{V}(\mu) = \left(\frac{\lambda\sigma^2}{n_1 + n_2\lambda}\right)$$

Q.68 Consider the rectangular distribution defined as :

$$f(x, \alpha, \beta) = \frac{1}{(\beta - \alpha)},\qquad \alpha < x < \beta$$

Then method of M.L.E. fails in this case and M.L.E.'s of α & β are given as :

$$\alpha = x_{(1)},\ \hat{\beta} = x_{(n)}$$

Q.69 Consider the following exponential distribution :

$f(x, \alpha, \beta) = y_0\; e{-}\beta\;(x - \alpha),\quad \alpha \leq x < \infty.$

where y_0 is a constant. Let $x_{(1)}, x_{(2)}$ ......, x(n) be an ordered sample of size n then MLE's of α & β are given as :

$$\alpha = x_{(1)} \text{ and } \beta = \left(\frac{1}{\overline{x} - x_{(1)}}\right)$$

Q.70 Let (t_1 & t_2) be two unbiased estimators of a parameter θ with variances (σ_1^2 and σ_2^2) and correlation ρ.

Then the linear combination $T = \sum_{i=1}^{2} l_i\, t_i$ is an unbiased estimator for θ if :

$(l_1 + l_2) = 1$

Q.71 In Q.70, the sampling variance of 'T' is given by :

$V(T) = l_1^2\, \sigma_1^2 + l_2^2\, \sigma_2^2 + 2 l_1\, l_2\, \rho\, \sigma_1\, \sigma_2.$

Q.72 In Q.70, the two constants l_1 and l_2 will be worked out as :

$$l_1 = \frac{\left(\sigma_2^2 - \rho\, \sigma_1\, \sigma_2\right)}{\left(\sigma_1^2 + \sigma_2^2\ -\ 2\rho\, \sigma_1\, \sigma_2\right)}$$

$$l_2 = \frac{\left(\sigma_1^2 - \rho\, \sigma_1\, \sigma_2\right)}{\left(\sigma_1^2 + \sigma_2^2\ -\ 2\rho\, \sigma_1\, \sigma_2\right)}$$

Q.73 In Q.70, the T is the "Best linear unbiased estimator" (combination) of θ.

Q.74 Let in Q.70, t_1 be a M.V.U.E. and t_2 is an other estimator of parameter θ with $V(t_2) = (\sigma^2/e)$. Then

Correlation $(t_1, t_2) = \sqrt{e}$

Q.75 The correlation coefficient between a most efficient estimator t_1' and any other estimator 't_2' with efficiency 'e' is given as :

correlation $(t_1, t_2) = \sqrt{e}$

Q.76 The most general form of the distribution differentiable in θ for which the sample mean '$\overline{x}_n$' is the M.L.E. of q, is written as :

$$f(x, \theta) = \text{constent Exp}\ \{\ (x - \theta)\ \frac{\partial \psi}{\partial \theta} + \psi(\theta) + 6(x)\ \}$$

Q.77 The most general form of continuous distribution for which the M.L.E. of the parameter θ is the "geometric mean" (G.M.) of the sample, is written as :

$$f(x, \theta) = \left(\frac{\alpha}{\theta}\right)^{\theta\frac{\partial\psi}{\partial\theta}} \text{ exp. } \{\psi(\theta) + 6\,(x)\}$$

where $\psi(\theta)$ and $6(x)$ are the arbitrary function of θ and x respectively.

Q.78 The most general form of the distribution for which the M.L.E. of θ is the "Harmonic mean" (H.M.) of the sample is given by :

$$f(x, \theta) = \text{constent. exp } [\,(x-\theta)\,\frac{\partial\psi}{\partial\theta} + \psi(\theta) + \xi(\mathrm{x})\,]$$

where $\psi(\theta)$ and $\xi(x)$ are the arbitrary function of θ & x.

Q.79 Consider the bivariate normal distribution as :

$f(x, \mu_1, \mu_2, \sigma_1^2, \sigma_2^2, \rho)$ and let μ_1 is known then M.L.E. of σ_1^2 is given as :

$$\sigma_1 = \frac{1}{n}\sum_{i=1}^{n}(x_i - \mu_1)^2$$

Q.80 In Q.79, if μ_2 is known then MLE of μ_2^2 is given by :

$$\mu_2^2 = \frac{1}{n}\sum_{i}^{n}(y_i - \mu_2)^2$$

Q.81 In Q.79, if (μ_1 & μ_2) are known and (σ_1^2, σ_2^2) are unknown then M.L.E. of p is given by :

$$\rho = \frac{\frac{1}{n}\sum^{n}(x_i - \mu_1)\,(y_i - \mu_2)}{\sigma_1\,\sigma_2}$$

Q.82 In Q.79, if (μ_1, μ_2) are unknown then their MLE's are given by :

$$\mu_1 = \frac{1}{n}\sum_{i}^{n} x_i \text{ and } \mu_2 = \frac{1}{n}\sum_{i}^{n} y_i$$

Q.83 The exponential distribution is given as :

$$f(x, \theta, p) = \frac{1}{\Gamma p} \exp (x - \theta) \; . \qquad (x - \theta)^{p-1}$$

$$\theta \leq x < \infty$$

$$p \geq 2$$

Let $x_{(1)}, x_{(2)}, ..., x_{(n)}$ be a ordered sample. Then smallest observation $x_{(1)}$ is neither sufficient statistic; nor the M.L.E. of θ.

Q.84 In Q.83, an estimator θ is unique and satisfies the condition that :

$$x_{(1)} - (p-1) \leq \theta < x_{(1)}$$

Q.85 Consider the following normal distribution :

$$f(x, \mu, \sigma^2) = \frac{1}{\sqrt{2\pi}\ \sigma} e^{-\frac{1}{2}\left(\frac{x-\mu}{\sigma}\right)^2}, -\sigma < x < \sigma.$$

Then the log normal distribution $y = e^x$ has the mean θ_1 given as

$$\theta_1 = e^{(\mu + \frac{\sigma^2}{2})}$$

Q.86 In Q.85, the variance v of the log normal distribution is given as :

$$v = e^{(2\mu + \sigma^2)} . (e^{\sigma^2} - 1)$$

Q.87 In Q.85, the M.L.E. of θ_1 is given as :

$$\theta_1 = e^{\bar{x} + \frac{s^2}{2}}$$

Q.88 In Q.85, the M.L.E. of the variance v is given as :

$$\hat{v} = e^{(2\bar{x} + s^2)} . (e^{s^2} - 1)$$

Q.89 Consider the bivariate normal distribution :

$$N (\mu_1, \mu_2, \sigma_1^2, \sigma_2^2, p).$$

Then estimating efficiency of the MLE ρ is given as :

$$e(\rho) = \left(\frac{1-\rho^2}{1+\rho^2}\right)^2$$

Q.90 The method of M.L.E. and method of moment do not lead to the "same estimator."

Q.91 Any unbiased estimator wih finite mean is "asymptotically consistent" estimator for parameter θ.

Q.92 Any consistent estimator with finite mean is "Asymptotically unbiased" for the parameter θ.

Q.93 The sampling variance (sample median x) for any distribution is given by the formula :

$$V(\tilde{x}) = \left(\frac{1}{4xy_0^2}\right)$$

where y_0 is the value of p.d.f. after putting the value of r.v.x equal to the value of population parameter θ.

Q.94 Consider the Cauchy distribution :

$$f(x, \theta) = \frac{1}{\pi[1+(x-\theta)^2]} \qquad -\infty < x < \infty$$

$$-\infty < \theta < \infty$$

Then using the general formula written in Q.93, the sampling variance of the sample median x is given as :

$$V(x) = \frac{\pi}{4n}$$

Q.95 Consider the normal distribution $N(x, \mu, \sigma^2)$ the the sampling variance of sample median x is given as :

$$V(x) = \left(\frac{\pi}{2}\right) \cdot \frac{\sigma^2}{n}$$

Q.96 The merit of an estimator is judged by the sampling distribution of the estimator.

Q.97 As the sample size n increases, the amount of sampling error = (Estimate – True value of parameter) decreases.

Q.98 In generl the acuracy of an estimator increases as the sample size n increases.

Q.99 In case of Cauchy distribution $f(x, \theta) = \frac{1}{\pi\{1+(x-\theta)^2\}}$, the accuracy of an estimator $t_n = \sum^n x_i$ does not dependes on the "sample size n"

becasue :

$$f(t_n, \theta) = \frac{1}{\pi\,[1+(t_n-\theta)^2]}$$

Q.100 The method of "Minimum variance estimation" is analogus to "Rao-Blackwell theorem."

Q.101 In case of normal distribution $N(x, \mu, \sigma^2)$, the M.V.U.E. does not exist for σ but it exists for σ^2.

Q.102 The method of minimum variance will give a solution only if there exists a sufficient estimator for parameter θ.

Q.103 Let $f(x, \theta)$, be a p.d.f. of r.v.x then we define

$$I = \int_{-\infty}^{\infty} \left(\frac{\partial}{\partial\theta} \log f\right)^2 f(x, \theta)\, dx.$$

and I is called "Intrinsic accuracy" of the distribution $f(x, \theta)$ for θ.

Q.104 In Q.103, sometime 'I' is called the "Amount of information" per member of the sample.

Q.105 For any estimator 't_n' we have the following relation :

Accuracy $(t_n) \leq$ Amount of information contained in the sample.

Q.106 Consider the following Trapezoidal distribution defined as :

$$f(x, \theta) = \frac{2x}{(2\theta + 1)}, \quad 0 \leq \theta < x \leq \theta + 1$$

Then the intrinsic accuracy of the distribution $f(x, \theta)$ for a sample of size n, is given as :

$$\text{Accuracy} = \frac{4n^2}{(2\theta + 1)^2}$$

Q.107 The "method of moment" for the estimation of population parameter θ has been developed by the statistician "Karl Pearson."

Q.108 In general the estimator obtained by applying the "method of moment" will be "less efficient" than the M.L.E.

Q.109 The method of moment will give solution identical with M.L. estimator only if the "parent distribution" is of a particular form.

Q.110 The most general form for the "method of moment" and "method of maximum likelihood estimation" to be identical is given as :

$f(x, \theta) = e\ \{a_0 + a_1\ x + a_2\ x_2 + \text{......}\ \text{......}\ \}$

where the constants a's depend upon θ.

Q.110* Let $m_r^{'}$ be the rth moment about the orign and $\hat{\theta}_r$ is the rth estimator of parameter θ_r (r = 1, 2, ..., k) respectively,

and $\theta_r = (m_1^{'}, m_2^{'}\ \text{..........}, m'_k)$. respectively then

$\theta_r = \theta_r\ (m_1^{'}, m_2^{'}\ \text{.......}, m_k^{'})$

Q.111 The sample moments are the "consistent estimators" of the corresponding population moments.

Q.112 Under quite general conditions, the estimates obtained by the "method of moment" are "Asymptotically normal" but in general not efficient.

Q.113 Prof. Karl Pearson has utilized the "method of moment" in fitting the "Pearsonian curves."

Q.114 The estimators obtained by the utilisation of "method of moments" are not in general "consistent and efficient."

Q.115 Consider the p.d.f. defined as :

$$f(x, \theta) = \frac{1}{\theta} \cdot e^{-\frac{x}{\theta}}, \quad 0 < x < \sigma.$$

Then for this distribution, the estimate obtained through the "method of moment" and "method of maximum likelihood" will be identical.

Q.116 In case of Poisson distribution $f(x, \lambda)$, the parameter λ can not be estimated through the "method of moment."

Q.117 Consider the rectangular distribution written as :

$$f(x, \theta) = 1 \text{ for } \theta - \frac{1}{2} \leq x < \theta + \frac{1}{2}$$

$$= 0, \text{ otherwise}$$

Then by the "method of moment" the parameter θ will be estimated as :

$$\theta = \bar{x}_n = \frac{1}{n} \sum^{n} x_i$$

Q.118 Consider the Pearson's type III distribution written as :

$$f(x, \alpha, \beta) = \frac{\beta^{\alpha}}{\overline{|\alpha}} x^{\alpha-1} e^{-\beta x}, \quad 0 \leq x < \infty$$

Let (μ'_1 & μ'_2) be the sample moments then (α, β) will be estimated as :

$$\alpha = \frac{\mu'^2_1}{(\mu'_2 - \mu'^2_1)}, \quad \beta = \frac{\mu'_1}{(\mu'_2 - \mu'^2_1)}$$

Q.119 Consider the double Poisson distribution defined as :

$$f(x, m_1, m_2) = \frac{1}{2}\left[\frac{e^{-m_1}\, m_1^x}{x!} + \frac{e^{-m_2}\, m_2^x}{x!}\right]$$

$$x = 0, 1, 2, \ldots\ldots\ldots$$

Then using the method of moment, the parameters (m_1, m_2) are estimated as :

$$m_1 = m_2 = \mu'_1 \pm \sqrt{\mu'_2 - \mu'_1 - \mu'^2_1}$$

Q.120 A statistic (estimator) does not depend upon any unknown parameter of the population.

Q.121 The sampling distribution of any statistic/estimator may very well depend upon unknown parameters of the parent population.

Q.122 Let ($x_1, x_2, \ldots.., x_n$) be a random sample of size n drawn from a parent population. Then $t = \sum_{i=1}^{n} x_i$ is a sufficient statistic for parameter θ in case of Poisson, Binomial, Gamma & Geometric distributions etc.

Q.123 An unbiased estimator of a parameter θ may not be unique.

Q.124 The method of minimum $\chi 2$ and modified minimum $\chi 2$ may be used when the observations are in the form of frequencies grouped into a finite numher of (say k) of mutually exclusive classes.

Q.125 A measure of discrepancy between the set of oberved and corresponding expected frequencies, is provided by Pearsonian Chi-square, defined as :

$$\theta^2 \quad = \sum_{i}^{k} [\{f_i - np_i(\theta)\}^2 / np_i(\theta)]$$

$$= \sum_{i}^{k} \frac{f_i^2}{np_i(\theta)} - n$$

Q.126 In Q.125, the measure of discrepancy is a function of the unknown parameters (say, θ_i, $i = 1, 2, \ldots.. k$).

Q.127 The method of minimum χ^2 uses those estimators of parameter θ which minimise χ^2.

Q.128 For large n(the number of observations), the minimum χ^2 equations and likelihood equations are identical.

Q.129 The method of minimum χ^2 is found more trouble some to apply in many cases and is not improvement over the M.L.E. method.

Q.130 In case of modified minimum χ'^2 method, we use those values of θ as estimates which minimises χ'^2 rather than χ^2 and is defined as :

$$\chi'^2 = \sum_{i}^{k} \frac{(np_i - f_i)^2}{f_i} = \sum_{i}^{k} \frac{n^2 p_i^2}{f_i} - n$$

Q.131 For large n, the estimates obtained through modified minimum χ'^2 method would be identical with the M.L. estimates.

Q.132 The least square estimates do not have any optimum property even "Asymptotic property" also is not possessed.

Q.133 The least square estimators obtained as linear function of sample value $(x_1, x_2, \ldots\ldots, x_n)$ will be M.V.U.E.

Q.134 The order statistics are used in testing experiment where certain ordered elements of a sample are known for testing and estimation purposes.

Q.135 The order statistic is ued in N.P. tests.

2.2 Fill in the Blanks

Q.1 From theoretical point of view, the most general method of estimation is the method of "...............................".

Q.2 The M.L.E. method was initially formulated by the statistician known as "...............................".

Q.3 The method of M.L.E. as a general method of estimation was first time introduced by Prof. "...............................".

Q.4 Let $x_1, x_2, \ldots\ldots, x_n$ be a random sample drawn from distribution $f(x, \theta)$, $\theta \in \Omega$. Then joint p.d.f. L(x, θ) is written as :

L (x, θ) = ..., $\theta \in \Omega\ \theta$

Q.5 The joint p.d.f. say L(x, θ), may be regarded as the function of "..............................".

Q.6 The joint p.d.f. $L(x, \theta)$ of random sample $x = (x_1, x_2, ..., x_n)$ is known as "..............................".

Q.7 Let U(x) be a non trivial function of random sample x. If $L\{U(x)/x\}$ is atleast as greate as L(x, 0), $\forall_\theta \in_\theta \Omega_\theta$, then U(x) is known as "...................." for parameter θ.

Q.8 The M.L.E. of θ is denoted by θ and is expressed as :

θ = ...

Q.9 The M.L.E. θ of parameter θ is the solution of

$$\frac{\partial}{\partial\theta}\log_e L(x,\theta)=0 \text{ and } \frac{\partial^2}{\partial\theta^2}\log(x,\theta) \text{ .... } < 0$$

Q.10 The equation $\frac{\partial}{\partial\theta}$. $\log L(x, \theta) = 0$ is usually known as "...................."

Q.11 In general, the M.L.E. say θ of parameter θ (....................) be unique.

Q.12 There is an assumption that for M.L.E. θ, the derivative $\frac{\partial}{\partial\theta}\log L(x,\theta)$ and $\frac{\partial^2}{\partial\theta^2}\log L(x, \theta)$ exist and are (........................) of θ in range R including the (..................) of θ for almost all x.

Q.13 For the existance of M.L.E., the following inequality hold :

$$\left|\frac{\partial}{\partial\theta}\log L(x,\theta)\right| < F_1(x) \text{ and}$$

$$\left|\frac{\partial^2}{\partial\theta^2}\log L(x,\theta)\right| < F_2(x)$$

where $F_1(x)$ and $F_2(x)$ are (........................) of x over (...............).

Q.14 For the existance of M.L.E., we assume that 3rd order derivative. $\frac{\partial^3}{\partial\theta^3}$ $\log L(x, \theta)$ exists such that :

$$\left| \frac{\partial^3}{\partial\theta^3} \log L(x,\theta) \right| < \text{...........}$$

where M(x) < k for k being a positive quantity.

Q.15 For every θ in R, we have

$$E\left[-\frac{\partial^3}{\partial\theta^3} \log L(x,\theta) \right] = \text{.............................} = I(\theta).$$

where $I_{(\theta)}$ is finite and non zero.

Q.16 For the existance of M.L.E. say$_{\theta}$, we assume that the range of integration is (......................) of parameter θ.

Q.17 In case of M.L.E., if the range of integration depends upon parameter θ, then $f(x, \theta)$ vanishes at the extreme depending upon (...............).

Q.18 The assumption made in Q.(17) is made to make the differentiation valid under. (......................).

Q.19 Cramer Rao theorem on consistency states that the likelyhood equation $\frac{\partial}{\partial\theta} \log L(x, \theta) = 0$, has a solution which converges in (......................). to the true value θ_0.

Q.20 M.L.E.'s are (..........................) estimators to the parameters.

Q.21 Hazoor Bajar theorem states that "Any consistent solution of likelihood equation provides a (..........). to the likelihood function with probability $p \rightarrow 1$ as sample sixe $n \rightarrow$ (..........).

Q.22 A consistent solution of likelihood equation is (....................) normally distributed about true value θ.

Q.23 Let $_{\theta}$ be the M.L.E. of parameter θ then we have $_{\theta}$ follows Asymptotically N (....., ...)

where θ_0 is the true value of θ.

Q.24 Let t(x) be a sufficient statistic (estimator) for parameter θ then M.L.E. is the function of (..........).

Q.25 Let $\frac{\partial}{\partial\theta} \log L(x, \theta) = \psi\,(t(x), \theta)$ then $t(x)$ is regarded as (..........) for parameter θ.

Q.26 If it is not possible to write $\frac{\partial}{\partial\theta} \log \mathrm{L}(x, \theta) = \{\psi(t(x), \theta)\}$, then there does not exist a (..........) t(x) for parameter θ.

Q.27 Let $\hat{\theta}$ be the M.L.E. of parameter θ. then the invarient property states that g($\hat{\theta}$) is the M.L.E. of (.....................).

Q.28 Let $\hat{\theta}$ be the M.L.E. of θ, then for large samples, $\hat{\theta}$ is most (..........) under the assumption that the sampling distribution of $\hat{\theta}$ tends to follow (..........).

Q.29 In case of multivariate distribution, let θ be the vector as $\theta = (\theta_1, \theta_2, ..., \theta_m)$. Then the necessary condition for the likelihood equation to have a (....................) is that

$$\frac{\partial}{\partial \theta_i} \log L(x, \theta) = 0 \ \forall \ i = 1, 2, ..., m.$$

Q.30 In Q.(29), a sufficient condition is that the matrix $\frac{\partial^2}{\partial \theta^2} \log L(x, \theta)$ is (..........) matrix.

Q.31 In Q.(29), the $E\left(\frac{\partial^2}{\partial \theta_i \, \partial \theta}\right) \log L(x, \theta)$ is called (....................).

Q.32 If the method of M.L.E. leads to a complicated equations, then the method adopted to get the solution is the (....................).

Q.33 The iterative method applied to get a solution in case of complicated likelihood equaution is also known as method of (.........).

Q.34 An estimator obtained through the application of method of scoring is often consistent but is less (..........).

Q.35 In case of Cauchy distribution $f(x, \theta) = \frac{1}{\pi} - \frac{1}{[1 + (x - \theta)^2]}$, the method of scoring is adopted to get (..........). of parameter θ.

Q.36 In Cauchy distribution, the sample median is the (..........) estimator of parameter θ.

Q.37 In Q.(35). the M.L.E. of θ is given by the (..........).

Q.38 Let t(x) be the sample median in Cauchy distribution $f(x, \theta) = \frac{1}{\pi}$

$\frac{1}{[1+(x-\theta)^2]}$, $-\infty < x < \infty$, and $\hat{\theta}$ is the M.L.E. of θ, then we have

$$\hat{\theta} = t(x) + \frac{4}{n}\sum_{i}^{n}\frac{(\theta_i - \theta)}{[1+(..........)^2]}.$$

Q.39 In Q.(38), the sampling variance of t(x) is given as :

$$V(t(x)) = \frac{1}{I_{(\theta)}} = ...$$

Q.40 Under some regularity conditions, if range of x is independent of θ then likelihood equation

.... = 0.

has an unique solution with probability 1 and that solution $\hat{\theta}$ will be (..........).

Q.41 If the regularity conditions for the efficiency of M.L.E. do not exist, even then the M.L.E. may have (..........) confered upon it by the (..........) of which, it is always a function.

Q.42 If the M.L.E. $\hat{\theta}$ is the root of $\frac{\partial}{\partial\theta}\log L(x,\theta) = 0$ for which $\hat{\theta} = \bar{x}_n$, then the most general form of the distribution (..........) in θ is expressed as : $f(x, \theta) = e^{\{A(\theta) + A'(\theta)(x-\theta) + B(x)\}}$

Q.43 In Q.(42), the sample mean $\bar{x}_n$ is (..........) for θ.

Q.44 In Q.(42), the $\hat{\theta}$ has minimum variance bound (M.V.B.) equal to (..........).

Q.45 Consider the binomial distribution B(x, n, p), then the consistent estimator t(x) for p(1–p) will be given as :

$$t(x) = \frac{r}{n}\,(..........) \text{ where } r = \sum_{i=1}^{n} x_i$$

Q.46 In Q.(45), the consistent estimator $t(x)$ for $\frac{1}{p}$ is given as :

$$t(x) = \frac{n}{r}.$$

Q.47 Let (T, U) be the consistent estimators for λ(θ). Then the statistic T* defined as :

T* = a T + b.U

with (a + b = ...) is also (..........) for $\lambda(\theta)$.

Q.48 If we add to M.L.E., a function of observations which tends in probability to zero then this modified estimator will have the (..........) property.

Q.49 Barkson has shown through sampling experiment that in some particular situation arised in "Bio-essay", (.....................) has smaller M.S.E. while M.L.E. sometimes has infinite (..........).

Q.50 The most general form of the distribution for which the simple geometric mean.

$$\text{G.M.}\ (x) = \left(\prod_{i}^{n} x_i\right)^{\frac{1}{n}}$$ is the (..........) of θ is of the form written as :

$$f(x, \theta) = \left(\frac{x}{\theta}\right)^{\theta A'(\theta)} \qquad e^{(A(\theta) + c(x))}$$

Q.51 The most general form of the p.d.f. of distribution is given as :

$$f(x, \theta) = e^{\left[\frac{1}{\theta}\{\theta A'(\theta) - A(\theta)\} - A'(\theta) + C(x)\right]}$$

for which the M.L.E. of θ is given by a statistic t(x) written as :

$$t(x) = \frac{n}{(..........)}$$

Q.52 A sufficient (though not necessary) condition that any of the stationary values (M.L.E.) say θ will be a (..........) is that

L" (x, θ) < 0

Q.53 If we find all the local maximas of L.F. in the manner written in Q.(52) and if there are more than one, then one must prefer the (..........) of them.

Q.54 If a M.V.B. estimator $t(x)$ exists for $\psi(\theta)$ and LF. has a solution θ then :

$t(x) = \psi$ (..........).

Q.55 If θ is an unbiased estimator for θ then in general we have :

$E(\psi(\theta))$ .... $\psi(E(\theta))$

Q.56 Any sequence of estimators say (θ_n) obtained through likelihood function is (..........).

Q.57 Let $x \sim N(\mu, \sigma^2)$ where σ^2 is known. Then M.L.E. of μ is given by an estimator $t_n(x)$ written as :

$t_n(x) = (..........)$

Q.58 In Q.(57), let μ be known, then M.L.E. of σ^2 is given as :

$\sigma^2 =$ ....

Q.59 Let in Q.(57), both (μ & σ^2) be unknown, then their M.L.E.'s are given respectively as :

$t_1 =$ ...., $t_2 =$ ....

Q.60 Let $x \sim P(x, \lambda)$, then M.L.E. of λ is given by a sample estimator written as :

$\lambda =$ ....

Q.61 Let $x \sim b(x, n, p)$ then M.L.E. of p is given by a sample estimator written as :

$p =$ ....

Q.62 In Q.(61), a consistent estimator for p^2 is given by a sample estimator given as :

$p^2 =$ ...., where $r = \sum_{i}^{n} x_i$

Q.63 Let $x \sim N(0, \theta^2)$ with p.d.f. written as :

$$f(x, 0, \theta^2) = \frac{1}{\sqrt{2\pi}\,\theta} \cdot e^{\frac{-1}{2}\left(\frac{x}{\theta}\right)^2}, \quad -\infty < x < \infty$$

Then M.L.E. of θ is given by a sample estimator θ written as :

$\theta = \sqrt{..........}$

Q.64 Consider the power series distribution written as :

$P(x = r) = a_r\, \theta^r / f(\theta)$, $r = 0, 1, 2,$ ....

Then M.L.E. of θ is given as :

$\theta = \theta\,(..........)$

Q.65 Consider two normal populations defined as :

$N(\mu\lambda, \sigma^2)$ and $N(\mu, \sigma^2)$ respectively. Then M.L.E. of μ is given as :

$$\mu = \left(\frac{n_1 \bar{x}_1 + n_2 \lambda \bar{x}_2}{\cdots\cdots\cdots}\right)$$

Q.66 In Q.(65), the sampling variance of the estimator μ is given as :

$V(\mu = (\ldots\ldots\ldots)$.

Q.67 Consider the rectangular distribution defined as :

$$f(x, \alpha, \beta) = \frac{1}{(\beta - \alpha)}, \quad \alpha < x < \beta$$

and let ($x_1 < x_2 <$ .... $< x_n$) be an ordered sample, the method of (..........) fails in this case and M.L.E.'s of (α, β) are given respectively as :

α = ... , β = ...

Q.68 Consider the following exponential distribution :

$f(x, \alpha, \beta) = y_0\ e{-}\beta(x - \alpha), \quad \alpha < x < \infty$

where y_0 is constant. Let $x_1, x_2, \ldots\ldots x_n$ be an ordered sample then M.L.E.'s of α and β are given as :

α = ... , β = (..........)

Q.69 Let (t_1, t_2) be two unbiased estimators of a parameters θ with sampling variances (σ_1^2 & σ_2^2) and correlation ρ. Then Linear combination

$T = \sum_{i=1}^{2} l_i\ t_i$ is an unbiased estimator for θ iff :

$l_1 + l_2 =$ ....

Q.70 In Q.(69), the sampling variance V(T) is given as :

$V(T) = l_1^2\ \sigma_1^2 + l_2^2\ \sigma_1^2 +$ ....

Q.71 In Q.(69), the two constant l_1 & l_2 will be worked out as :

$$l_1 = \frac{(\ldots\ldots\ldots)}{\left(\sigma_1^2 + \sigma_2^2 - 2\rho\,\sigma_1\sigma_2\right)},\ l_2 = \frac{(\ldots\ldots\ldots)}{\left(\sigma_1^2 + \sigma_2^2 - 2\rho\,\sigma_1\sigma_2\right)}$$

Q.72 In Q.(69), the estimator T is the "Best linear ....) estimator for θ.

Q.73 Let in Q.(69), t_1 be a minimum variance unbiased estimator and t_2 be an other estimator of θ with $V(t_2) = (\sigma^2/e)$. then we have :

Correlation $(t_1, t_2) =$ ....

Q.74 The correlation coefficient (ρ) between a most efficient estimator t_1, and any other estimator t_2 with efficiency 'e' is given as :

$\rho\ (t_1, t_2) =$ ....

Q.75 Let a most general form of the distribution, differentiable in θ is given as :

$$f(x, \theta) = \text{constant Exp } \{ (x-\theta) \frac{\partial\psi}{\partial\theta} + y(q) + 6(x) \}.$$

Then the M.L.E. θ of θ is given as :

$\theta =$ .... (Sample values)

Q.76 The most general form of the continuous distribution is given as :

$$f(x, \theta) = \left(\frac{\alpha}{\theta}\right)^{\theta\frac{\partial\psi}{\partial\theta}} . \quad \exp \{ \psi(\theta) + 6(x) \}$$

where $\psi(\theta)$ and $6(x)$ are the arbitrary function of θ and x respectively. Then M.L.E. of θ say θ is given as :

$\theta =$ .... (Sample value)

Q.77 The most general form of a distribution is given as :

$$f(x, \theta) = \text{constant . exp } [(x-\theta) \frac{\partial\psi}{\partial\theta} + \psi(\theta) + 6(x)]$$

where $\psi(\theta)$ & $6(x)$ are the arbitrary function of θ and x. Then M.L.E. of θ say θ is given as :

$\theta =$ .... (Sample values).

Q.78 Consider the bivariate normal distribution as f $(x, y, \mu_1, \mu_2, \sigma_1^2, \sigma_2^2, \rho)$ and let μ_1 be known, then M.L.E. of σ_1^2 is given as :

$\hat{\sigma}_1^2 =$ ....

Q.79 If in Q.(78), the μ_1 is known then M.L.E. of σ_2^2 is given as $\hat{\sigma}_2^2 =$ ....

Q.80 In Q.(78), if (μ_1, μ_2) are known and (σ_1^2, σ_2^2) are unknown then M.L.E. of ρ say ρ is given as :

$$\hat{\rho} = \frac{\ldots\ldots\ldots\ldots\ldots}{\hat{\sigma}_1 \hat{\sigma}_2}$$

Q.81 If in Q.(78), the (μ_1, μ_2) are unknown then their MLE's are given as :

μ_1 = ..., μ_2 = ...

Q.82 Consider the normal distribution $N(x, \mu, \sigma^2)$. Then the log normal distribution $y = e^x$ has the mean θ_1 given as :

θ_1 = ...

Q.83 In Q.(82), the variance 'V' of log normal distribution $y = e^x$ is given as :
$V = e^{(2\mu + \sigma 2)}$. (..........)

Q.84 In Q.(82), the M.L.E. of θ_1 is given as :

$$\theta_1 = e^{\ldots\ldots + \frac{\$^2}{2}}$$

where $\$^2$ has usual meaning.

Q.85 In Q.(82), the M.L.E. of variance "V" of log normal distribution $y = e^x$ is given as :

$\lozenge = e^{(2\bar{x} + \$^2)}$. (..........)

Q.86 Consider the bivariate normal distribution as

$N(x, y, \mu_1, \mu_2, \sigma_1^2, \sigma_2^2, \rho)$ then the estimating efficiency of the MLE. ρ is given as

e (ρ) = $(\ldots\ldots\ldots)^2$

Q.87 The method of M.L.E. and method of moment do not lead to the (..........).

Q.88 Any unbiased estimator with finite mean is (..........) estimator for parameter θ.

Q.89 Any consistent estimator with finite mean is (...........) estimator for parameter θ.

Q.90 The sampling variance of sample median (x) for any distribution is given by the expression :

$$V(\tilde{x}) = \left(\frac{1}{\ldots\ldots\ldots\ldots\ldots y_0^2}\right)$$

where y_0 is the value of p.d.f. after putting the value of r.v.'x' equal to the value of population parameter θ_0.

Q.91 For Cauchy distribution $f(x, \theta) = \dfrac{1}{\pi\,[1 + (x - \theta)^2\,]}$, with the help of formula in Q.(90), the $V(x)$ is given as :

$V(x) = (\ldots\underline{\pi}\ldots\ldots\ldots\ldots)$

Q.92 In case of normal distribution $N(x, \mu, \sigma^2)$ the $V(x)$ is given as :

$$V(x) = (\ldots\ldots\ldots\frac{\sigma^2}{n})$$

Q.93 The merit of an estimator is judged by the sampling variance of the (.........).

Q.94 As the sample size n increases, the amount of samling error (...........) decreases.

Q.95 In case of Cauchy distribution $f(x, \theta) = \dfrac{1}{\pi\{1 + (x - \theta)^2\}}$ the accuracy of an estimator $t_n = \sum^{n} x_i$ does not depend on (..........). because

$$f(t_n, \theta) = \frac{1}{\pi\,[1 + (t_n - \theta)^2\,]}$$

Q.96 The method of "Minimum variance estimation" is analogus to (....................) theorem.

Q.97 In case of normal distribution $N(x, \mu, \sigma^2)$, the M.V.U.E. does not exist for (..........) but it exists for (..........).

Q.98 The method of "Minimum variance" will give a solution only if there exists a (..........) for parameter θ.

Q.99 In general, the accuracy of an estimator increases as the sample size n ....

Q.100 For the p.d.f. $f(x, \theta)$, we define :

$$I_\theta = \int_{-\infty}^{\infty} \left(\frac{\partial}{\partial \theta} \log f \right)^2 f(x, \theta)\, dx$$

and I_θ is known as (..........) for θ.

Q.101 In Q.(100), some time I_θ is called as the amount of (..........) per member of the sample.

Q.102 For any estimator t_n, we have the following relation Accuracy (t_n) .... Amount of information contained in the sample.

Q.103 Consider the Trapezoidal distribution given as :

$$f(x, \theta) = \frac{2x}{2\theta + 1}, \quad 0 < \theta < x < \theta + 1.$$

Then intrinsic accuracy of $f(x, \theta)$ for sample size n, is given as :

$$\text{Accuracy } (f(x, \theta)) = \frac{4n^2}{(..........)}$$

Q.104 The (method of moment) for estimation of population parameter θ, has been developed by statistician (..........).

Q.105 In general, the estimator obtained through "Method of moment" will be (..........) than the M.L.E.

Q.106 The method of moment will give a solution identical with M.L. estimator only if the (..........) is of a particular form.

Q.107 The sample moments are the (..........) for corresponding population moments.

Q.108 Under quite general conditions, the estimates obtained through "Method of moment" are (..........) normal but in general not efficient.

Q.109 Prof. Karl Pearson has utilized the "Method of moment" to fit the curves known as (..........).

Q.110 The estimators obtained through "Method of Moment" are not in general ".......... and ....) estimators.

Q.111 Consider the p.d.f. as $f(x, \theta) = \frac{1}{\theta} e^{-\frac{x}{\theta}}, 0 < x < \infty$

Then the estimate for θ obtained through "Method of moment" and through "Method of M.L.E." will be (..........).

Q.112 In case of Poisson distribution $f(x, \lambda)$ the parameter λ can not be estimated through method of (..........).

Q.113 Consider the rectangular distribution :

$$f(x, \theta) = 1, (\theta - \frac{1}{2}) \leq x < (\theta + \frac{1}{2})$$

Then by the method of (.........), the θ will be estimated as :

$$\theta = \frac{1}{n} \sum_{i}^{n} x_i$$

Q.114 Consider the Pearson's type III distribution written as :

$$f(x, \alpha, \beta) = \frac{\beta^\alpha}{\Gamma\alpha} x^{\alpha-1} e^{-\beta x}, \quad 0 < x < \infty$$

Let (μ_1' & μ_2') be the sample moments then (α, β) will be estimated as :

$$\alpha = \frac{\mu_1'^2}{(..........)}, \quad \beta = \frac{\mu_1'^2}{(..........)}$$

Q.115 Consider the double Poisson distribution defined as :

$$f(x, m_1, m_2) = \frac{1}{2}\left[\frac{e^{-m_1} m_1^x}{x!} + \frac{e^{-m_2} m_2^x}{x!}\right]$$

Then through method of moment, the parameters (m_1, m_2) are estimated as :

$$m_1 = \mu_1' + \sqrt{\mu_2' - \mu_1' - \mu_1'^2}, \quad m_2 = \mu_1' \,.........\, \sqrt{....................}$$

Q.116 A statistic (estimator) does not depend upon any unknown (..........).

Q.117 The sampling distribution of any statistic/estimator may very well depend upon (..........) of the parent population.

Q.118 The method of minimum χ^2 and modified minimum χ^2, may be used when the observations are in the form of frequencies grouped into finite number of (say k) .... classes.

Q.119 The method of minimum χ^2 uses those estimates of parameter θ which (..........) χ^2.

Q.120 For large n (the sample size), the minimum χ^2 equations and likelihood equations are (..........).

Q.121 The method of minimum χ^2 is found more troublesome to apply in many cases and is not improvement over (..........). method.

Q.122 In case of modified minimum χ'^2 method, we use those, values of θ as estimates which (..........) rather than χ^2.

Q.123 For large n, the estimates obtained through modified minimum χ^{12} method would be identical with (..........).

Q.124 The least square estimates do not have any optionum property even (..........) also is not possessed.

Q.125 The least square estimators obtained as linear function of sample values will be minimum variance (..........). estimates.

2.3 Multiple Choice Questions and Answers

Q.1 From theoretical point of view, the most general method of estimation is known as method of :

(a) Maximum likelihood (b) Point estimation
(c) Interval estimation (d) Moment.

Q.2 M.L.E. method was initially formulated by the statistician known as :

(a) Prof. R.A. Fisher (b) Prof. Mahalanobis
(c) C.F. Gauss (d) Prof. Karl Pearson

Q.3 The method of M.L.E. as a general method was first time introduced by :

(a) Prof. R.A. Fisher (b) Prof. Karl Pearson
(v) Prof. J. Nemom (d) Prof. C.R. Rao

Q.4 Let $x_1, x_2, \ldots, xn$ be a random sample drwn from a distribution $f(x, \theta)$, $\theta \in \Omega\ \theta$. Then their joint p.d.f. say $L(x, \theta)$ is given by :

(a) $L(x, \theta) = \prod_{i}^{n} f(x_i, \theta)$ (b) $L(x, \theta) = \sum_{i}^{n} f(x_i, \theta)$

(c) $L(x, \theta) = \int_{-\infty}^{\infty} f(x_i, \theta)\ dx_i$ (d) None

Q.5 The joint p.d.f. $L(x, \theta)$ may be regarded as the function of :

(a) Parameter θ alone (b) θ and an estimator t_n
(c) θ and a statistic (tx) (d) None

Q.6 The joint p.d.f. $L(x, \theta)$ of random sample $x_1, x_2, \ldots, x_n$, is known as :

(a) Likelihood equation (b) Likelihood function

(c) M.L.E. (d) Probability law

Q.7 Let U(x) be a non trivial function of random sample x. If L{U(x) / x } is as great as L(x) $\forall$ $\theta \in \Omega$ θ, then U(x) ks known as :

(a) Sufficient estimator for θ.

(b) Maximum likelihood estimator for θ.

(c) Consistent estimator for θ.

(d) Minimum variance estimator for θ.

Q.8 Let MLE of θ be denoted by θ and is expressed as :

(a) $\theta = U'(x)$ (b) $\theta = U(x)$

(c) $\theta = \dfrac{U(x)}{n}$ (d) None

Q.9 The MLE θ of θ is the solution of :

(a) $\dfrac{\partial}{\partial\theta}\log L(x, \theta) = 0$ and $\dfrac{\partial^2}{\partial\theta^2}\log L(x, \theta) > 0$

(b) $\dfrac{\partial}{\partial\theta}\log L(x, \theta) = 0$ and $\dfrac{\partial^2}{\partial\theta^2}\log L(x, \theta) < 0$

(c) $\dfrac{\partial}{\partial\theta}\log L(x, \theta) = 0$ and $\dfrac{\partial^2}{\partial\theta^2}\log L(x, \theta) = 0$

(d) $\dfrac{\partial}{\partial\theta}\log L(x, \theta) \geq 0$ and $\dfrac{\partial^2}{\partial\theta^2}\log L(x, \theta) \geq 0$

Q.10 The equation $\dfrac{\partial}{\partial\theta}\log L(x, \theta) = 0$, is usually known as :

(a) An estimator (b) Likelihood equation

(c) Consistent solution (d) Consistent equation

Q.11 In general, the MLE say θ of parameter θ.

(a) Will be unique (b) Will not be unique

(c) May be unique (d) May or may not be unique

Q.12 There is an assumption for MLE say θ that the derivatives $\dfrac{\partial}{\partial\theta}$ log L(x, θ) and $\frac{\partial^2}{\partial\theta^2}$ log L(x, θ) exist and are (..........) of θ in range R including the (..........) of θ for almost all x.

(a) Continuous function, true value.

(b) True value, continuous function

(c) Dependent, true value

(d) Independent, continuous function

Q.13 For the existance of MLE say θ, the following inequalities are true :

(a) $\left|\frac{\partial}{\partial\theta}\log L(x,\theta)\right| < F_1(x)$. and $\left|\frac{\partial^2}{\partial\theta^2}\log L(x,\theta)\right| < F_2(x)$

(b) $\left|\frac{\partial}{\partial\theta}\log L(x,\theta)\right| > F_1(x)$ and $\left|\frac{\partial^2}{\partial\theta^2}\log L(x,\theta)\right| > F_2(x)$

(c) $\left|\frac{\partial}{\partial\theta}\log L(x,\theta)\right| > F_1(x)$ and $\left|\frac{\partial^2}{\partial\theta^2}\log L(x,\theta)\right| < F_2(x)$

(d) $\left|\frac{\partial}{\partial\theta}\log L(x,\theta)\right| < F_1(x)$. and $\left|\frac{\partial^2}{\partial\theta^2}\log L(x,\theta)\right| > F_2(x)$

Q.14 For the existance of M.L.E., we assume that 3rd order derivative $\frac{\partial^3}{\partial\theta^3}$ log L(x, θ) exists and for M(x) < k, lies as :

(a) >M(x)

(b) $\geq$ M(x)

(c) M(x)

(d) < M(x)

Where k is a positive quantity :

Q.15 For every θ is R and $I_{(\theta)}$ being finite and non zero, we have :

(a) $E\left(-\frac{\partial^2}{\partial\theta^2}\log L(x,\theta)\right) > \int_{-\infty}^{\infty}\frac{\partial^2}{\partial\theta^2}\log L(x,\theta)\,dx$

(b) $E\left(-\frac{\partial^2}{\partial\theta^2}\log L(x,\theta)\right) = \int_{-\infty}^{\infty}\frac{\partial^2}{\partial\theta^2}\log L(x,\theta)\,dx$

(c) $E\left(-\frac{\partial^2}{\partial\theta^2}\log L(x,\theta)\right)=0$

(d) $E\left(-\frac{\partial^2}{\partial\theta^2}\log L(x,\theta)\right)<\int_{-\infty}^{\infty}(-\frac{\partial^2}{\partial\theta^2}\log L(x,\theta))\,dx$

Q.16 For the existence of M.L.E. say θ, we assume that the range of integration is :

(a) Dependent of θ. (b) Independent of θ.

(c) Not dependent of θ. (d) Book (b) and (c)

Q.17 In case of M.L.E., if the range of integration depends on θ, then $f(x, \theta)$ vanishes at the :

(a) Extreme depending on θ. (b) Extreme independent of θ.

(c) Extreme > θ. (d) Extreme < θ.

Q.18 The assumption made in Q.(17) is made to make the differentiation valid under :

(a) Differentiation sign. (b) Integral sign.

(c) Summation sign. (d) Both (b) & (c).

Q.19 Cramer Rao theorem on "consistency property" of an estimator θ states that the equation $\frac{\partial}{\partial\theta}\log L(x,\theta)=0$, has a solution which converges to the true value θ_0.

(a) In probability (b) With probability 1.0

(c) With probability < 0.5 (d) With probability > 0.5

Q.20 M.L.E.'s are (..........) to the parameters θ's.

(a) Unbiased estimators (b) Efficient estimators

(c) Consistent estimators (d) Biased estimators

Q.21 Hazoor Bajar theorem states that any consistent solution of likelihood equation provides a (..........) to the likelihood function with probability p $\to 1$ as sample size $n \to$ (..........).

(a) Maximum, ∞. (b) Minimum, ∞.

(c) Maximum, 100 (d) Minimum, ∞

Q.22 A consistent solution of the likelihood equation about the true value θ_0 is :

(a) Normally distributed

(b) Asymptotically normally distributed
(c) Continuously distributed
(d) Uniformally distributed

Q.23 Let θ be the M.L.E. of θ with true value θ_0 then we have θ distributed as :

(a) Asymptotically $N(\theta_0, I_\theta)$ (b) Asymptotically $N(I_{(\theta)}, \theta_0)$
(c) Asymptotically $N(\mu, \sigma^2)$ (d) Asymptotically $N(\mu_0, \sigma_0^2)$

Q.24 Let $t(x)$ be a sufficient statistic (estimator) for parameter θ, then M.L.E. is the function of :

(a) θ (b) $t(x)$
(c) $t^2(x)$ (d) $\sqrt{t(x)}$

Q.25 Let $\frac{\partial}{\partial\theta}\log L(x, \theta) = \psi(t(x), \theta)$, then $t(x)$ is regarded for parameter θ as :

(a) An unbiased estimator (b) A consistent estimator
(c) A sufficient estimator (d) Linear estimator

Q.26 If it is not possible to write $\frac{\partial}{\partial\theta}\log L(x, q) = \{\psi(t(x), \theta\}$, then there does not exist for parameter θ a.

(a) Sufficient statistic (b) Consistent estimator
(c) Efficient estimator (d) M.V.U.E.

Q.27 Let θ be the MLE of θ, then invariant property states that $g(\theta)$ is the :

(a) MLE of $g(\theta)$ (b) MLE of $g(\theta^2)$
(c) Unbiased estimator of $g(\theta)$ (d) Biased estimator of $g(\theta)$

Q.28 Let θ be the MLE of θ, then for large samples, the sampling distribution of θ tends to follow the :

(a) Binomial distribution (b) Normal distribution
(c) Uniform distribution (d) Poisson distribution

Q.29 Let $\theta = (\theta_1, \theta_2, ... , \theta_m)$ be a vector of parameters. Then necessary condition for the likelihood equation to have a absolute maxima is that :

(a) $\frac{\partial}{\partial \theta_i} \log L(x, \theta) > 0$

(b) $\frac{\partial}{\partial \theta_i} \log L(x, \theta) < 0$

(c) $\frac{\partial}{\partial \theta_i} \log L(x, \theta) = 0$

(d) $\frac{\partial}{\partial \theta_i} \log L(x, \theta) \geq 0$

for all i = 1, 2, ..., m

Q.30 In Q.(29), a sufficient condition is that the matrix $\frac{\partial^2}{\partial \theta^2} \log L(x, \theta)$ is

(a) Positive definite (b) Negative definite
(c) Singular (d) Diagonal

Q.31 In Q.(29), the matrix $E\left(\frac{\partial^2 \log}{\partial \theta_i \, \partial \theta_j} L(x, \theta)\right)$ is called :

(a) Dispersion matrix (b) Correlation matrix
(c) Diagonal matrix (d) Asymetrical matrix

Q.32 If the method of MLE leads to complicated equations, then the method adopted to get the solution is the :

(a) Method of iteration (b) Method of moment
(c) Method of minimum $\chi 2$ (d) Method of least square

Q.33 The iterative method applied to get a solution in case of complicated likelihood equation is also known as the :

(a) Method of moment
(b) Method of minimum χ^2
(c) Method of scoring
(d) Method of least square

Q.34 An estimator obtained through method of scoring is often consistent but is

(a) Less efficient (b) Less sufficient
(c) More efficient (d) Less biased

Q.35 In case of Cauchy distribution $f(x, \theta) = \frac{1}{\pi} \frac{1}{1+(x-\theta)^2}$ the method of scoring is used to get :

(a) Unbiased estimator for θ.0 (b) Sufficient estimator for θ.

(c) M.L.E. for θ. (d) Efficient estimator for θ.

Q.36 In Cauchy distribution $f(x, \theta) = \frac{1}{\pi} \frac{1}{1+(x-\theta)^2}$, the parameter θ is estimated through :

(a) Sample statistic (b) Sample median x.

(c) Sample estimator. (d) Sample correlation

Q.37 In Q.(36), the M.L.E. of θ is given by the :

(a) Sample mean. (b) Sample median.

(c) Sample variance s^2. (d) Sample regression coefficient

Q.38 An estimator for parameter θ through method of scoring is often :

(a) Efficient estimator (b) Consistent estimator

(c) Unbiased estimator (d) Minimum variance estimator

Q.39 Let t(x) be the sample median of Cauchy distribution $f(x, \theta) = \frac{1}{\pi} \frac{1}{1+(x-\theta)^2}$ then V (t(x)) is equal to

(a) $\frac{2}{n}$ (b) $\frac{n}{2}$

(c) $\frac{2}{3}n$ (d) 2 n

Q.40 Under some regularity conditions, if range of x is independent of parameter θ then the solution obtained from the likelihood equation will be :

(a) Consistent estimator (b) M.V.B. estimator

(c) Sufficient estimator (d) None.

Q.41 If the regularity conditions for the efficiency of M.L.E. do not exist, even then the M.L.E. may have minimum variance which is conferred by a statistic known as :

(a) Efficient statistic (b) Sufficient statistic

(c) Consistent statistic (d) Unbiased statistic

Q.42 If the MLE $\theta = \overline{x}_n$, then most general form of the distribution differentiable in θ is expressed as :

(a) $f(x, \theta) = e^{\{ A(\theta) + A'(\theta)(x-\theta) + B(x) \}}$

(b) $f(x, \theta) = e^{\{ A(\theta)(x-\theta) + A'(\theta) + B(x) \}}$

(c) $f(x, \theta) = e^{\{ A(\theta) + A''(\theta)(x-\theta) + B(x) \}}$

(d) $f(x, \theta) = e^{\{ A(\theta) . B(x) + A'(\theta)(x-\theta) \}}$

Q.43 In Q.(42), the sample mean $\overline{x}_n$ is

(a) Consistent statistic for θ.

(b) Efficient statistic θ.

(c) Sufficient statistic for θ.

(d) Minimum variance estimator for θ.

Q.44 In case of $b(x, n, p)$, the consistent estimator for p (1–p) will be given as :

(a) $t(x) = \left(\frac{r}{n}\right)^2$ (b) $t(x) = \frac{r}{n}\left(1 - \frac{r}{n}\right)^2$

(c) $t(x) = \frac{r}{n}\left(1 - \frac{r}{n}\right)$ (d) $t(x) = \frac{n}{r}\left(1 - \frac{r}{n}\right)$

Q.45 In Q.(42), the MLE θ has "Minimum variance bound" equal to

(a) $V(\theta) = \frac{1}{n A''(\theta)}$ (b) $V(\theta) = n A''(\theta)$

(c) $V(\theta) = \frac{1}{n . A'(\theta)}$ (d) $V(\theta) = \frac{n}{A''(\theta)}$

Q.46 In case of $b(x, n, p)$, the consistent estimator for p is given as $t(x)$ written as :

(a) $t(x) = \frac{r}{n}$ (b) $t(x) = \frac{n}{r}$

(c) $t(x) = \frac{(n-r)}{n}$ (d) $t(x) = \frac{r}{n}(1 - \frac{r}{n})$

Q.47 Let (t, u) be consistent estimators for λ(θ). Then the statistic t* = (at + bu), will also be consistent statistic for λ(θ) for :

(a) a + b = 0 (b) a + b = 1

(c) a + b > 0 (d) a + b < 0

Q.48 If we add to M.L.E., a function of observations which tends in probability to zero, then this modified estimator will have the :

(a) Asymptotic property (b) Sufficiency property

(c) Unbiasedness property (d) Efficiency property

Q.49 Barkson has shown through sampling experiment that in some particular situation arised in "Bioessay" the M.L.E. some time has :

(a) Infinite M.S.E. (b) Finite M.S.E.

(c) M.S.E. > 100 (d) M.S.E. < zero.

Q.50 Consider the following most general form of distribution written as :

$$f(x, \theta) = \left(\frac{x}{\theta}\right)^{\theta} A'(\theta)\ e^{(A(\theta) + C(x))}$$

The M.L.E. of θ will be given by a statistic t(x) defined as :

(a) Arithmatic mean of sample.

(b) Geometric mean of sample.

(c) Harmonic mean of sample

(d) Sample median x.

Q.51 Consider the most general form of the pdf given as :

$$f(x, \theta) = e[\tfrac{1}{\theta}\ \{\theta A'(\theta) - A(\theta)\}\ A'(\theta) + C(x)\ ].$$

Then the M.L.E. of θ will be given by a statistic $t(x)$ defined as :

(a) $t(x) = \frac{1}{n}\sum^{n} x_i$ (b) $t(x) = \left(\prod_{i}^{n} x_i\right)^{\frac{1}{n}}$

(c) $t(x) = \left(\frac{n}{\sum^{n}\frac{1}{x_i}}\right)$ (d) $t(x)$ = sample median x

Q.52 A sufficient (though not necessary) condition that any of the stationary values (M.L.E.) say θ will be a "Local maxima" is that :

(a) L''(x, θ) = 0 (b) L''(x, θ) < 0

(c) $L''(x, \theta) \geq 0$ (d) $L''(x, \theta) = 1.0$

Q.53 If we find more than one local maximas from the equation $L'(x, \theta) = 0$ and $L''(x, \theta) < 0$, then one must prefer the :

(a) Largest one (b) Smallest one

(c) A.M. of all maximas (d) Median of maximas

Q.54 Let $t(x)$ be the M.V.B. estimator for $\psi(\theta)$ and L.F. has a solution as θ then we have :

(a) $t(x) = \psi(\theta_0)$ (b) $t(x) = \psi(\theta)$

(c) $t(x) = \psi(\theta^2)$ (d) $t(x) = (\psi(\theta))^2$

Q.55 Let θ be an unbiased estimator for θ then in general we have the result :

(a) $E(\psi(\theta)) = \psi(E(\theta))$ (b) $E(\psi(\theta)) \neq \psi(E(\theta))$

(c) $E(\psi'(\theta)) > \psi(E(\theta))$ (d) $E(\psi(\theta)) \leq \psi(E(\theta))$

Q.56 Any sequence of estimators say (θ_n) obtained through likelihood equation is

(a) Not consistent estimator

(b) Consistent estimator

(c) Efficient estimator

(d) Unbiased estimator

Q.57 Let x ~ $N(\mu, \sigma^2)$ where σ^2 is known. Then M.L.E. of μ is given by an estimator t(x) defined as :

(a) $t(x) = \left(\prod^{n} x_i\right)^{\frac{1}{n}}$ (b) $t(x) = \frac{1}{n}\sum^{n} x_i$

(c) $t(x) = \frac{n}{\sum^{n} \frac{1}{x_i}}$ (d) $t(x)$ = sample median

Q.58 In Q.(57), let μ be known then M.L.E. σ^2 is given as :

(a) $\sigma^2 = \frac{1}{n}\sum^{n}(x_i - \bar{x})^2$ (b) $\sigma^2 = \frac{1}{n-1}\sum^{n}(x_i - \mu)^2$

(c) $\sigma^2 = \frac{1}{n}\sum^{n}(x_i - \mu)^2$ (d) $\sigma^2 = \frac{1}{n-1}\sum^{n}(x_i - \bar{x})^2$

Q.59 Let in Q.(57), both (μ, σ^2) are unknown then their MLE's are given respectively as :

(a) $\bar{x}_n, \frac{1}{n}\sum^{n}(x_i - \bar{x}_n)^2$ (b) $\bar{x}_n, \frac{1}{n}\sum^{n}(x_i - \mu)^2$

(c) $\bar{x}, \frac{1}{n-1}\sum^{n}(x_i - \bar{x}_n)^2$ (d) $\bar{x}_n, \frac{1}{n-1}\sum^{n}(x_i - \mu)^2$

Q.60 Let $x \sim P(x, \lambda) = \frac{e^{-\lambda}\lambda^x}{x!}$, then M.L.E. of λ is given as :

(a) λ = sample median $\bar{x}$ (b) $\lambda = \frac{1}{n}\sum_{i}^{n} x_i$

(c) $\lambda = \left(\prod_{i=1}^{n} x_i\right)^{\frac{1}{n}}$ (d) $\lambda = \frac{n}{\sum^{n}\frac{1}{x_i}}$

Q.61 Let x ~ $b(x, n, p)$ and r = $\sum^{n} x_i$ then M.L.E. of p is given as :

(a) $p = \frac{n}{r}$ (b) $p = \left(\frac{r}{n}\right)^2$

(c) $p = \frac{r}{n}$ (d) $p = \sqrt{\frac{r}{n}}$

Q.62 In Q.(61), a consistent estimator for p^2 is given by a sample estimator t_n defined as :

(a) $t_n = \frac{r}{n}$ (b) $t_n = \left(\frac{r}{n}\right)^2$

(c) $t_n = \frac{n}{r}$ (d) $t_n = \frac{r}{\sqrt{n}}$

Q.63 Let $x \sim N(0, \theta^2)$ with p.d.f. given as :

$$f(x, 0, \theta^2) = \frac{1}{\sqrt{2\theta}\ \theta} e^{-\frac{1}{2}\left(\frac{x}{\theta}\right)^2}, \quad -\infty < x < \infty$$

Then MLE θ of θ is given as :

(a) $\theta = \frac{1}{n}\sum_{i}^{n} x_i^2$ (b) $\theta = \left(\frac{1}{n}\sum^{n} x_i^2\right)^{\frac{1}{2}}$

(c) $\theta = \frac{1}{n}\sum^{n} x_i$ (d) $\theta = \left(\frac{1}{n}\sum^{n} x_i\right)^{\frac{1}{2}}$

Q.64 Consider the power series distribution wirtten as :

$P(x = r) = a_r\, \theta^r / f(\theta), \quad r = 0, 1, 2, \ldots$

Then M.L.E. θ of parameter θ is given as :

(a) $\theta = \frac{\theta . f(\theta)}{f'(\theta)}$ (b) $\theta = \frac{\theta' f'(\theta)}{f(\theta)}$

(c) $\theta = \frac{f(\theta)}{\theta\, f'(\theta)}$ (d) $\theta = \frac{\theta . f'(\theta)}{f(\theta)}$

Q.65 Consider two normal populations defined as $N(\mu, \lambda\, \sigma^2)$ and $N(\mu, \sigma^2)$ respectively. Then MLE μ of μ is given as :

(a) $\mu = \frac{(n_1\, \overline{x}_1 + n_2\, \lambda\, \overline{x}_2)}{(n_1 + n_2\, \lambda)}$ (b) $\mu = \left(\frac{n_1\, \overline{x}_1 + n_2\, \lambda\, \overline{x}_2}{n_1\, \lambda + n_2}\right)$

(c) $\frac{\mu}{2} = \left(\frac{n_1\, \lambda\, \overline{x}_1 + n_2\, \overline{x}_2}{n_1 + n_2\, \lambda}\right)$ (d) $\mu = \left(\frac{n_1\, \lambda\, \overline{x}_1 + n_2\, \overline{x}_2}{n_1\, \lambda + n_2}\right)$

Q.66 In Q.(65), the sampling variance V(p) is given as :

(a) $V(\mu)=\left(\dfrac{\sigma^2}{n_1+n_2\lambda}\right)$ (b) $V(\mu)=\left(\dfrac{\lambda\sigma^2}{n_1+n_2\lambda}\right)$

(c) $V(\mu)=\left(\dfrac{\lambda\sigma^2}{n_1\lambda+n_2}\right)$ (d) $V(\mu)=\left(\dfrac{\sigma^2}{n_1\lambda+n_2}\right)$

Q.67 Consider the rectangular distribution defined as :

$$f(x,\alpha,\beta)=\frac{1}{(\beta-\alpha)},\quad \alpha<x<\beta$$

and let $(x_1<x_2<\ldots\ldots<x_n)$ be an ordered sample from $f(x,\alpha,\beta)$ then M.L.E.'s of (α,β) are respectively given as :

(a) $\alpha=x_n,\ \beta=x_1$ (b) $\alpha=x_1,\ \beta=x_n$

(c) $\alpha=\dfrac{x_1}{2},\dfrac{x_n}{2}$ (d) $\alpha=x_1^2,\ \beta=x_n^2$

Q.68 Consider the following exponential distribution

$f(x,\alpha,\beta)=y_0\ e{-}\beta(x{-}\alpha),\quad \alpha<x<\infty$

where y_0 is a constant. Let $(x_1, x_2, \ldots\ldots, x_n)$ be an ordered sample from $f(x,\alpha,\beta)$. Then MLE's of (α,β) are respectively given as :

(a) $\alpha=x_1,\ \beta=x_n$ (b) $\alpha=\dfrac{1}{(\bar{x}-x_1)},\ \beta=x_n$

(c) $\alpha=x_1,\ \beta=\dfrac{1}{(\bar{x}-x_1)}$ (d) $\alpha=\dfrac{1}{(x_1-\bar{x})},\ \beta=x_n$

Q.69 Let (t_1, t_2) be two unbiased estimators of a parameter θ with sampling variances (σ_1^2,σ_2^2) respectively and correlation (t_1, t_2) as ρ. Then linear combination $t=\sum_{i=1}^{2} l_i\, t_i$ is an unbiased estimator of θ if :

(a) $l_1+l_2=0$ (b) $l_1^2+l_2^2=1$

(c) $l_1+l_2=1$ (d) $l_1+l_2\geq 1$

Q.70 In Q.(69) the sampling variance $v(t)$ is given as :

(a) $V(t)=l_1\ \sigma_1^2+l_2\ \sigma_2^2+2\,l_1\,l_2\,\rho.\ \sigma_1\ \sigma_2$

(b) $V(t) = l_1 \sigma_1^2 + l_2 \sigma_2^2 + 2 \rho . \sigma_1 \sigma_2$

(c) $V(t) = l_1^2 \sigma_1^2 + l_2^2 \sigma_2^2 + 2 l_1 l_2 \rho \sigma_1 \sigma_2$

(d) $V(t) = l_1^2 \sigma_1^2 + l_2^2 \sigma_2^2 + 2 \rho \sigma_1 \sigma_2$

Q.71 In Q.(69) the constant l_1 will be worked out as :

(a) $l_1 = \left(\dfrac{\sigma_2^2 - \rho \sigma_1 \sigma_2}{\sigma_1^2 + \sigma_2^2 - 2 \rho \sigma_1 \sigma_2} \right)$

(b) $l_1 = \left(\dfrac{\sigma_1^2 - \rho \sigma_1 \sigma_2}{\sigma_1^2 + \sigma_2^2 - 2 \rho \sigma_1 \sigma_2} \right)$

(c) $l_1 = \left(\dfrac{\sigma_1^2 - \rho \sigma_1 \sigma_2}{\sigma_1^2 + \sigma_2^2 - 2 \rho \sigma_1 \sigma_2} \right)$

(d) $\dfrac{\sigma_1^2 - \rho \sigma_1 \sigma_2}{(\sigma_1 \sigma_2)}$

Q.72 In Q.(69), the estimator t is the :

(a) Best linear unbiased estimator for θ.

(b) Consistent estimator for θ.

(c) Unbiased estimator for θ.

(d) Both (a) & (c)

Q.73 Let in (69), t_1 be M.V.U.E. and t_2 be an other estimator for θ with $v(t_2) = \dfrac{\sigma^2}{e}$, then we have :

(a) Correlation $(t_1, t_2) = e$

(b) Correlation $(t_1, t_2) = \sqrt{e}$

(c) Correlation $(t_1, t_2) = (1 - e)$

(d) Correlation $(t_1, t_2) = (1 - e^2)$

Q.74 The correlation coefficient p between a most efficient estimator t_1 and any other estimator t_2 with efficiency 'e' is given as :

(a) $\rho(t_1, t_2) = e$

(b) $\rho(t_1, t_2) = e^2$

(c) $\rho(t_1, t_2) = \sqrt{e}$

(d) $\rho(t_1, t_2) = (1 - \sqrt{e})$

Q.75 Let a most general form of a distribution, differentiable in θ is given as :

$$f(x, \theta) = \text{constant exp. } \{ (x - \theta) \frac{\partial \psi}{\partial \theta} + \psi(\theta) + 6(x) \}$$

Then MLE θ of θ is given as :

(a) G.M. of sample values.

(b) H.M. of sample values.

(c) A.M. of sample values.

(d) Mediam of sample values

Q.76 Consider a most general form of a continuous distribution written as :

$$f(x, \theta) = \left(\frac{\alpha}{\theta}\right)^{\theta \frac{\partial \psi}{\partial \theta}} . \exp \{ \psi(\theta) + 6(x) \}$$

where ψ(θ) and 6(*x*) are the arbitrary function of θ and *x* respectively. Then MLE θ of θ is given as :

(a) θ = A.M. of sample values. (b) G.M. of sample values.

(c) Median of sample values. (d) H.M. of sample values.

Q.77 The most general form of a distribution is given as $f(x, \theta)$ = constant exp

$$[(x - \theta) \frac{\partial \psi}{\partial \theta} + \psi(\theta) + 6(x)]$$

where ψ(θ) and 6(*x*) are the arbitrary functions of θ & *x*. Then MLE. θ of θ is given as :

(a) A.M. of sample values. (b) G.M. of sample values.

(c) Median of sample values. (d) H.M. of sample values.

Q.78 Consider bivariate normal distribution $f(x, y, \mu_1, \mu_2, \sigma_1^2, \sigma_2^2, \rho)$ Let μ_1 be known then MLE $\hat{\sigma}_1^2$ of σ_1^2 is given as :

(a) $\hat{\sigma}_1^2 = \frac{1}{n} \sum_i^n (x_i - \mu_1)^2$ (b) $\hat{\sigma}_1^2 = \frac{1}{n} \sum^n (x_i - \overline{x})^2$

(c) $\hat{\sigma}_1^2 = \frac{1}{n-1} \sum_i^n (x_i - \mu_1)^2$ (d) $\hat{\sigma}_1^2 = \frac{1}{n-1} \sum^n (x_i - \overline{x})^2$

Q.79 Let in Q.(78), μ_2 is known then MLE $\hat{\sigma}_2^2$ of σ_2^2 is given as :

(a) $\frac{1}{n-1}\sum_i^n (y_i - \mu_2)^2$ (b) $\frac{1}{n-1}\sum_i^n (y_i - \bar{y})^2$

(c) $=\frac{1}{n}\sum_i^n (y_i - \bar{y})^2$ (d) $\frac{1}{n}\sum^n (y_i - \mu_2)^2$

Q.80 Let in Q.(78), μ_1 & μ_2 are unknow then their M.L.E's are respectively given as :

(a) $\mu_1 = \frac{1}{n}\sum^n y_i$, $\mu_2 = \frac{1}{n}\sum_{i=1}^n x_i$ (b) $\mu_1 = \frac{1}{n}\sum^n x_i$, $\mu_2 = \frac{1}{n}\sum^n y_i$

(c) $\mu_1 = \frac{1}{n}\sum_i^n x_i^2$, $\mu_2 = \frac{1}{n}\sum_i^n y_i^2$ (d) $\mu_1 = \frac{1}{n}\sum^n y_i^2$, $\mu_2 = \frac{1}{n}\sum_i^n x_i^2$

Q.81 Let in Q.(78), (μ_1, μ_2) are known and (σ_1^2, σ_2^2) are unknown then MLE ρ of ρ is given as :

(a) $\rho = \frac{1}{n}\sum^n (x_i - \mu_1)(y_i - \mu_2) / \hat{\sigma}_1 \hat{\sigma}_2$

(b) $\rho = \frac{1}{n}\sum_i^n (y_i - \mu_2)(x_i - \mu_1) / \sigma_1 \sigma_2$

(c) $\rho = \frac{1}{n}\sum^n (x_i - \mu_1)(y_i - \mu_2) / \sigma_1 \sigma_2$

(d) Both (a) & (b)

Q.82 Consider normal distribution $N(x, \mu, \sigma^2)$. Then log normal distribution $y = e^x$ has the mean θ_1 given as :

(a) $\theta_1 = e^{(\mu + \sigma^2)}_0$ (b) $\theta_1 = e^{(\mu + \frac{\sigma^2}{2})}$

(c) $\theta_1 = e^{(\mu + \sigma)}$ (d) $\theta_1 = e^{(\sigma^2 + \mu^2)}$

Q.83 In Q.(82), the variance 'v' of log normal distribution $y = e^x$ is given as :

(a) $v = (e^{\sigma} - 1)$ (b) $v = (1 - e^{\sigma})$

(c) $v = (e^{\sigma^2} - 1)$ (d) $v = (1 - e^{\sigma^2})$

Q.84 In Q.(82), the MLE θ_1 of θ_1 is given as :

(a) $\theta_1 = e^{(\bar{x} + \hat{s}^2)}$ (b) $\theta_1 = e^{(\bar{x} + \hat{s}^2/2)}$

(c) $\theta_1 = e^{(\bar{x} + \hat{s})}$ (d) $\theta_1 = e^{(\bar{x} + \hat{s}/2)}$

Where $s^2 = \frac{1}{n-1} (\sum^n (x_i - \bar{x})^2$

Q.85 In Q.(83), the MLE. of variance 'V' of log normal distribution $y = e^x$ is given as :

(a) $v = e^{(2\bar{x} + \hat{s}^2)} . (e^{\hat{s}^2} - 1)$ (b) $v = e^{(\bar{x} + \hat{s}^2)} . (e^{\hat{s}^2} - 1)$

(c) $v = (1 - e^{\hat{s}^2})\ e(\bar{x} + \hat{s}^2)$ (d) $v = (1 - e^{\hat{s}})\ (e^{(\bar{x} - \hat{s}^2)})$

Q.86 Consider the bivariate normal distribution $N(\mu_1, \mu_2, \sigma_1^2, \sigma_2^2, \rho)$ the estimating efficiency of MLE ρ is given as :

(a) $e(\rho) = \left(\frac{1-\rho^2}{1+\rho^2}\right)^2$ (b) $e(\rho) = \left(\frac{1-\rho^2}{1+\rho^2}\right)$

(c) $e(\rho) = \left(\frac{1-\rho}{1+\rho}\right)$ (d) $e(\rho) = \left(\frac{1-\rho}{1+\rho}\right)^2$

Q.87 Method of M.L.E. and method of moment lead to the :

(a) Same estimators (b) Different estimators

(c) Unbiased estimators (d) None.

Q.88 Any unbiased estimator with finite mean for parameter θ is

(a) Consistent estimator (b) Biased estimator

(c) Asymptotically consistent (d) Sufficient estimator

Q.89 Any consistent estimator with finite mean for parameter θ is

(a) Unbiased estimator

(b) Asymptotically unbiased estimators

(c) Asymptotically consistentestimator

(d) Biased estimator

Q.90 Let y_0 be the value of p.d.f. of a distribution after putting the value of r.v. (x) equal to the value of population parameter θ_0. The sampling variance 'v' of sample median x from the distribution is given as :

(a) $v(x) = \frac{\pi}{4n}$ (b) $v(x) = \frac{\pi}{2n}$

(c) $v(x) = \frac{2\pi}{3n}$ (d) $v(x) = \frac{2\pi^2}{3n}$

Q.91 In case of Cauchy distribution $f(x, \theta) = \frac{1}{\pi\{1+(x-\theta)^2\}}$, the variance of sample median x is given as :

(a) $v(x) = \frac{\pi}{2n}$ (b) $v(x) = \frac{\pi}{4n}$

(c) $v(x) = \frac{2}{3}\frac{\pi}{n}$ (d) $v(x) = \frac{2\pi^2}{3n}$

Q.92 In case of normal distribution N (μ, σ^2), the sampling variance of sample median x is given as :

(a) $v\ (x) = \frac{\pi}{2}\frac{\sigma^2}{n}$ (b) $v\ (x) = \frac{\pi}{4}\frac{\sigma^2}{n}$

(c) $v\ (x) = \frac{2\pi}{3}.\frac{\sigma^2}{n}$ (d) $v\ (x) = \frac{\pi}{4}\frac{\sigma}{n}$

Q.93 The merit of an estimator is judged by the sampling variance of :

(a) Estimator (b) MVBE.

(c) Efficient estimator (d) Biased estimator,

Q.94 As the sample size n increases, the amount of sampling error :

(a) Increases (b) Remains constant

(c) Decreases (d) None

Q.95 In case of Cauchy distribution $f(x, \theta) = \frac{1}{\pi}\frac{1}{\{1+(x-\theta)^2\}}$ the accuracy of an estimator $t_n = \sum^{n} x_i$:

(a) Depends on n. (b) Does not dependent n.
(c) Is independent of n. (d) Both (b) & (c)

Q.96 The method of "Minimum variance estimation" is analoguas to :
(a) Factorization theorem (b) N.P. Lemma.
(c) Rao-Blackwell theorem (d) Minimum χ^2 method

Q.97 In case of $N(\mu, \sigma^2)$, the M.V.U.E. does not exist for :
(a) σ^2 (b) σ
(c) $\frac{\sigma}{n}$ (d) Both (b) & (c)

Q.98 The method of "Minimum variance" will give a solution only if for parameter θ, there exists :
(a) Sufficient statistic (b) Unbiased estimator
(c) M.L.E. for θ. (d) None.

Q.99 In general, the accuracy of an estimator increases as the sample size n :
(a) Decreases (b) Increases
(c) Remains unaffected (d) None.

Q.100 For the p.d.f. $f(x, \theta)$, we define

$$I_\theta = \int_{-\infty}^{\infty} \left(\frac{\partial}{\partial\theta} \log f\right)^2 f(x,\theta)\, dx$$

then I_θ is known as :
(a) Error for θ. (b) Information for θ.
(c) Efficiency of θ. (d) Biasness for θ.

Q.101 In Q.(100), some times I_θ is also known as :
(a) Intrinsic accuracy for θ. (b) Biasness of θ.
(c) Efficiency of θ. (d) None.

Q.102 For any estimator t(n), we have the relation :
(a) Accuracy (t_n) = Amount of information
(b) Accuracy $(t_n) >$ Amount of information
(c) Accuracy $(t_n) \geq$ Amount ofo information
(d) Accuracy $(t_n) \leq$ Amount of information

Q.103 Consider the Trapezoidal distribution given as :

$$f(x, \theta) = \frac{2x}{2\theta + 1}, \quad 0 \leq \theta \leq x \leq \theta + 1.$$

Then intrinsic accuracy for $f(x, \theta)$ for a sample of size n is given as :

(a) Accuracy $(f(x, \theta)) = \dfrac{4n^2}{(2\theta + 1)^2}$

(b) Accuracy $(f(x, \theta)) = \left(\dfrac{2n}{2\theta + 1}\right)^2$

(c) Accuracy $(f(x, \theta)) = \dfrac{2n}{2\theta + 1}$

(d) Both (a) & (b)

Q.104 The "Method of moment" for estimation of parameter θ has been developed by the statistician :

(a) Prof. Mahalanobis (b) Prof. Karl Pearson

(c) Prof. J. Neymam (d) Prof. C.R. Rao

Q.105 In general the efficiency of an estimator obtained through the "Method of moment" will be :

(a) Equal to that of M.L.E.

(b) Less than that of M.L.E.

(c) Greater than that of M.L.E.

(d) Greater than or equal to that of M.L.E

Q.106 The "Method of moment" and "Method of M.L.E." will give an identical solution to parameter θ only if :

(a) The parent population is of a particular form.

(b) Parent population is continuous.

(c) Prarent population is discrete.

(d) Parent population is of exponential form.

Q.107 The population moments are estimated through corresponding the :

(a) Samplel moment (b) Sample median

(c) Sample A.M. (d) Sample G.M.

Q.108 Under quite general conditions, the estimators obtained through "Method of moment" are :

(a) Normally distributed

(b) Asymptotically normally distributed
(c) Efficient
(d) Unbiased

Q.109 Prof. Karl Pearson has utilized the "Method of moment" to fit the curves known as :

(a) Pearsonian curves. (b) Exponential curves.
(c) Logrithmic curves (d) Non linear curves

Q.110 The estimators obtained through the "Method of moment" in general are :

(a) Not consistent (b) Not efficient
(c) Consistent (d) Both (a) & (b)

Q.111 Consider the p.d.f. as $f(x, \theta) = \frac{1}{\theta} \cdot e^{-\frac{x}{\theta}}, \; 0 < x < \infty$

Then the estimators obtained through "Method of moment" and "Method of M.L.E." are :

(a) Identical (b) Efficient
(c) Consistent (d) Both (b) & (c)

Q.112 In case of Poisson distribution $\rho(x, \lambda) = \frac{e^{-\lambda} \lambda^x}{x!}$, the parameter λ can not be estimated through :

(a) Method of M.L.E. (b) Method of minimum variance
(c) Method of moment (d) Neyman Pearson Lemma

Q.113 Consider the rectangular distribution $f(x, \theta)$ as :

$$f(x, \theta) = 1, \quad (\theta - \frac{1}{2}) \leq x \leq (\theta + \frac{1}{2})$$

Then by the "Method of moment", the parameter q will be estimated as :

(a) $\theta = \frac{1}{n} \sum^{n} x_i^2$ (b) $\theta = \frac{1}{n} \sum^{n} (x_i - \bar{x}_n)^2$

(c) $\theta = \frac{1}{n} \sum^{n} x_i$ (d) None.

Q.114 Consider the Pearson's type III distribution written as :

$$f(x, \alpha, \beta) = \frac{\beta^{\alpha}}{\Gamma\alpha} x^{\alpha-1} \cdot e^{-\beta x}, \; 0 < x < \infty$$

Let (μ'_1 & μ'_2) be the sample moments then (α, β) wil be estimated as :

(a) $\alpha = \dfrac{\mu'^2_1}{(\mu'_2 - \mu'^2_1)}$, $\beta = \dfrac{\mu'^2_1}{(\mu'_2 - \mu'^2_1)}$

(b) $\alpha = \dfrac{\mu'^2_1}{(\mu'_2 - \mu'_1)}$, $\beta = \dfrac{\mu'^2_1}{(\mu'_2 - \mu'_1)}$

(c) $\alpha = \dfrac{\mu'_1}{(\mu'_2 - \mu'_1)}$, $\hat{\beta} = \dfrac{\mu'_2}{(\mu'_2 - \mu'_1)}$

(d) $\alpha = \left(\dfrac{\mu'_1}{\mu'_2 - \mu'_1}\right)^2$ $\beta = \left(\dfrac{\mu'_2}{\mu'_2 - \mu'_1}\right)^2$

Q.115 Consider the double Poisson distribution defined as :

$$f(x, m_1, m_2) = \frac{1}{2}\left[\frac{e^{-m_1} m_1^x}{x!} + \frac{e^{-m_2} m_2^x}{x!}\right]$$

Then through "Method of moment", the parameters (m_1, m_2) are estimated as :

(a) $m_1 = \mu'_1 + \sqrt{\mu'_2 - \mu'_1 - \mu'^2_1}$, $m_2 = \mu'_1 - \sqrt{\mu'_2 - \mu'_1 - \mu'^2_1}$

(b) $m_1 = \mu'_1 - \sqrt{\mu'_2 - \mu'_1 - \mu'^2_1}$, $m_2 = \mu'_1 + \sqrt{\mu'_2 - \mu'_1 - \mu'^2_1}$

(c) Both (a) & (b) (d) None

Q.116 Any statistic (estimator) does not depend upon :

(a) Unknown parameter (b) Population mean

(c) Population variance (d) Population correlation

Q.117 The sampling distribution of any statistic (estimator) may very well depend upon :

(a) Population correlation (b) Population variance

(c) Population mean (d) Population parameter

Q.118 The method of minimum χ^2 and modified minimum χ^2, may be used when the values are in the form of frequencies grouped into number of (say k) :

(a) Continuous classes (b) Mutually exclusive classes

(c) Mutually disjoint classes (d) Both (b) & (c)

Q.119 Method of minimum χ^2 uses those estimates of parameter θ which :

(a) Maximises χ^2 (b) Minimises χ^2

(c) Give average χ^2 (d) None

Q.120 For large sample size n, the minimum χ^2 and likelihood equations are :

(a) Not identical (b) Different

(c) Identical (d) None

Q.121 The method of minimum χ^2 is found more troublesome to apply in many cases and is not improvement over :

(a) MLE method (b) Minimum variance method

(c) N.P. Lemma (d) Factorization method

Q.122 In case of modified minimum χ^2 method, we use those value of θ as estimates which :

(a) Maximises χ' rather then χ^2.

(b) Minimises χ'^2 rather than χ^2.

(c) Gives less value χ'^2 than χ^2.

(d) Both (b) & (c).

Q.123 For large samplel size n, the estimates obtained through modified minimum χ^{*2} method would be identical with :

(a) Sample moment (b) Sample median

(c) M.L.E. (d) Sample mean

Q.124 The least square estimates do not have any optimum property even :

(a) Asymptotic property is not met.

(b) Asymptotic property is met.

(c) Continuity property is met.

(d) Monotonicity property is met.

Q.125 The least square estimators obtained as linear function of sample values will be :

(a) Minimum variance unbiased estimators

(b) Biased estimators

(c) Sufficient estimators

(d) None

2.4 Key Answers

Write True/False

Q.	Ans.	Q.	Ans.	Q.	Ans.	Q.	Ans.
1.	True	32.	True	63.	True	94.	True
2.	True	33.	True	64.	True	95.	True
3.	True	34.	True	65.	True	96.	True
4.	True	35.	True	66.	True	97.	True
5.	True	36.	True	67.	True	98.	True
6.	True	37.	True	68.	True	99.	True
7.	True	38.	True	69.	True	100.	True
8.	True	39.	True	70.	True	101.	True
9.	True	40.	True	71.	True	102.	True
10.	True	41.	True	72.	True	103.	True
11.	True	42.	True	73.	True	104.	True
12.	True	43.	True	74.	True	105.	True
13.	True	44.	True	75.	True	106.	True
14.	True	45.	True	76.	True	107.	True
15.	True	46.	True	77.	True	108.	True
16.	True	47.	True	78.	True	109.	True
17.	True	48.	True	79.	True	110.	True
18.	True	49.	True	80.	True	111.	True
19.	True	50.	True	81.	True	112.	True
20.	True	51.	True	82.	True	113.	True
21.	True	52.	True	83.	True	114.	True
22.	True	53.	True	84.	True	115.	True
23.	True	54.	True	85.	True	116.	True
24.	True	55.	True	86.	True	117.	True
25.	True	56.	True	87.	True	118.	True
26.	True	57.	True	88.	True	119.	True
27.	True	58.	True	89.	True	120.	True
28.	True	59.	True	90.	True	121.	True
29.	True	60.	True	91.	True	122.	True
30.	True	61.	True	92.	True	123.	True
31.	True	62.	True	93.	True	124.	True
						125.	True

Fill in the Blanks

Q.	A.
1.	Maximum likelihood
2.	C.F. Gauss
3.	Prof. R.A. Fisher
4.	$\prod_{i}^{n} f(x_i, \theta)$
5.	θ alone
6.	Likelihood function
7.	M.L. estimator
8.	$\theta = U(x)$
9.	$\log_e L(x, \theta)$
10.	Likelihood equation
11.	May or may not be
12.	Continuous function, True value
13.	Integrable function, $(-\infty, \infty)$
14.	M(x)
15.	$\int_{-\infty}^{\infty} (-\frac{\partial^2}{\partial\theta^2} \log L(x, \theta)\, dx$
16.	Independent
17.	θ
18.	Integral sign
19.	Probability
20.	Constant
21.	Maximum, $n \to \infty$
22.	Assymptotically
23.	$(\theta_0, \frac{1}{I_{(\theta_0)}})$
24.	$t(x)$
25.	Sufficient statistic
26.	Sufficient statistic
27.	$g(\theta)$
28.	Efficient, Normal distribution
29.	Absolute maxima
30.	Negative definite
31.	Dispersion matrix
32.	Interative method
33.	Method of scoring
34.	Efficient
35.	M.L. estimator
36.	Consistent
37.	Sample median
38.	$1 + (x_i - \theta)^2$
39.	$\frac{2}{n}$
40.	$\frac{\partial}{\partial\theta} \log L(x, \theta) = 0$, M.V.B.E.
41.	Maximum variance, sufficient statistic
42.	Differentiable in θ
43.	Sufficient statistic
44.	$\frac{1}{(n A''(\theta))}$
45.	$\left(1 - \frac{r}{n}\right)$
46.	$\frac{n}{r}$
47.	= 1, consistent estimator
48.	Assymptotic property
49.	Minimum c2 estimator, M.S.E.
50.	M.L.E. of θ.

51. $\sum_{i}^{n} \frac{1}{x}$

52. Local maxima

53. Largest one

54. θ

55. $\neq$

56. Consistent

57. $\frac{1}{n}\sum^{n} x_i$

58. $\frac{1}{n}\sum^{n} (x_i - \mu)^2$

59. $\bar{x}, \frac{1}{n}, \sum^{n} (x_i - \bar{x})^2$

60. $\frac{1}{n}\sum^{n} x_i$

61. $\frac{r}{n} = \frac{1}{n}\sum^{n} x_i$

62. $\left(\frac{r}{n}\right)^2$

63. $\sqrt{\frac{1}{n}\sum^{n} x_i^2}$

64. $\frac{f'(\theta)}{f(\theta)}$

65. $(n_1 + n_2 \lambda)$

66. $\frac{\lambda \sigma^2}{(n_1 + n2\lambda)}$

67. M.L.E., $x_{(1)}, x_{(n)}$

68. $x_{(1)}, \frac{1}{(\bar{x} - x_{(1)})}$

69. $= 1$

70. $2\, l_1\, l_2\, \rho\, \sigma_1\, \sigma_2$

71. $(\sigma^2_2 - \rho\, \sigma_1\, \sigma_2$

72. Unbiased

73. $\sqrt{e}$

74. $\sqrt{e}$

75. A.M. (sample values)

76. G.M. (sample values)

77. H.M. (sample values)

78. $\frac{1}{n}\sum^{n} (x_i - \mu_1)^2$

79. $\frac{1}{n}\sum^{n} (x_i - \mu_2)^2$

80. $\frac{1}{n}\sum^{n} (x_i - \mu_1)(y_i - \mu_2)$

81. $\frac{1}{n}\sum^{n} x_i, \frac{1}{n}\sum^{n} y_i$

82. $e\,(\mu + \frac{\sigma^2}{2})$

83. $(e^{\sigma^2} - 1)$

84. $\overline{x}$

85. $(e^{\sigma^2} - 1)$

86. $\dfrac{(1-\rho)^2}{(1+\rho^2)}$

87. Same estimator

88. Asymptotically consistent

89. Asymptotically imbiased

90. 4n

91. 4n

92. $\dfrac{\pi}{2}$

93. Estimator

94. Decreases

95. Sample size n

96. Rao-Blackwell

97. σ^2

98. Sufficient statistic

99. Increases

100. Information

101. Intrinsic accuracy.

102. $\leq$

103. $(2\theta + 1)^2$

104. Karl Pearson

105. Less efficient

106. Parent distribution

107. Consistent estimator

108. Asymptotically

109. Pearsonian curves

110. Consistent & efficient

111. Identical

112. Moment

113. Moment

114. $\left(\mu'_2 - \mu'^2_1\right), \left(\mu'_2 - \mu'^2_2\right)$

115. $-\sqrt{\mu'_2 - \mu'_2 - \mu^2_1}$

116. Parameter

117. Parameter

118. Mutually exclusive

119. Minimisses

120. Identical

121. M.L.E.

122. Minimise modified minimum χ'^2

123. M.L. Estimators

124. Asymptotic property

125. Unbiasedness property

Multiple choice Questions and Answers

Q.	A.	Q.	A.	Q.	A.	Q.	A.
1.	(a)	32.	(a)	63.	(b)	94.	(c)
2.	(c)	33.	(c)	64.	(d)	95.	(d)
3.	(a)	34.	(a)	65.	(a)	96.	(c)
4.	(a)	35.	(c)	66.	(b)	97.	(d)
5.	(a)	36.	(b)	67.	(b)	98.	(a)
6.	(b)	37.	(b)	68.	(c)	99.	(b)
7.	(b)	38.	(b)	69.	(c)	100.	(b)
8.	(b)	39.	(a)	70.	(c)	101.	(a)
9.	(b)	40.	(b)	71.	(a)	102.	(d)
10.	(b)	41.	(b)	72.	(d)	103.	(d)
11.	(d)	42.	(a)	73.	(b)	104.	(b)
12.	(a)	43.	(c)	74.	(c)	105.	(b)
13.	(a)	44.	(c)	75.	(c)	106.	(a)
14.	(d)	45.	(a)	76.	(b)	107.	(a)
15.	(b)	46.	(b)	77.	(d)	108.	(b)
16.	(d)	47.	(b)	78.	(a)	109.	(a)
17.	(a)	48.	(a)	79.	(c)	110.	(d)
18.	(b)	49.	(a)	80.	(b)	111.	(a)
19.	(a)	50.	(b)	81.	(d)	112.	(c)
20.	(c)	51.	(c)	82.	(b)	113.	(c)
21.	(a)	52.	(b)	83.	(c)	114.	(a)
22.	(b)	53.	(a)	84.	(b)	115.	(c)
23.	(b)	54.	(b)	85.	(a)	116.	(a)
24.	(b)	55.	(b)	86.	(a)	117.	(d)
25.	(c)	56.	(b)	87.	(b)	118.	(a)
26.	(a)	57.	(b)	88.	(c)	119.	(b)
27.	(a)	58.	(c)	89.	(b)	120.	(c)
28.	(b)	59.	(a)	90.	(a)	121.	(a)
29.	(c)	60.	(b)	91.	(b)	122.	(a)
30.	(b)	61.	(c)	92.	(b)	123.	(c)
31.	(a)	62.	(b)	93.	(a)	124.	(a)
						125.	(a)

❑❑❑

CHAPTER - 3

Interval Estimation

3.1 Write True/False

Q.1 The estimates obtained though various methods of point estimation, in general, do not concide with the true values of the parameters.

Q.2 The statistician is interested in finding an 'interval' within which the true value of the parameter θ is expected to lie with certain 'confidence coefficient', say $(1 - \alpha)$.

Q.3 The statistician want to find two values (t_1, t_2) being the function of sample values $(x_1, x_2, \ldots\ldots, x_n)$ from a p.d.f. $f(x, \theta)$ such that :

$P(t_1 < \theta < t_2) = (1 - \alpha)$

Then, the interval (t_1, t_2) is known as confidence interval for the parameter θ.

Q.4 In Q.3, the coefficient "$(1 - \alpha)$" is known as the "Degree of confidence or confidence coefficient."

Q.5 The degree of confidence "$(1 - \alpha)$" is measured in term of probability.

Q.6 In interval estimation, the risk of making an incorrect decision about the parameter "θ" can be controlled.

Q.7 An interval $(t_1 < \theta < t_2)$ is correct in long run of statistical experience and is measured with confidence coefficient '$(1 - a)$'.

Q.8 The limits (t_1, t_2) are known as the 'lower' and 'upper' limits of confidence interval.

Q.9 The quantities (t_1, t_2) depend only on the confidence coefficient '$(1 - \alpha)$' and the sample size "n".

Q.10 For a fixed "α" the totality of the "(t_1, t_2)" from different samples determines a field within which the 'θ' is asserted to lie.

Q.11 The field described in Q.10, is known as "confidence belt or Region of acceptance."

Q.12 Consider a p.d.f. P(t, θ) of t and assume two functions A(θ) & B(θ). Such that :

$$P\,(A_{(\theta)} < t < B_{(\theta)}) = \int_{A(\theta)}^{B(\theta)} P(t, \theta)\, dt = (1 - \alpha)$$

Then A(θ) and B(θ) be the functions of parameter θ for fixed "α".

Q.13 In Q.12, let us assume that :

$P(t < A(\theta)) = \alpha_1$

$P(t > b(\theta)) = \alpha_1$

such that $\alpha_1 + \alpha_2 = (1-\alpha)$

Then A(θ) and B(θ) will depend on the sufficient statistic "t".

Q.14 In Q.13 if $\alpha_1 = \alpha_2 = \frac{1}{2}(1 - \alpha)$ then the interval (A(θ), B(θ)) is called "central inverval."

Q.15 In Q.13, if $\alpha_1 \neq \alpha_2 \neq \frac{1}{2}(1 - \alpha)$ then the interval (A(θ), B(θ)) is known as "Non central" interval.

Q.16 Let $x_1, x_2, \ldots, x_n$ be a random sample from N(μ, σ^2) where σ^2 is known. Then confidence interval for the population mean μ is given as :

$$(\bar{x} - t_{\alpha/2} \frac{\sigma}{\sqrt{n}} < \mu < \bar{x} + t_{\alpha/2} \frac{\sigma}{\sqrt{n}})$$

with confidence coefficient (1 – α).

Q.17 Let $x_1, x_2 \ldots, x_n$ be a random sample from N(μ, σ^2) where μ & σ^2, both are unknown. Then the confidence interval for population mean μ is given as :

$$(\bar{x} - t_{\alpha/2} \frac{\$}{\sqrt{n}} \leq \mu \leq \bar{x} + t_{\alpha/2} \frac{\$}{\sqrt{n}})$$

where s^2 is the sample variance.

Q.18 Consider the normal distribution N(μ, σ^2) where μ is known but σ^2 is unknown. Then confidence interval for σ^2 is given as :

$$\sum_{i}^{n} \frac{(x_i - \mu)^2}{\chi^2_{\alpha/2,\, n-1}} \leq \sigma^2 \leq \sum_{i-1}^{n} \frac{(x_i - \mu)^2}{\chi^2_{1-\alpha/2,\, n-1}}$$

Q.19 Consider normal distribution N(μ, σ^2) where (μ & σ^2) are unknown. Then pair of confidence limits for σ^2 with confidence coefficient (1 – α) is given as :

$$\sum_{i}^{n} \frac{(x_i - \bar{x})^2}{\chi^2_{\alpha-/2, n-1}} \leq \sigma^2 = \frac{(n-1)\, s^2}{\chi^2_{\alpha/2, n-1}}$$

$$\text{and } \sum_{i}^{n} \frac{(x_i - \bar{x})^2}{\chi^2_{1-\alpha/2, n-1}} = \frac{(n-1)\, s^2}{\chi^2_{1-\alpha/2, n-1}}$$

Q.20 The simple method of obtaining the confidence intervals for the parameter 'θ' fails if no function of a sufficient statistic 'T' say $\psi(T, \theta)$ exists whose sampling distribution is independent of θ.

Q.21 Let t_1 & t_2 be the two limits such that
$P(t_1 \leq \theta < t_2) = 1 - a$, is satisfied. ....(1)

The confidence interval (t_1', t_2') also satisfies the result (1) and if $(t_2 - t_1) \leq (t_2', t_1')$ $\forall\ \theta \in \Omega_\theta$. Then the shortest confidence interval (t_1, t_2) is known as "uniformly shortest confidence interval" for θ.

Q.22 Consider the confidence interval $(t_{\alpha_2}, t_{\alpha_1})$ such that

$$P_\theta\left[\bar{x} - t_{\alpha_1} \frac{\sigma}{\sqrt{n}} \leq \theta \leq \bar{x} + t_{\alpha_2} \frac{\sigma}{\sqrt{n}}\right] = (1-\alpha)$$

where $\alpha_1 + \alpha_2 = \alpha$. Then the length $(t_{\alpha_2} - t_{\alpha_1})$ of confidence interval will be minimum when $(\alpha_1 = \alpha_2 = \alpha/2)$.

Q.23 In some situations, the length of confidence interval may depend on (i.e. be the function of) observations.

Q.24 The confidence interval for which the expected length $E_\theta\,(t_2 - t_1)$ is minimum, may be called as "Interval with the shortest expected length."

Q.25 Consider the normal distribution $N(\mu, \sigma^2)$ with known μ. Also consider the inequality given as :

$$P_{\sigma^2}\left[\sum_{i}^{n} \frac{(x_i - \mu)}{\sigma^2} < \chi^2_{1-\alpha_1, n}\right] = \alpha_1$$

$$\text{and } P_{\sigma^2}\left[\sum_{i}^{n} \frac{(x_i - \mu)^2}{\sigma^2} < \chi^2_{\alpha_2, n}\right] = \alpha_2$$

Then above inequalities lead to the following result :

$$P_{\sigma^2}\left[\sum_{i}^{n}\frac{(x_i-\mu)^2}{\chi^2_{\alpha_2,n}} \le \sigma^2 \le \sum_{i=1}^{n}\frac{(x_i-\mu)^2}{\chi^2_{1-\alpha,n}}\right] = (1-\alpha)$$

Q.26 The confidence interval defined in Q.25, has the length given as :

$$(t_2 - t_1) = n\sigma^2\left[\frac{1}{\alpha_{1-\alpha_1,n}} - \frac{1}{\chi^2_{\alpha_2,n}}\right]$$

Q.27 The probability density function for the interval defined in Q.25 is the p.d.f. of chi square (χ^2) with n degree of freedom.

Q.28 In case of discrete random variables, we can get a confidence interval for each α $(0 < \alpha < 1)$ with confidence coefficient exactly equal to $(1 - \alpha)$.

Q.29 To avoid the problem faced in Q.28. the one way is to consider only that confidence coefficient which is atleast equal to $(1 - \alpha)$.

Q.30 A finite or infinite interval where at least one of the end points is a r, v, is known as "Random interval."

Q.31 Let $T_1 = t_1(x_1, x_2, x_3, \ldots, x_n)$ be a statistic for which $P(t_1 < \theta) > (1 - \alpha)$, $\forall\ \theta \in \Omega\ \theta$. Then T_1 is known as "Loower confidence bound" for the parameter θ with confidence level $(1 - \alpha)$.

Q.32 A random interval $[T_1, \infty]$ is known as "Lower one sided $(1 - \alpha)$ confidence interval for parameter θ.

Q.33 If $T_2 = t_2(x_1, x_2, \ldots, x_n)$ is a statistic for which $P(\theta < t_2) \ge (1 - \alpha)$ $\forall\ \theta \in \Omega\ \theta$. Then t_2 is known as "upper confidence bound for θ" with confidence level equal to $(1 - \alpha)$.

Q.34 A random interval defined as $(-\infty, T)$ is known as an "Upper one sided $(1 - \alpha)$ confidence interval for parameter θ.

Q.35 Let (T_1, T_2) be an interval such that $P[T_1 < \theta < T_2] > (1 - \alpha)$. Then the random interval (T_1, T_2) is known as two sided $(1 - \alpha)$ confidence interval for θ.

Q.36 If a $(1 - \alpha)$ two sided C.I. (T_1, T_2) for θ is available and if $\psi(\theta)$ is strictly monotone function of θ then a $(1 - \alpha)$ two osided C.I. for $\psi(\theta)$ is $(\psi(T_1), (\psi(T_2))$ or $(\psi(T_2), \psi(T_1)$ according whether '$\psi(\theta)$' is strictly increasing or decreasing function of θ.

Q.37 If a $(1 - \alpha)$ one sided C.I. (lower confidence bound) say $[T_1, \infty)$ for parameter θ, and if $\psi(\theta)$ is strictly monotone function of θ then a one sided $(1 - \alpha)$ C.I. for $\psi(\theta)$ is $[\psi(T_1), \infty\ E)$.

Q.38 If a $(1 - \alpha)$ one sided C.I. (upper confidence bound) say $(-\infty, T_2)$ for parameter θ and if $\psi(\theta)$ is strictly monotone function of θ then one sided $(1 - \alpha)$ C.I. for $\psi(\theta)$ is $(-\infty, \psi(T_2)]$.

Q.39 Confidence intervals are not unique because different statistics are available to constitute $(1-\alpha)$ confidence interval.

Q.40 Confidence interval can not be unique because different multiple of the some statistic in a two sided C.I. can be utilized.

Q.41 The minimization of the length $l = (t_2 - t_1)$ of C.I. (t_1, t_2) is one of the criterion of selecting C.I.

Q.42 A wearness of the minimization criteria for C.I. is that in some situations, $(t_2 - t_1)$ may be a random variable.

Q.43 One of the criterion to select the C.I. is to minimize the expected length $E(t_2 - t_1)$.

Q.44 A weekness of the minimization criterion for $E(l)$ is that this criteria depends on the unknown true value of the parameter.

Q.45 In large sample theory, the confidence intervals based on $\left(\frac{\partial}{\partial\theta} \log L\right)$ has minimum expected length, where L is the likelihood function of sample values.

Q.46 A random interval (t_1, t_2) is said to be unbiased for θ if :

$$P(t_1 \le \theta' \le t_2) \quad \begin{array}{l} \ge 1 - \alpha, \text{ if } \theta = \theta' \\ \le 1 - \alpha, \text{ if } \theta \ne \theta' \end{array}$$

Q.47 For a $(1 - \alpha)$ lower confidence bound for θ, the statistic "t_1" is called uniformaly most accurate if :

$$P(t_1 \le \theta') < P(t^* \le \theta') \ \forall \ \theta' < \theta$$

where 't^*' is any $(1 - \alpha)$ arbitrary lower confidence bound for θ.

Q.48 For a $(1 - \alpha)$ upper confidence bound for θ, the statistic 't_2' is called uniformally most accurate if :

$$P(t_2 > \theta') < P(t^* \ge \theta') \ \forall \ \theta' < \theta.$$

where t^* is any $(1-\alpha)$ upper confidence bound for θ.

Q.49 Suppose $f(x, \theta)$, $\theta \in \Omega\ \theta$ denotes the family of density and have M.L.E. in (Tx). Also suppose that the distribution function $F(t, \theta)$ of $T = T(x)$ is a continuous function of t for each θ. Then the exists a unifirmaly most accurate lower confidence bound $\underset{\sim}{\theta}(x)$ for θ. at each confidence level $(1 - a)$.

Q.50 If x denotes the observed value of X and $t = t(x)$ and if the equation F(t,

$\theta) = (1 - \alpha)$ has a solution $\theta = \underset{\sim}{\theta}$ in Ω_θ then this solution is unique and $\underset{\sim}{\theta}(x) = \underset{\sim}{\theta}$.

Q.51 If $f(x, \theta) = \dfrac{g(x)}{h(\theta)}$, $a(\theta) \le x \le b(\theta)$ where $b(\theta)$ is a monotone decreasing function of a (θ) then a single sufficient statistic for θ is :

$\underset{\sim}{\theta} = \text{Min}, (a^{-1}(x_1), b^{-1}(x_n))$

Q.52 Let $\psi = \dfrac{h(\theta)}{h(\theta)}$. Then $P\{\alpha^{\frac{1}{n}} \le \psi \le 1\} = (1 - \alpha)$ from which a confidence interval for θ can be obtained. Then this interval is shorter than other intervals based on the distribution of ψ.

Q.53 Let $f(x, \theta) = \dfrac{g(x)}{h(\theta)}$, $a(\theta) < x < b(\theta)$.

where $a(\theta)$ is monotone increasing function and $b(\theta)$ is monotone decreasing function of parameter θ. Then :

$$P\left[h^{-1}\left(\frac{h(\hat{\theta})}{\alpha^{\frac{1}{n}}}\right) \le \theta \le \hat{\theta}\right] = (1 - \alpha)$$

where $\underset{\sim}{\theta} = \text{Min}\,(a^{-1}(x_1), b^{-1}(x_n))$.

Q.54 For large samples, the central confidence interval with confidence coefficient $(a = 0.95)$, for the mean λ of the Poisson distribution is given as :

$$\left[\bar{x} \mp \left(\frac{1.92}{\sqrt{n}} + \sqrt{\frac{3.82\,\bar{x}}{n} + \frac{3.69}{n}}\right)\right]$$

Q.55 Consider the p.d.f., $f(x, \theta) = \theta.\, e^{-\theta x}$, $0 \le x < \infty$, the central confidence interval for the parameter θ with confidence coefficient $\alpha = 0.95$ is given as :

$$\left[\left(\frac{1}{\bar{x}} - \frac{1.96}{\sqrt{n}\,\bar{x}}\right) \le \theta \le \left(\frac{1}{\bar{x}} + \frac{1.96}{\sqrt{n}\,\bar{x}}\right)\right]$$

3.1.1 Theory of Confidence Sets

Q.56 There are infinitely many ways to obtain a confidence interval for a given parametric function $\psi(\theta)$ even when the confidence coefficient $(a - \alpha)$ is fixed.

Q.57 The choice of statistic 'T' on the basis of which the confidence coefficient is obtained, is not unique.

Q.58 The division of level of significance (α) into two parts say (α_1 & α_2) corresponding to two tails areas of sampling distribution of the statistic 'T' to get a confidence interval is not unique.

Q.59 Let S be a set of the parametric space. Then a set "$S_0(\theta)$" to cover the parameter 'θ' is constituted such that

$$S_0(\theta) \Longleftrightarrow \theta \in S.$$

Q.60 A family of sets S(x) for varying $x \in$ X of the parametric spce $\Omega\ \theta$, is said to be a family of confidence sets at the confidence coefficient $(1-\alpha)$ if:

$$P [S(x_1, x_2, ..., x_n) \subset \theta] = (1 - \alpha)\ \forall\ \theta \in \Omega\ \theta.$$

Q.61 'A' family of sets $S_0(X)$ for warying $x\ \varepsilon$ X, of the parametric space $\Omega\ \theta$, is said to constitute a family of uniformaly most accurate (most selective or smallest) confidence set if we have :

$$P_\theta [S_0 (x_1, x_2, ... , x_n) \subset \theta] = (1 - \alpha),\ \forall\ \theta \in \Omega\ \theta \qquad ...(2)$$

and

$$P_{\theta'} [S_0 (x_1, x_2, ... , x_n) \subset \theta] \leq P_\theta [S(x_1, x_2, ..., x_n) \subset \theta] \qquad ...(2)$$

where S(x) is an arbitrary set satisfying (2).

Q.62 The set $S_0(x)$ defined in Q.61, is the smallest confidence set of level of significance 'α'.

Q.63 In most of the cases, a family of uniformaly most accurate (U.M.A.) sets can not be obtained.

Q.64 A family of sets S(x), for varying $x \in$ X, of parametric space $\Omega\ \theta$, is said to make a family of "Unbiased confidence set" of confidence coefficient $(1 - \alpha)$ if we have

$$P_{\theta'} [S(x_1, x_2, ..., x_n) \subset \theta] \leq (1 - \alpha).\ \forall\ \theta, \theta^1 \in \Omega\ \theta, \theta \neq \theta'.$$

Q.65 S(x) is a family of unbiased sets iff :

the probability for $S(x_1, x_2, ..., x_n)$ to cover θ, when some alternative value say θ', is true, does not exceed the same probability for the case when θ itself is the true value.

Q.66 An unbiasedness is the desirable feature of the family of confidence set.

Q.67 A family $S_0(x)$ of sets for varying $x \in \chi$ of $\Omega\ \theta$, is said to constitute a family of "Uniformly most accurate unbiased (U.M.A.U.) confidence set of confidence coefficient $(1 - \alpha)$ if :

$$P_\theta [\ S_0(x_1, x_2, \ldots.., x_n) \subset \theta\ ] = (1 - \alpha)\ \forall\ \theta \in \Omega\ \theta. \qquad \ldots(4)$$

$$P_{\theta'} [\ S_0(x_1, x_2, \ldots.., x_n) \subset \theta \leq] \leq (1 - \alpha)\ \forall\ \theta \in \Omega\ \theta, \theta \neq \theta' \qquad \ldots(5)$$

and

$$P_\theta [\ S_0(x_1, x_2, \ldots.., x_n) \subset \theta\ ] \leq P_{\theta'} [\ S(x_1, x_2, \ldots.., x_n) \subset \theta\ ]$$

$$\forall\ \theta, \theta' \in \Omega\ \theta, \theta \neq \theta'.$$

for the arbitrary set S(x), satisfying (4) & (5).

Q.68 For each $\theta_0 \in \Omega\ \theta$, let $A_0(\theta)$ be the acceptance region of a level of significance 'α' for testing $H_0 : \theta = \theta_0$. and for each $x \in \chi$. Let S(x) denotes the set of parametric values defined as :

$S(x) = \{\ \theta / x \in A(\theta)\ \}$.

Then $S(x)$ is a family of confidence sets for θ at the confidence coefficient $(1 - \alpha)$.

Q.69 For each θ_0 let $A_0(\theta_0)$ be the acceptance region at the level of significance α. Then to test the hypothesis $H_0 : \theta = \theta_0$ against all alternatives, the set S0 (x) defined as :

$S_0(x)\ \{\ \theta / x \in A_0(\theta_0)\ \}$

for varying $x \in X$, is a family of U.M.A. confidence set at the confidence level $(1 - \alpha)$.

Q.70 A family of "Uniformly most accurate unbiased" (U.M.A.U) confidence sets may be obtained by starting with a family of U.M.A.U. tests.

Q.71 Let $A_0(\theta_0)$ be the acceptance region at the level of significance α, for U.M.P.U. test for $H0 : \theta = \theta_0$ V.S. all alternatives. Let

$S_0(x) = \{\ \theta / x \in A_0(\theta)\ \}$

The the sets $S_0(x)$ for varying $x \in \chi$, are a family of U.M.A.U. confidence sets at the confidence level $(1 - \alpha)$.

Q.72 Let $x \sim N(\mu, \sigma^2)$. Then U.M.P.U. test for $H_0 : \mu = \mu_0$ vs all the alternatives at the level of significance 'α' have the C.R. given as :

$$W_0(\mu_0) = \{\ x / \frac{(\bar{x} - \mu_0)\sqrt{n}}{\sigma} > t_{\alpha/2}$$

Q.73 In case of the problem discribed in Q.72, the acceptance region is given as :

$$A_0 (\mu_0) = \{ x / \frac{(\bar{x} - \mu_0)\sqrt{n}}{\sigma} \leq t\alpha/2 \}$$

Q.74 In case of problem described in Q.73, for given μ, $x \in A_0 (\mu_0)$ iff :

$$\left(\bar{x} - t_{\alpha/2} \frac{\sigma}{\sqrt{n}} \leq \mu \leq \bar{x} + t_{\alpha/2} \frac{\sigma}{\sqrt{n}} \right)$$

Q.75 In situation described in Q.73, for given x, an U.M.P.U., a confidence set of confidence level (1 – α) is provided as :

$$S_0(x) = \left\{ \mu / \left(\bar{x} - t_{\alpha/2} \frac{\sigma}{\sqrt{n}} \leq \mu \; \bar{x} + t_{\alpha/2} \frac{\sigma}{\sqrt{n}} \right) \right\}$$

Q.76 Let x~ N(μ, σ^2) when μ is known but σ^2 is unknown. Then the U.M.P.U. test for $H_0 : \sigma^2 = \sigma_0^2$ v.s. all alternatives at the level of signifcance 'α' has the C.R. given as :

$$W_0 (\sigma^2) = \left\{ x / \sum_i^n \frac{(x_i - \mu)^2}{\sigma_0^2} < \chi_1^2 \; or > \chi_2^2 \right\}$$

where $\left(\chi_1^2 \; \& \; \chi_2^2\right)$ are such that :

$$\int_{\chi_1^2}^{\chi_2^2} f(\chi^2) \, d\,(\chi^2) = (1 - \alpha) \text{ and } \chi_1^2 \, f(\chi_1^2) = \chi_2^2 \, f(\chi_2^2)$$

Here f is the p.d.f. of chi-square statistic with n.d.f.

Q.77 In Q.76, the acceptance region is provided as :

$$A_0 (\sigma_0^2) = \left\{ x / \chi_1^2 \leq \sum_i^n \frac{(x_i - \mu)^2}{\sigma_0^2} \leq \chi_2^2 \right\}$$

Q.78 In problem described in Q.76, for given σ^2, $x \in A_0 (\sigma^2)$ if :

$$\left(\sum_i^n \frac{(x_i - \mu)^2}{\chi_2^2} \leq \sigma^2 \leq \sum_i^n \frac{(x_i - \mu)^2}{\chi_1^2} \right)$$

Q.79 In case of the problem discussed in Q.76, an uniformly most accurate unbiased (U.M.A.U.) confidence set at the confidence level $(1-\alpha)$ for given x, is provided as :

$$S_0(x) = \left\{ \sigma^2_{,} / \sum_{i-1}^{n} \frac{(x_i-\mu)^2}{\chi_2^2} \leq \sigma^2 \leq \sum_{i=1}^{n} \frac{(x_i-\mu)^2}{\chi_1^2} \right\}$$

and it is an interval.

Q.80 Suppose (6 & η) denote the sets of moments of parameter θ to be estimated and that of nuisance parameter respectively. Then a family of sets S(x), for satisfying $x \in \chi$ of the space 6, is said to be a family of confidence sets (for 6) at the confidence level $(1-\alpha)$, if :

$P6, \eta\ [\ S(x_1, x_2, \ldots\ldots, x_n) \subset 6\ ] = (1-\alpha)\ \forall\ 6\ \&\ \eta$ ...(6)

Q.81 A family of sets S0(x) for varying $x \in \chi$, of the space 6 is is said to provide a family of uniformaly most accurate (U.M.A.) confidence sets if:

$P6, \eta\ [\ S_0(x_1, x_2, \ldots\ldots, x_n) \subset 6\ ] = (1-\alpha)\ \forall\ 6\ \&\ \eta$ ...(7)

and

$P6', \eta\ [\ S_0(x_1, x_2, \ldots\ldots, x_n) \subset 6\ ] \leq P6, \eta\ [\ S_0(x_1, x_2, \ldots\ldots, x_n) \subset 6\ ]$

for what ever the other family of sets S(x) satisfying 6 may be.

Q.82 A family of sets S(x) for varying $x \in \chi$ of the space 6 is said to provide a family of unbiased confidence sets of level (1 – a) if :

$P6', \eta\ [\ S_0(x_1, x_2, \ldots\ldots, x_n) \subset 6\ ] = (1-\alpha)\ \forall\ 6, 6, \eta$ ...(8)

Q.83 A family of sets S_0(x) for varying $x \in$ X of the space 6 is said to provide a family of uniformly most accurate unbiased confidence sets for 6 of level $(1-\alpha)$ if :

$P6', \eta\ [\ S_0(x_1, x_2, \ldots\ldots, x_n) \subset 6\ ] = (1-\alpha)\ \forall\ 6, \eta$ ...(9a)

$P6', \eta\ [\ S_0(x_1, x_2, \ldots\ldots, x_n) \subset 6\ ] = (1-\alpha)\ \forall\ 6, 6', \eta$ ...(9b)

and

$P6', \eta\ [\ S_0(x_1, x_2, \ldots\ldots, x_n) \subset 6\ ] \leq P6', \eta\ [\ S(x_1, x_2, \ldots\ldots, x_n) \subset 6\ ]$

$\forall\ 6, 6', \eta$ ...(9c)

for whatever other family of sets S(x) satisfying 9(a) & 9(b) may be.

Q.84 One can state that there exists a relationship between the concept of "confidence sets" and the concept of "Similar tests" in the presence of "nuisance parameters."

Q.85 Let $A(6_0)$ be the acceptance region for each 6_0, at a level α similar test for the hypothesis H_0; $6 = 6_0$ and for each $x \in \chi$, let S(x) be the set of 6 values defined by :

$S(x) = \{ 6/x \in A(6) \}$.

Then the set S(x) is said to be a family of confidence sets for 6 at the level $(1 - \alpha)$.

Q.86 Let $A_0(6_0)$ be the acceptance region for each 6_0, of the level α, U.M.P. similar test for $H_0 : 6 = 6_0$. vs all alternatives. Then the set $S_0(x)$ defined as :

$S_0(x) = \{ 6/ x \in A_0 (6) \}$

for varying $x \in \chi$, is a family of "U.M.A. confidence sets" for 6 at the level $(1 - \alpha)$.

Q.87 Let $A_0(6_0)$ for each 6_0 be the acceptance region of the confidence level $(1 - \alpha)$ U.M.P.U. similar test for H0 : $6 = 6_0$ VS. all alternatives. Then the set $S_0(x)$ defined as :

$S_0(x) = \{ 6/ x \in A_0 (6) \}$

for varying $x \in \chi$, is a family of U.M.A.U. confidence sets for 6 at the confidence level $(1 - \alpha)$

Q.88 Let x~ $N(\mu, \sigma^2)$, where μ & σ^2 are both unknow. Then an U.M.A.U. similar test to test the hypothesis $H_0 : \mu_0 = \mu$ v.s. all alternatives at level α has the C.R. given as :

$$W_0 (\mu_0) = \left\{ x / \frac{|\overline{x} - \mu_0| \sqrt{n}}{s} > t_{\alpha/2, (n-1)} \right\}$$

Q.89 In Q.(88), the acceptance region $A_0(\mu_0)$ is given as :

$$A0(\mu_0) = \left\{ x / \frac{|\overline{x} - \mu_0| \sqrt{n}}{s} \le t_{\alpha/2, (n-1)} \right\}$$

Q.90 In Q.(89), for given μ, the observation x belongs to the acceptance region $A_0(\mu)$ iff :

$$\left(\overline{x} - t_{\alpha/2, n-1} \frac{s}{\sqrt{n}} \le \mu \le \overline{x} + t_{\alpha/2, n-1} \cdot \frac{s}{\sqrt{n}} \right)$$

Q.91 In Q.(88), for given x, an U.M.A.U. confidence set $S_0(x)$ for μ with confidence coefficient $(1 - \alpha)$ is given as :

$$S_0(x) = \left\{ \mu \,/\, (\bar{x} - t_{\alpha/2, n-1} \frac{s}{\sqrt{n}}) \leq \mu \leq (\bar{x} + t_{\alpha/2, n-1} \cdot \frac{s}{\sqrt{n}}) \right\}$$

Q.92 In Q.(91) the confidence set S_0(x) is a confidence interval with confidence coefficient $(1 - \alpha)$.

Q.93 If in Q.(88), we have to test the hypothesis H_0: $\sigma^2 = \sigma_0^2$ vs all alternatives then for an U.M.P.U. test of level α, the C.R. is given as :

$$W_0(\sigma_0^2) = \left\{ x \,/\, \frac{(n-1)s^2}{\sigma_0^2} < \chi_1^2 \text{ or } \frac{(n-1)s^2}{\sigma_0^2} > \chi_2^2 \right\}$$

where χ_1^2 & χ_2^2 are with df equal to $(n - 1)$.

Q.94 In Q.(93), if $\chi_1^2 f(\chi_1^2) = \chi_2^2 \, f(\chi_2^2)$ then :

$P(\chi^2 < \chi_1^2) + P(\chi^2 > \chi_2^2) = \alpha \cdot$

Q.95 In Q.(93), the acceptance region is given as :

$$A_0(\sigma_0^2) = \left\{ x \,/\, \chi_1^2 \leq \frac{(n-1)\, s^2}{\sigma_0^2} \leq \chi_2^2 \right\}$$

Q.96 In Q.(93), for given σ^2, $x \in A_0(\sigma^2)$ iff :

$$\frac{(n-1)\, s^2}{\chi_2^2} \leq \sigma^2 \leq \frac{(n-1)\, s^2}{\chi_1^2}$$

Q.97 In Q.(93). for given x, a U.M.A.U. confidence set S_0(x) for σ^2 with confidence coefficient $(1–\alpha)$ is given as :

$$S_0(x) = \left\{ \sigma^2 \,/\, \frac{(n-1)\, s^2}{\chi_2^2} \leq \sigma^2 \leq \frac{(n-1)\, s^2}{\chi_1^2} \right\}$$

Q.98 The confidence set $S_0(x)$ provided in Q.(97) is a confidence interval for σ^2 with confidence coefficient $(1 - \alpha)$.

Q.99 In Q.(93),j for given x, 9. U.M.A.U. confidence set for σ with confidence coefficient $(1 - \alpha)$ is given as :

$$S_0^*(x) = \left\{ \sigma \,/\, \frac{(n-1)\, s}{\chi_2} \leq \sigma^2 \leq \frac{(n-1)\, s}{\chi_1} \right\}$$

3.2 Fill in the Blanks

Q.1 The estimates obtained through various methods of point estimation,in general, do not coincide with the true value of (..........).

Q.2 The statistician is interested in finding an (..........) within which the true value of parameter θ, is expected to lie with certain (..........) coefficient, say $(1 - \alpha)$.

Q.3 The statistician want to find two values (t_1, t_2) as the function of sample values (x). from a population $f(x, \theta)$ such that :

P (.......... $\leq \theta \leq$ ....) $= (1 - \alpha)$.

Q.4 In Q.(3), the coefficient (1 – a) is known as degree of (..........) or confidence (..........).

Q.5 The degree of confidence '$(1 - \alpha)$' is measured in term of (..........).

Q.6 In interval estimation, the risk of making an incorrect decision about the (..........) θ. can be controlled.

Q.7 An interval $(t_1 \leq \theta \leq t_2)$ is correct in long run of statistical experience and is measured with confidence coefficient equal to (..........).

Q.8 The limits (l_1, l_2) are known as the (.......... and ....) limits of confidence interval.

Q.9 The quantities (t_1, t_2) depend only on the confidence coefficient equal to (..........) and the sample size n.

Q.10 For fixed "α" the totality of the (t_1, t_2) obtained from all possible samples determines a field within which the θ is expected to lie is known as (..........).

Q.11 Consider a p.d.f. $P(t, \theta)$ of t and assume two functions $A(\theta)$ & $B(\theta)$ such that :

$P (A(\theta) < t < B(\theta) = (1 - \alpha)$

then the $(1 - \alpha)$ is known as confidence (..........).

Q.12 In Q.(11), let us assume that :

$P (t < A(\theta)) = \alpha_1$ and $P (t > B(\theta)) = \alpha_2$

such that $(\alpha_1 + \alpha_2) =$ (..........), the confidence coefficient.

Q.13 In Q(12), the $A(\theta)$ and $B(\theta)$ will depend on the (..........) statistic.

Q.14 In Q.(12), if $\alpha_1 = \alpha_2 = \frac{1}{2}(1 - \alpha)$, then the interval $(A(\alpha), B(\alpha))$ is called (..........).

Q.15 If in Q.(12), $\alpha_1 \neq \alpha_2 \neq \frac{1}{2}(1-\alpha)$, then the interval (A (θ), B(θ)) is known as (..........).

Q.16 Let $\bar{x}$ be a sample mean from $N(\mu, \sigma^2)$ for σ^2 being known then the confidence interval for μ is given as :

$$(\bar{x} - t_{\alpha/2} \ldots\ldots < \mu < \bar{x} + \ldots\ldots\ldots \frac{\sigma}{\sqrt{n}})$$

with confidence coefficient (1 – α).

Q.17 Let ($\bar{x}$ & σ^2) be sample mean and sample variance from $N(\mu, \sigma^2)$, where (μ & σ^2) are unknown. Then confidence interval for μ is given as :

$$(\bar{x} - t_{\alpha/2} \ldots\ldots < \mu < \bar{x} + \ldots\ldots\ldots \frac{s}{\sqrt{n}})$$

Q.18 Let $x \sim N(\mu, \sigma^2)$ where μ is known. The confidence interval for σ^2 is given as :

$$\frac{\sum_{i}^{n}(x_i - \mu)^2}{\chi^2_{\alpha/2, n-1}} \leq \sigma 2 \leq \frac{\sum^{n}(x_i - \mu)^2}{\ldots\ldots\ldots}$$

Q.19 Let in Q.(18), both (μ, σ^2) be unknown. Then pair of confidence limits for σ^2 with confidence coefficient (1 – α) is given as :

$$\frac{\sum_{i}^{n}(x_i - \bar{x})^2}{\chi^2_{\alpha/2, n-1}} = \frac{(n-1)s^2}{(\ldots\ldots\ldots)}$$

and

$$\frac{\sum^{n}(x_i - \bar{x})^2}{\chi^2_{1-\alpha/2, n-1}} = \frac{(n-1)s^2}{(\ldots\ldots\ldots)}$$

Q.20 The simple method for obtaining confidence intervals for 'θ' fails if no function of a (..........) exists whose sampling distribution is independent of θ.

Q.21 Let (t_1, t_2) be confidence interval for θ such that :
$P(t_1 < \theta < t_2) = (1 - \alpha)$, is satisfied. ...(1)

Let the confidence interval $(t_1^*, t_2^*) \forall \theta \in \Omega\, \theta$. Then (t_1, t_2) is known as uniformly (..........) for θ.

Q.22 Consider a confidence interval $\left(t_{\alpha_1}, t_{\alpha_2}\right)$ for θ such that :

$$P\left(\bar{x} - t_{\alpha_1}\frac{\sigma}{\sqrt{n}} \leq \theta \leq \bar{x} + t_{\alpha_2}\frac{\sigma}{n}\right) = (1 - \alpha)$$

where $(\alpha_1 + \alpha_2) = \alpha$, Then the length $(t_{\alpha_2} - t_{\alpha_1})$ of confidence interval will be (..........) when $(\alpha_1 = \alpha_2 = \alpha/2)$.

Q.23 The confidence interval (t_1, t_2) for which the expected length $E\theta\,(t_2 - t_1)$ is minimum and is known as "Interval with shortest ....".

Q.24 Let $x \sim N(\mu, \sigma^2)$ with known μ. Also consider the following inequalities given as :

$$P_{\sigma^2}\left[\sum^{n}\left(\frac{x_i - \mu}{\sigma}\right)^2 < \chi^2_{(1-\alpha),n}\right] = \alpha_1$$

and

$$P_{\sigma^2}\left[\sum_{i}^{n}\left(\frac{x_i - \mu}{\sigma}\right)^2 < \chi^2_{\alpha_2,n}\right] = \alpha_2$$

Then, above inequalities lead to the following result :

$$P_{\sigma^2}\left[\frac{\sum^{n}(x_i - \mu)^2}{\chi^2_{\alpha_2,n}} \leq \sigma^2 \leq \frac{\sum^{n}(x_i - \mu)^2}{\chi^2_{1-\alpha,\, n}}\right] = (...........).$$

Q.25 The confidence interval defined in Q.(24), has the length given as :

$$(t_2 - t_1) = \eta\,\sigma^2\left[\frac{1}{(..........)} - \frac{1}{(..........)}\right]$$

Q.26 The p.d.f. for the interval defined in Q.(24) will be the p.d.f. of (..........) distribution with n.d.f.

Q.27 In case of discrete r.v., we can not get a confidence interval for each α, $(0 < \alpha < 1)$ with confidence coefficient exactly equal to (..........).

Q.28 To avoide the problem faced in Q(27), one way is to considere only that confidence coefficient which is atleaslt equal to (..........).

Q.29 A finite or infinite interval where atleast one of the end points, is a r.v., is known as (..........).

Q.30 Let $t_1 = t_1(x)$ be a statistic for which $P(t_1 \leq \theta) \geq (1-\alpha) \forall \theta \in \Omega\, \theta$. Then t_1 is known as (..........) bound for parameter θ with confidence level equal to (.........).

Q.31 A random interval $[(t_1, \infty)]$ is known as (..........) $(1-\alpha)$ confidence interval for parameter θ.

Q.32 Let $t_2 = t_2(x)$ be a statistic for which :

$P(\theta < t_2(x)) \geq (1-\alpha), \forall \theta \in \Omega\, \theta.$

Then t_2 is known as (.........) bound for θ with confidence level equal to $(1-\alpha)$.

Q.33 A random interval defined as $[-\infty, t]$ is known as (..........) $(1-\alpha)$ confidence interval for (..........) θ.

Q.34 Let (t_1, t_2) be an interval such that $P[t_1 < \theta < t_2] > (1-\alpha)$. Then the random interval (t_1, t_2) is known as (..........) sided $(1-\alpha)$ confidence interval for θ.

Q.35 If a $(1-\alpha)$ two sided confidence interval (t_1, t_2) for θ is available and $\psi(\theta)$ is strictly monotone function of θ then a $(1-\alpha)$ two sided C.I. for $\psi(\theta)$ is $[\psi(t_1), \psi(t_2)]$ or $[\psi(t_2), \psi(t_1)]$ according whether $\psi(\theta)$ is strictly (..........) or (..........). function of θ.

Q.36 If a $(1-\alpha)$ one sided C.I. (Lower confidence bounded says that (t_1, ∞) is for parameter θ and if $\psi(\theta)$ is strictly monotone function of θ, then one sided $(1-\alpha)$ C.I. for $\psi(\theta)$ is [.........., ...).

Q.37 If a $(1-\alpha)$ one side C.I. (Upper confidence bounded) say $(-\infty, t_2]$ exists for parameter θ and if $\psi(\theta)$ is strictly monotone function of θ then one sided $(1-\alpha)$ C.I. for $\psi(\theta)$ is (..........].

Q.38 Confidence intervals are not unique because different statistics are available to constitute (..........) confidence (...........).

Q.39 Confidence interval can not be unique because different multiple of the some (...........) in a two sided C.I. can be utilized.

Q.40 The minimization of the length l = (..........) of C.I. (t_1, t_2) is one of the criterion of selecting (..........).

Q.41 A weakness of the minimization criteria for C.I. is that in some situations, the length l = (..........). may be a r.v.

Q.42 One of the criterion to select the C.I. is to minimise the (..........) say E $(t_2 - t_1)$.

Q.43 One of the minimization criterion for E(l) is that this criteria depends upon unknown value of (..........).

Q.44 In large sample theory, the confidence intervals based on $\left(\frac{\partial}{\partial\theta}\log L\right)$ has (........) expected length.

Q.45 A random interval (t_1, t_2) is said to be unbiased for θ if :

$$P[t_1 < \theta' < t_2] \geq (..........), \text{ if } \theta' = \theta$$
$$\leq (..........), \text{ if } \theta' \neq \theta.$$

Q.46 For a (1 – a) lower confidence bound for θ, the statistic 't_1' is known uniformly most accurate if :

$P(t_1 \text{ .... } \theta') < P(t^* \leq \theta') \ \forall \ \theta' < \theta.$

where t* is any (1 – α) arbitrary lower confidence bound for ....

Q.47 For a (1 – α) upper confidence bound for θ, the statistic 't' is known as Uniformly most accuracte if :

$P(t_2 \text{ .... } \theta') < P(t^* \text{ ..... } \theta') \ \forall \ \theta' < \theta.$

where t* is any (1 – α) upper (..........) bound for θ.

Q.48 Suppose $f(x, \theta)$, $\theta \in \Omega\ \theta$ denotes the family of density and have M.L.E. in T(x). Also suppose that the distribution function F(t, θ) of T = $t(x)$ is a continuous function of $t(x)$ for each θ. Then there exists a uniformly most accurate lower confidence bound θ(x) for θ at each (...........) leval (1 – α).

Q.49 If x denotes the observed value of X, $t = t(x)$ and if the equation F(t, θ) = (1 – α) has a solution $\theta = \underset{\sim}{\theta}$ in $\Omega\ \theta$ then this solution is (...........) and $\underset{\sim}{\theta}(x)$ = ...

Q.50 Let $f(x, \theta) = \frac{g(x)}{h(\theta)}$, $a(\theta) < x < b(\theta)$ where b(θ) is a montone decreasing function of a(θ) then a single sufficient statistic for θ is θ given as :

θ = Min. (.........., ...).

Q.51 Let $f(x, \theta) = \frac{g(x)}{h(\theta)}$, $a(\theta) \leq x \leq b(\theta)$

where $a(\theta)$ is monotone increasing function and $b(\theta)$ is monotone decreasing function of θ. Then :

$$P\left[h-1\left(\frac{h(\theta)}{\alpha^{\frac{1}{n}}}\right) \leq \theta \leq \theta \quad\right] = (1-\alpha)$$

and θ = Min. {.........., ...}.

Q.52 For large samples, the central confidence interval with confidence coefficient $(1-\alpha = 0.95)$, for the mean λ of Poisson distribution is given as :

$$(t_1, t_2) = \left[\bar{x} \quad \left(\frac{1.92}{\sqrt{\ldots\ldots}} + \sqrt{\frac{3.82\,\bar{x} + 3.69}{\ldots\ldots}}\right)\right]$$

Q.53 Consider the p.d.f., $f(x, \theta) = \theta\, e^{-\theta x}$, $0 \leq x < \infty$. The central confidence interval for the parameter θ with confidence coefficient $(1-\alpha) = 0.95$ is given as :

$$\left[\left(\frac{1}{\bar{x}} - \frac{1.96}{\sqrt{n}\,\bar{x}}\right) \leq \theta \leq (\ldots\ldots\ldots.)\right]$$

Q.54 There are infinitely many ways to obtain a confidence interval for a given (..........) ψ(θ) even when the confidence coefficient (..........) is fixed.

Q.55 The division of level of significance 'α', into two parts say (α_1 & α_2) corresponding to (..........) areas of (..........) distribution of the statistic $t(x)$ to get the confidence interval is not unique.

Q.56 A family of sets say S(x) for varying $x \in \chi$ of the parametric space Ω_θ, is said to be a family of (..........) sets at the confidence coefficient (1 – α) if :

$P\,[\,S\,(x_1, x_2, \ldots\ldots, x_n) \subset \theta\,] = (\ldots\ldots\ldots.)\ \forall\ \theta \in \Omega_\theta$.

Q.57 Let $S_0(x)$ be a family of uniformly most accurate (Most selective or smallest) confidence set for varying $x \in X$. Then $S_0(x)$ is the (..........) confidence set of level of significance '..........'.

Q.58 A family of sets S(x) for varying $x \in X$, of parametric space Ω_θ, is said to make a family of an (..........) of confidence coefficient (1 – α) if we have :

$P_{\theta'}\,[\,S\,(x_1, x_2, \ldots\ldots, x_n) \subset \theta\,] < (1-\alpha)\ \forall\ \theta' \in \Omega\,\theta.\ \theta' \neq \theta$.

Q.59 The S(x) is a family of (..........) sets if :

The probability for $S(x_1, x_2, \ldots, x_n)$ to cover θ, does not exceed the probability for the case when θ is itself the true value.

Q.60 An (..........) is the desirable property of the family of confidence sets.

Q.61 For each $\theta_0 \in \Omega_\theta$, let $A(\theta_0)$ be a acceptance region of a confidence level $(1 - \alpha)$ for testing $H_0 : \theta = \theta_0$ and for each $x \in X$. Let $S(x)$ denotes the set of parametric values defined as :

$S(x) = \{ \theta / x \in A(\theta) \}$.

Then S(x) is a family of (..........) for θ at the confidence level $(1 - \alpha)$.

Q.62 Let $A_0(\theta_0)$ be the acceptance region at the level of significance 'α' for U.M.P.U. test to test $H_0 : \theta = \theta_0$ vs. all alternatives. Let sets $S_0(x)$ be as :

$S_0(x) = \{ \theta \mid x \in A_0(\theta) \}$

Then the sets $S_0(x)$ for varying $x \in X$, is a family of (..........) confidence sets at the confidence level $(1 - \alpha)$.

Q.63 Let $x \sim N(\mu, \sigma^2)$ where σ^2 is known. Then U.M.P.U. test for $H_0 : \mu = \mu_0$ v.s. all alternatives at the level of significance 'α' have the C.R. given as :

$$W_0 (\mu_0) = \left\{ X / \frac{(\bar{x} - \mu_0)\sqrt{\ldots\ldots}}{\ldots\ldots\ldots} > t_{\alpha/2} \right\}$$

Q.64 In case of problem described in Q(63), the acceptance region is given as :

$$A_0 (\mu_0) = \left\{ X / \frac{(\ldots\ldots\ldots)\sqrt{\;}}{\sigma} \leq t_{\alpha/2} \right\}$$

Q.65 In case of problem described in Q.(64), for given μ, we have $X \in A_0(\mu_0)$ if:

$$(\bar{x} - t_{\alpha/2} \frac{\sigma}{\sqrt{n}} \leq \mu \leq \ldots\ldots\ldots)$$

Q.66 In situation described in Q.(63) for given X, an U.M.P.U., test, a confidence set $S_0(x)$ of confidence level $(1 - \alpha)$ is provided as :

$$S_0(x) = \{ \mu / (\bar{x} - t_{a/2} \frac{\sigma}{\sqrt{n}} \leq \mu \leq (\ldots\ldots\ldots) \}$$

Q.67 Let X ~ N (μ, σ^2) when μ is known as $\underset{0}{\mu}$ but σ^2 is unknow. Then an U.M.P.U. test for $H_0 : \sigma^2 = \underset{0}{\sigma^2}$ vs. all alternatives at the level of significance 'a' has the C.R. given as :

$$W_0 (\underset{0}{\sigma^2}) = \left\{ X / \sum_{i}^{n} (..........)^2 < \chi_1^2 \text{ or } > \chi_2^2 \right\}$$

where (χ_1^2 or χ_2^2) are Chi-squares with n.d.f.

Q.68 In Q.(67), the acceptance region is provided as :

$$A_0 (\underset{0}{\sigma^2}) = \left\{ X / \chi_1^2 \le \sum^{n} (..........)^2 \le \chi_2^2 \right\}$$

Q.69 In problem described in Q.(67), for given σ^2, $x \in A_0 (\sigma^2)$ if :

$$\left(\sum_{i=1}^{n} \frac{(..........)^2}{\chi_2^2} \le \sigma^2 \le \sum^{n} \frac{(..........)^2}{\chi_1^2} \right)$$

Q.70 In case of problem discussed in Q.(67), an uniformly most accurate unbiased (U.M.A.U.) confidence set for σ^2 at the confidence level $(1 - \alpha)$ for given x is provided as :

$$S_0(x) = \left\{ \sigma^2_{,} \Big/ \sum^{n} \frac{(..........)^2}{\chi_2^2} \le \sigma^2 \le \sum^{n} \frac{(x_i - \mu)^2}{\chi_1^2} \right\}$$

which is an interval.

Q.71 A family of sets S(x) for varying $x \in$ X. with the space 6 is said to provide a family of (.........) sets of confidence level $(1 - \alpha)$ if :

$P6', \eta\, [\, S_0 (x_1, x_2, ..., x_n) \subset 6 \,] \le (1 - \alpha) \;\forall\; 6, 6', \eta$.

Q.72 One can state that there exists a relationship between the concept of "Confidence sets" and the concept of "Similar sets" in the presence of (..........).

Q.73 Let A(6_0) be the acceptance region for each 6_0, at a level of significance, α, a similar test for the hypothesis $H_0 : 6 = 6_0$ and for each $x \in$ X. Let S(x) be the set of 6 defined as :

$S(x) = \{ 6 / x \in A(6) \}$

Then the set S(x) is said to be a family of (..........) for θ at the confidence level (..........).

Q.74 Let $A_0(\theta_0)$ be the acceptance region for each θ_0 of the level α, U.M.P. similar test for $H_0 : \theta = \theta_0$ V.S. all alternatives. Then the set S_0(x). defined as :

$S_0(x) = \{ \theta / x \in A_0(\theta) \}$

is a family of uniformly most accurate (..........) for θ at confidence level (..........).

Q.75 Let $A_0(\theta_0)$ for each θ_0 be the acceptance region of confidence level (1 – a), U.M.P.U. similar test for $H_0 : \theta = \theta_0$ vs all alternatives. Then the set $S_0(x)$ defined as :

$S_0(x) = \{ \theta / x \in A_0(\theta) \}$

for varying x $\in$ X, is the family of (..........) confidence sets for θ at the confidence a level $(1 - \alpha)$.

Q.76 Let $X \sim N(\mu, \sigma^2)$ with (μ, σ^2) being unknown. Then an U.M.A.U. similar test to test $H_0 : \mu = \mu_0$ vs all alternatives at level α has the C.R. given as :

$$W_0(\mu_0) = \left\{ x \,/\, \frac{|..........|\sqrt{n}}{..........} > t_{\alpha/2,\,(n-1)} \right\}$$

Q.77 In Q.(76), the acceptance region $A_0(\mu_0)$ is given as :

$$A_0(\mu_0) = \left\{ x \,/\, \frac{|..........|\,.....}{s} \le t_{\alpha/2,\,(n-1)} \right\}$$

Q.78 In Q.(77), for given μ, the observation x belongs to the acceptance region $A_0(\mu)$ if :

$(\bar{x} - .... < m < \bar{x} + ...)$

Q.79 In Q.(77), for given x, an U.M.A.U. confidence set S_0(x) for μ with confidence coefficient $(1 - \alpha)$ is given as :

$S_0(x) = \{ \mu / (..........) < \mu < (..........) \}$

Q.80 In Q.(79), the confidence set $S_0(x)$ is a confidence interval with (..........) equal to $(1 - \alpha)$.

Q.81 In Q.(76), to test $H_0 : \sigma^2 = \sigma_0^2$ v.s. all alternatives, for an U.M.P.U. test of level α, the C.R. is given as :

$$W_0(\sigma_0^2) = \left\{ x \Big/ \frac{(n-1)\,s^2}{\sigma_0^2} < \chi_1^2 \text{ or } (\text{..........}) > \chi_2^2 \right\}$$

where (χ_1^2, χ_2^2) are with d.f. equal to $(n-1)$.

Q.82 In Q.(81) if $\chi_1^2 f(\chi_1^2) = \chi_2^2\, f(\chi_2^2)$ then :

$$P(\chi_2 < \chi_1^2) + P(\text{..........}) = \alpha$$

Q.83 In Q.(81), the acceptance region is given as :

$$A_0(\sigma_0^2) = \{x / \chi_1^2 \leq \text{.........} \leq \chi_2^2\}$$

Q.84 In Q.(81), for given σ^2, $x \in A_0(\sigma^2)$ if :

$$\frac{(n-1)\,s^2}{\chi_2^2} \leq \sigma^2 \leq . \frac{(\text{..........})\,\ldots}{\chi_1^2}$$

Q.85 In Q.(81), for given x, an U.M.A.U. confidence set $S_0(x)$ for σ^2 with confidence coefficient (1 – a) is given as :

$$S_0(x) = \left\{ \sigma^2 \Big/ \frac{(\text{..........})\,\ldots}{\chi_2^2} \leq \sigma^2 \leq \frac{(\text{..........})\,\ldots}{\chi_1^2} \right\}$$

Q.86 The confidence set $S_0(x)$ provided in Q.(85) is a confidence interval for (...........), with confidence coefficient equal to (..........).

Q.87 In Q.(81), for given x, a U.M.A.U. confidence set for σ' with confidence coefficient (..........), is given as :

$$S_0^*(x) = \left\{ \sigma \Big/ \frac{(n-1)\,s}{\chi_2} \leq \sigma \leq \frac{(\text{..........})\,\ldots}{\chi_1} \right\}$$

3.3 Multiple Choice Questions and Answers

Q.1 The estimates obtained through various methods of point estimation, in general, do not conside with the true value of :

(a) Population parameter (b) Population variance

(c) Population constant (d) Both (a) & (c)

Q.2 The statistician is interested in finding an (..........) within which the true value of parameter θ, is expected to lie with certain confidence coefficient (1 – α).

(a) Estimate (b) Interval
(c) Estimator (d) Extended value

Q.3 Let 'α' be the level of significance, then confidence coefficient is equal to :

(a) α (b) 2α
(c) $(1 - \alpha)$ (d) $\alpha/2$

Q.4 In Q.(3), the coefficient (1 – a) is known as :

(a) Degree of confidence (b) Confidence coefficient
(c) Both (a) & (b) (d) None

Q.5 The degree of confidence $(1 - \alpha)$ is measured in term of :

(a) Probability (b) Association
(c) Coefficient of correlation (d) Coefficient of determines

Q.6 In an interval estimation, the risk of making an incorrect decision about (..........) θ, can be controlled.

(a) True value of. (b) Parameter θ.
(c) Both (a) & (b) (d) None

Q.7 In an interval $(t_1 \leq \theta < t_2)$, the limits (t_1, t_2) are known as the :

(a) Upper & lower limits of the interval.
(b) Lower & upper limits of the interval.
(c) Length of the interval.
(d) None.

Q.8 The quantities (t_1, t_2) of a confidence interval $(t_1 < \theta < t_2)$ depend on confidence coefficient equal to :

(a) a, (b) $\left(1 - \frac{\alpha}{1}\right)$
(c) $(1 - \alpha)$ (d) 2α

Q.9 For fixed 'α' the totality of the (t_1, t_2) obtained from all possible samples determines a field within which θ is expected to lie is known as :

(a) Confidence belt (b) Region of acceptance
(c) Both (a) & (b) (d) None

Q.10 Consider a p.d.f., $P(t, \theta)$ of 't' and assume two functions $A(\theta)$ & $B(\theta)$ such that :

$P [A(\theta) < t < B(\theta)] = (1 - \alpha)$

then $(1 - \alpha)$ is known as :

(a) Confidence coefficient (b) Level of significance

(c) Coefficient of determination (d) None

Q.11 In Q.(10), let us assume that :

$P(t < A(\theta)) = \alpha_1$ and $P(t > B(\theta)) = \alpha_2$

Then $(\alpha_1 + \alpha_2)$ will be equal to :

(a) α, the level of significance.

(b) Confidence coefficient $(1 - \alpha)$

(c) Confidence level.

(d) Both (b) & (c).

Q.12 In Q.(11), the function $A(\theta)$ & $B(\theta)$ will depend on the :

(a) Unbiased estimator (b) Biased estimator

(c) Sufficient statistic (d) None

Q.13 In Q.(11), if $\alpha_1 = \alpha_2 = \frac{(1-\alpha)}{2}$, then the interval $(A(\theta), B(\theta)$ is known as :

(a) Confidence interval (b) Central interval

(c) Extreame interval (d) None

Q.14 In Q.(11), $\alpha_1 = \alpha_2 \neq \frac{1}{2}(1 - \alpha)$, then the interval $(A(\theta), B(\theta)$ is known as :

(a) Central interval (b) Non central interval

(c) Confidence level (d) Level of significance

Q.15 Let $\bar{x}$ be the sample mean from $N(\mu, \sigma^2)$ where σ^2 is known and sample size n is large. Then confidence interval for μ is given as :

(a) $(\bar{x} \quad t_{\alpha/2} \frac{s}{\sqrt{n}})$ (b) $(\bar{x} \pm t_{\alpha/2} \frac{s}{\sqrt{n}})$

(c) $(\bar{x} \quad Z_{\alpha/2} \cdot \frac{\sigma}{\sqrt{n}})$ (d) $(\bar{x} \pm Z_{\alpha/2} \frac{\sigma}{\sqrt{n}})$

Q.16 Let $(\bar{x}, s^2)$ be the sample mean and sample variance from $N(\mu, \sigma^2)$ where (μ, σ^2) are unknown. Then confidence interval for μ is given as :

(a) $(\bar{x} \pm t_{\alpha/2} \frac{s}{\sqrt{n}})$ (b) $(\bar{x} \quad t_{\alpha/2} \frac{s}{\sqrt{n}})$

(c) $(\bar{x} \pm Z_{\alpha/2} \frac{\sigma}{\sqrt{n}})$ (d) $(\bar{x} \quad Z_{\alpha/2} \frac{\sigma}{\sqrt{n}})$

Q.17 Let x ~ N (μ, σ^2) where μ is known. Then confidence interval for σ^2 is given as :

(a) $$\left(\frac{\sum^{n} (x_i - \mu)^2}{\chi^2_{\alpha/2,\, n-1}}, \frac{\sum^{n} (x_i - \mu)^2}{\chi^2_{1-\alpha/2,\, n-1}} \right)$$

(b) $$\left(\frac{\sum^{n} (x_i - \mu)^2}{\chi^2_{(1-\alpha/2),\, n-1}}, \frac{\sum^{n} (x_i - \mu)^2}{\chi^2_{\alpha/2,\, n-1}} \right)$$

(c) $$\left(\frac{\sum^{n} (x_i - \bar{x})^2}{\chi^2_{\alpha/2,\, n-1}}, \frac{\sum^{n} (x_i - \bar{x})^2}{\chi^2_{(1\alpha/2),\, n-1}} \right)$$

(d) $$\left(\frac{\sum^{n} (x_i - \bar{x})^2}{\chi^2_{\alpha/2,\, n}}, \frac{\sum^{n} (x_i - \bar{x})^2}{\chi^2_{(1-\alpha/2),\, n}} \right)$$

Q.18 Let x~ N(μ, σ^2) where both (μ, σ^2) are unknown. Then confidence limits for σ^2 with confidence coefficient $(1 - \alpha)$ is given as :

(a) $\left(\dfrac{\sum\limits^{n}(x_i-\mu)^2}{\chi^2_{\alpha/2,\,n-1}},\ \dfrac{\sum\limits^{n}(x_i-\mu)^2}{\chi^2_{(1-\alpha/2),\,n-1}}\right)$

(b) $\left(\dfrac{\sum\limits^{n}(x_i-\bar{x})^2}{\chi^2_{\alpha/2,\,n-1}},\ \dfrac{\sum\limits^{n}(x_i-\bar{x})^2}{\chi^2_{(1-\alpha/2),\,n-1}}\right)$

(c) $\left(\dfrac{\sum\limits^{n}(x_i-\mu)^2}{\chi^2_{\alpha/2,\,n}},\ \dfrac{\sum\limits^{n}(x_i-\mu)^2}{\chi^2_{(1-\alpha/2),\,n-1}}\right)$

(d) $\left(\dfrac{\sum\limits^{n}(x_i-\bar{x})^2}{\chi^2_{\alpha/2,\,n}},\ \dfrac{\sum\limits^{n}(x_i-\bar{x})^2}{\chi^2_{(1-\alpha/2),\,n}}\right)$

Q.19 The simple method to get a confidence interval for parameter θ fails if no function of a (..........). exists whose sampling distribution is independent of θ.

(a) Biased estimator (b) M.L.E.

(c) Efficient estimator (d) Sufficient statistic

Q.20 The simple method to get a confidence interval for parameter θ fails if no function of a sufficient statistic exists and sampling distribution is :

(a) Dependent of θ. (b) Independent of θ.

(c) Not dependent of θ. (d) Both (b) & (c)

Q.21 Let (t_1, t_2) and (t_1^*, t_2^*) be two confidence interval for 'θ' such that :

$P(t_1 \le \theta \le t_2) = (1-\alpha)$ and $P(t_1^* \le \theta \le t_2^*) = (1-\alpha)$ are satisfied. Let $(t_2 - t_1) < (t_2^* - t_1^*)\ \forall\ \theta \in \Omega\ \theta$. Then (t_1, t_2) is known as :

(a) Uniformally shortest confidence interval.

(b) Bounded shortest confidence interval.

(c) Uniformally longest confidence interval.

(d) Both (b) & (c)

Q.22 Let $(t_{\alpha_1}, t_{\alpha_2})$ be confidence interval for θ such that :

$$P\left(\bar{x} - t_{\alpha_1}\frac{s}{\sqrt{n}} \le \theta \le \bar{x} + t_{\alpha_2}\frac{s}{\sqrt{n}}\right) = (1-\alpha)$$

where $\alpha_1 + \alpha_2 = \alpha$. Then the length $(t_{\alpha_2} - t_{\alpha_1})$ of confidence interval will be minimum if :

(a) $\alpha_1 = \alpha_2 = \alpha$ (b) $\alpha_1 = \alpha_2 = \frac{\alpha}{2}$

(c) $\alpha_1 \ne \alpha 2 \ne \frac{\alpha}{2}$ (d) $\alpha_1 = \alpha_2 = (1-\alpha)$

Q.23 The confidence interval (t_1, t_2) for which the expected length $E_\theta(t_2 - t_1)$ is minimum is known as :

(a) Interval with shortest expected length.

(b) Interval with median expected length.

(c) Interval with least expected length.

(d) Both (a) & (c)

Q.24 Let x ~ N (μ, σ^2) with known μ. Also consider the following inequality written as :

$$P_{\sigma^2}\left[\sum^{n}\left(\frac{x_i - \mu}{\sigma}\right)^2 < \chi^2_{(1-\alpha_1),\, n}\right] = \alpha_1$$

and $$P_{\sigma^2}\left[\sum_{i}^{n}\left(\frac{x_i - \mu}{\sigma}\right)^2 > \chi^2_{2\alpha_2),\, n}\right] = \alpha_2$$

$$\text{Then } P_{\sigma^2}\left[\sum^{n}\frac{(x_i-\mu^2}{\chi^2_{\alpha_2,n}} \le \sigma^2 \le \frac{\sum^{n}(x_i-\mu)^2}{\chi^2_{(1-a),n}}\right]$$

is equal to :

(a) α (b) 2α

(c) $\frac{\alpha 1}{2}$ (d) $(1-\alpha)$

Q.25 The confidence interval defined in Q.(24) has the length $(t_2 - t_1)$ equal to :

(a) $n\,\sigma^2\left(\frac{1}{\chi^2_{(1-\alpha_1),n}} - \frac{1}{\chi^2_{\alpha_2,n}}\right)$

(b) $n\,\sigma^2\left(\frac{1}{\chi^2_{\alpha_2,n}} - \frac{1}{\chi^2_{(1-\alpha_1),n}}\right)$

(c) $n\,\sigma^2\left(\frac{1}{\chi^2_{\alpha_1,n}} - \frac{1}{\chi^2_{\alpha_2,n}}\right)$

(d) $n\,\sigma^2\left(\frac{1}{\chi^2_{\alpha_1,n-1}} - \frac{1}{\chi^2_{\alpha_2,n-1}}\right)$

Q.26 The p.d.f. for the interval defined in Q.(24) will be the p.d.f. of a statistic known as :

(a) Z statistic (b) F statistic

(c) χ^2 with n d.f. (d) χ^2 with $(n-1)$ d.f.

Q.27 In case of diserete r.v., we can not get a confidence interval for each α, $(0 < \alpha < 1)$ with confidence coefficient equal to :

(a) $(1-\alpha)$ (b) α

(c) $\frac{\alpha}{2}$ (d) $\left(1-\frac{\alpha}{2}\right)$

Q.28 To avoid the problem faced in Q.(27), one has to consider only that confidence coefficient which is equal to :

(a) α (b) $\left(1-\frac{\alpha}{2}\right)$

(c) $(1-\alpha)$ (d) 2α

Q.29 A finite or infinite interval in which atleast one of the end points, is a r.v., is known as :

(a) Random interval (b) Confidence interval

(c) Central interval (d) Shortest interval

Q.30 Let $t_1 = t_1(x)$ be a statistic for θ for which $P(t_1 \le \theta) > (1-\alpha)\ \forall\ \theta \in \Omega_\theta$. Then for parameter θ, the t_1 is known as :

(a) Lower confidence bound.

(b) Upper confidence bound

(c) Central confidence bound

(d) Confidence interval

Q.31 A random interval $[\,t_1, \infty\,]$ for parameter θ with confidence coefficient $(1-\alpha)$ is known as confidence interval with :

(a) Lower one sided (b) Upper one sided

(c) Central one sided (d) Central two sided

Q.32 Let $t_2 = t_2(x)$ be a statistic for parameter θ for which $P\,(\,\theta < t_2\,(x)\,) \ge (1-\alpha)$, $\forall\ \theta \in \Omega_\theta$ then for parameter θ, the $t_2(x)$ is known as :

(a) Lower confidence bound (b) Upper confidence bound

(c) Central confidence bound (d) Confidence interval

Q.33 A random interval $(-\infty, t]$ for parameter θ with confidence coefficient $(1-\alpha)$ is known as confidence interval with :

(a) Lower one sided (b) Upper one sided

(c) Central two sided (d) Central one sided

Q.34 Let (t_1, t_2) be an interval such that $P\,(\,t_1 \le \theta \le t_2] \ge (1-\alpha)$. Then the random interval (t_1, t_2) is known as :

(a) Two sided with confidence coefficient $(1-\alpha)$.

(b) One sided with confidence coefficient $(1-\alpha)$

(c) Two sided with confidence coefficient $(1-2\alpha)$.

(d) Two sided with confidence coefficient $(1-\alpha/2)$.

Q.35 Let (t_1, t_2) be a two sided confidence interval for θ, and $\psi(\theta)$ be strictly monotone function of θ. Then two sided C.I. for $\psi(\theta)$ is $[\,\psi(t_1), \psi(t_2)\,]$ if

$\psi(\theta)$ is strictly :

(a) Increasing function of θ. (b) Decreasing function of θ.

(c) Monotone function of θ. (d) An unbiased estimator for θ.

Q.36 Let $[\, t_1, \infty\,]$ be a $(1 - \alpha)$ one sided C.I. (Lower confidence bound) for parameter θ. and $\psi(\theta)$ is strictly monotone function of θ. Then one sided $(1 - \alpha)$ C.I. for $\psi(\theta)$ is :

(a) $[\, \psi(t_1), \infty\,)$ (b) $[\, \infty\, , \psi(t_1)\,)$

(c) $[-\infty\, , \psi(t_1)\,)$ (d) $[\, \psi(t), -\infty\,)$

Q.37 Let $(-\infty, t_2]$ be a $(1 - \alpha)$ one sided C.I. (Upper confidence bound for parameter θ and $\psi(\theta)$ is a strictly monotone function of θ then one sided $(1 - \alpha)$ C.I. for $\psi(\theta)$ is :

(a) $(\, -\infty, \psi(t_2)\,]$ (b) $(\, \psi(t_2), -\infty\,]$

(c) $(\, \psi(t_2), \infty\,]$ (d) $(\, \infty, \psi(t_2)\,]$

Q.38 Confidence intervals are not unique because (..........) are available to constitute $(1 - \alpha)$ confidence interval.

(a) Different biased estimators.

(b) Different parameters.

(c) Different statistics

(d) Different M.L.E.'s

Q.39 Confidence interval can not be unique because in two sided C.I., the different multiple of the same (..........) can be utilized.

(a) Estimators (b) Statistics

(c) MLE's (d) Regression coefficients

Q.40 The minimization of the length $(t_2 - t_1)$ of C.I. (t_1, t_2) is one of the criterion for selecting the :

(a) Confidence interval (b) Confidence coefficient

(c) Level of significance (d) Sufficient statistic

Q.41 A weakness of the minimization criteria for C.I. is that in some situations, the length $l = (t_2 - t_1)$ may be the :

(a) Fixed value (b) Statistic

(c) Random variable (d) Constant value

Q.42 One of the criterion to select the C.I. is to :

(a) Minimize the length, $E\,(t_2 - t_1)$

(b) Minimize the length, $E\,(t_1 - t_2)$

(c) Maximize the length, $E\,(t_2 - t_1)$

(d) Maximize the length, $E(t_1 - t_2)$

Q.43 One of the minimization criterion for expected length of C.I. is that this criteria depends upon unknown value of ....

(a) Estimator (b) Parameter

(c) Statistic (d) Estimated

Q.44 In large sample theory, the confidence interval based on $(\frac{\partial}{\partial\theta} \log L)$ has :

(a) Maximum expected length

(b) Minimum expected length

(c) Average expected length

(d) Moderate expected length

Q.45 A random interval (t_1, t_2) is said to be unbiased for θ if $P(t_1 \leq \theta' \leq t_2)$ is :

(a) $\geq (1 - \alpha)$ if $\theta' = \theta$. (b) $\leq (1 - \alpha)$ if $\theta' \neq \theta$.

(c) Both (a) & (b) (d) None

Q.46 For a $(1 - \alpha)$ lower confidence bound for θ, the statistic 't_1' is known uniformally most accurate if $P(t_1 \leq \theta')$ is :

(a) $< P(t^* \leq \theta') \ \forall \ \theta' < \theta$.

(b) $> P(t^* \leq \theta') \ \forall \ \theta' < \theta$.

(c) $\geq P(t^* \leq \theta') \ \forall \ \theta' < \theta$.

(d) $\leq P(t^* \geq \theta') \ \forall \ \theta' < \theta$.

Q.47 For a $(1 - \alpha)$ confidence bound for θ, the statistic 't_2' is known as uniformally most accurate for θ. if $P(t_2 \geq \theta')$ is :

(a) $> P(t^* \geq \theta') \ \forall \ \theta' < \theta$.

(b) $< P(t^* \geq \theta') \ \forall \ \theta' < \theta$.

(c) $\geq P(t^* \geq \theta') \ \forall \ \theta' < \theta$.

(d) $\geq P(t^* \leq \theta') \ \forall \ \theta' < \theta$.

Q.48 Let $f(x, \theta), \theta \in \Omega_\theta$ be the family of density functions and have MLE in $T(x)$. Also let $F(t(x), \theta))$ be a continuous distribution function for each θ. Then, there exists a uniformally most accurate :

(a) Upper confidence bound $\underset{\sim}{\theta}(x)$ for θ.

(b) Central confidence bound $\underset{\sim}{\theta}(x)$ for θ.

(c) Lower confidence bound $\underset{\sim}{\theta}(x)$ for θ.

(d) Moderate confidence bound $\underset{\sim}{\theta}(x)$ for θ.

Q.49 Let x be the observed vaule of X, $t = t(x)$ be a statistic. If the equation $F(t, \theta) = (1 - \alpha)$ has a solution $\theta = \underset{\sim}{\theta}$ in Ω θ, then this solution is :

(a) Unique and $\underset{\sim}{\theta}(x) = \theta$.

(b) Random variable and $\underset{\sim}{\theta}(x) = \theta$.

(c) Unbiased and $\underset{\sim}{\theta}(x) = \theta$.

(d) BL.UE. and $\underset{\sim}{\theta}(x) = \theta$.

Q.50 Let $f(x, \theta) = \dfrac{g(x)}{h(\theta)}$, $a(\theta) \le x \le b(\theta)$ is a monotone decreasing function of $a(\theta)$, then a single sufficient statistic for θ is θ given as :

(a) Max $\{ a^{-1}(x_1), b^{-1}(x_n) \}$ (b) Min. $\{ a^{-1}(x_1), b^{-1}(x_n) \}$

(c) $\dfrac{1}{2}(a(x_1) + b(x_n))$ (d) $a(x_1) + b(x_n)$

Q.51 Let $f(x, \theta) = \dfrac{g(x)}{h(\theta)}$, $a(\theta) < x < b(\theta)$ where $a(\theta)$ and $b(\theta)$ are monotone increasing and decreasing function of θ respectively. Let

$$P\left[h{-}1\left(\frac{h(\theta)}{\alpha^{\frac{1}{n}}} \right) \le \theta \le \theta \right] = (1 - \alpha)$$

Then θ will be equal to :

(a) Min $\{ a^{-1}(x_1), b^{-1}(x_n) \}$ (b) Max $\{ a^{-1}(x_1), b^{-1}(x_n) \}$

(c) Min $\{ a(x_1), b(x_n) \}$ (d) Max $\{ a_1(x_1), b(x_n) \}$

Q.52 For large samples, the central confidence interval with confidence coefficient $(1 - \alpha) = 0.95$, for mean λ of Poisson distribution is given as :

(a) $(t_1, t_2) = \bar{x} \left(\dfrac{1.92 + \sqrt{3.82\,\bar{x} + 3.69}}{\sqrt{n}} \right)$

(b) $(t_1, t_2) = \bar{x} \pm \left(\dfrac{1.92 + \sqrt{3.82\,\bar{x} + 3.69}}{\sqrt{n}} \right)$

(c) $(t_1, t_2) = \overline{x} \quad \left(\dfrac{1.92 + \sqrt{3.82\,\overline{x} + 3.69}}{n} \right)$

(d) $(t_1, t_2) = \overline{x} \pm \left(\dfrac{1.92 + \sqrt{3.82\,\overline{x} + 3.69}}{n} \right)$

Q.53 Consider the p.d.f., $f(x, \theta) = \theta\, e^{-\theta x}$, $0 < x < \infty$, The central confidence interval for the parameter θ with confidence coefficient $(1 - \alpha) = 0.95$ is given as :

(a) $\left(\dfrac{1}{\overline{x}} - \dfrac{1.96}{\sqrt{n}\,\overline{x}} \right) \le \theta \le \left(\dfrac{1}{\overline{x}} + \dfrac{1.96}{\sqrt{n}\,.\,\overline{x}} \right)$

(b) $\dfrac{1}{\overline{x}} \left(1 - \dfrac{1.96}{\sqrt{n}\,\overline{x}} \right) \le \theta \le \dfrac{1}{\overline{x}} \left(1 + \dfrac{1.96}{\sqrt{n}} \right)$

(c) $\dfrac{1}{\overline{x}} \left(1 - \dfrac{1.96}{n} \right) \le \theta \le \dfrac{1}{\overline{x}} \left(1 + \dfrac{1.96}{n} \right)$

(d) Both (a) & (b)

Q.54 There are infinitely many ways to obtain a confidence interval for a given (..........). $\psi(\theta)$ of θ even when the confidence coefficient $(1 - \alpha)$ is fixed.

(a) Non parametric function (b) Parametric function

(c) Monotone function (d) Decreasing function

Q.55 The division of level of significance (α) into two parts say (α_1, α_2) corresponding to two tailed areas of (..........) distribution of the statistic $t(x)$ to the confidence interval is not unique.

(a) Population (b) Sampling

(c) Continuous (d) Dicrease

Q.56 A family of sets say $S(x)$ for varying $x \in X$. of the parameter space Ω_θ, is said to be a family of (..........) sets at the confidence level $(1 - \alpha)$ if :

$P\,[\,S\,(x_1, x_2, \ldots.., x_n) \subset \theta\,] = (1 - \alpha)\ \forall\ \theta \in \Omega\,\theta.$

(a) Confidence (b) Monotone

(c) Discrete (d) Continuous

Q.57 Let $S_0(x)$ be a family of uniformally most accurate (Most selective or smallest) confidence set for varying $x \in X$. Then $S_0(x)$ is the (..........) confidence set of level of significance α.

(a) Central (b) Largest

(c) Smallest (d) Average

Q.58 A family of sets S(x) for varying x ∈ X, of parametric space Ω θ, is said to make a family of an (..........) of confidence coefficient (1 – a) if we have :

$P\theta' [S(\underset{\sim}{x}) \subset \theta] \leq (1 - \alpha) \ \forall \ \theta' \in \Omega\ \theta, \theta' \neq \theta.$

(a) Efficient confidence set (b) Unbiased confidence set

(c) Sufficient confidence set (d) Consistant confidence set

Q.59 If the probability for a confidence set S(x). to cover θ, does not exceed the same probability for the case when θ is itself the true value. Then the set S(x) is known as a family of :

(a) Unbiased set (b) Efficient set

(c) Consistent set (d) Sufficient set

Q.60 The desirable property of the family of confidence sets is :

(a) Consistancy (b) Unbiasedness

(c) Sufficiency (d) Efficiency

Q.61 Let $A(\theta_0)$ be a acceptance region of a confidence level (1 – α) for each $\theta_0 \in \Omega_\theta$. for testing $H_0 : \theta = \theta_0$ for each $x \in X$. Let S(x) denotes the set of parametric values defined as :

$S(x) = \{ \theta / x \in A(\theta) \}$

Then for parameter θ, the S(x) is a family of :

(a) Confidence set (b) Confidence interval

(c) Sufficient set (d) Sufficient interval

Q.62 Let $A_0 (\theta_0)$ be the acceptance region at the level of significance 'α', for U.M.P.U. test to test $H_0 : \theta = \theta_0$ VS all alternatives. Let sets $S_0(x)$ be as :

$S_0(x) = \{ \theta / x \in A_0(\theta) \}$

Then the sets $S_0(x)$ for varying $x \in X$, are a family of (..........) confidence set at confidence level (1 – α) :

(a) Uniformly most accurate unbiased

(b) Uniformly most sufficient

(c) Uniformly most efficient

(d) Uniformly most consistant

Q.63 Let $x \sim N(\mu, \sigma^2)$ and σ^2 is known and sample size $n > 30$. Then U.M.P.U. test for $H_0 : \mu = \underset{0}{\mu}$ vs all altlernatives at the level of significance 'α' have the C.R. given as :

(a) $W_0(\mu_0) = \left\{ x \Big/ \frac{(\bar{x} - \mu_0)\sqrt{n}}{\sigma} < t_{\alpha} \right\}$

(b) $W_0(\mu_0) = \left\{ x \Big/ \frac{(\bar{x} - \mu_0)\sqrt{n}}{\sigma} > t_{\alpha} \right\}$

(c) $W_0(\mu_0) = \left\{ x \Big/ \frac{(\bar{x} - \mu_0)\sqrt{n}}{\sigma} < t_{\alpha/2} \right\}$

(d) $W_0(\mu_0) = \left\{ x \Big/ \frac{(\bar{x} - \mu_0)\sqrt{n}}{\sigma} > t_{\alpha/2} \right\}$

Q.64 In case of problem discussed in Q(63), the acceptance region is given as :

(a) $A_0(\mu_0) = \left\{ x \Big/ \frac{(\bar{x} - \mu_0)\sqrt{n}}{\sigma} \geq t_{\alpha/2} \right\}$

(b) $A_0(\mu_0) = \left\{ x \Big/ \frac{(\bar{x} - \mu_0)\sqrt{n}}{\sigma} < t_{\alpha} \right\}$

(c) $A_0(\mu_0) = \left\{ x \Big/ \frac{(\bar{x} - \mu_0)\sqrt{n}}{\sigma} \leq t_{\alpha/2} \right\}$

(d) $A_0(\mu_0) = \left\{ x \Big/ \frac{(\bar{x} - \mu_0)\sqrt{n}}{\sigma} \geq t_{\alpha} \right\}$

Q.65 In case of problem discussed in Q.(64), for given μ_0 we have $x \in A_0(\mu_0)$

if:

(a) $(\bar{x} - t_{\alpha/2} \frac{\sigma}{\sqrt{n}} \leq \mu \leq \bar{x} + t_{\alpha/2} \frac{\sigma}{\sqrt{n}})$

(b) $(\bar{x} + t_{\alpha/2} \frac{\sigma}{\sqrt{n}} \leq \mu \leq \bar{x} - t_{\alpha/2} \frac{\sigma}{\sqrt{n}})$

(c) $(\bar{x} - t_{\alpha} \cdot \frac{\sigma}{\sqrt{n}} \leq \mu \leq \bar{x} + t_{\alpha} \frac{\sigma}{\sqrt{n}})$

(d) $(\bar{x} + t_{\alpha} \cdot \frac{\sigma}{\sqrt{n}} \cdot \leq \mu \leq \bar{x} - t_{\alpha} \cdot \frac{\sigma}{\sqrt{n}})$

Q.66 In situation described in Q.(63), for given X, an U.M.P.U. test, a confidence set $S_0(x)$ of confidence level $(1 - \alpha)$ is provided as :

(a) $S_0(x) = \{ \mu / (\bar{x} + t_{\alpha/2} \frac{\sigma}{\sqrt{n}}) \leq \mu \leq (\bar{x} - t_{\alpha/2} \frac{\sigma}{\sqrt{n}}) \}$

(b) $S_0(x) = \{ \mu / (\bar{x} - t_{\alpha/2} \frac{\sigma}{\sqrt{n}}) \leq \mu \leq (\bar{x} + t_{\alpha/2} \frac{\sigma}{\sqrt{n}}) \}$

(c) $S_0(x) = \{ \mu / (\bar{x} - t_{\alpha} \cdot \frac{\sigma}{\sqrt{n}}) \leq \mu \leq (\bar{x} + t_{\alpha} \cdot \frac{\sigma}{\sqrt{n}}) \}$

(d) $S_0(x) = \{ \mu / (\bar{x} + t_{\alpha} \cdot \frac{\sigma}{\sqrt{n}}) \leq \mu \leq (\bar{x} - t_{\alpha} \cdot \frac{\sigma}{\sqrt{n}}) \}$

Q.67 Let $x \sim N(\mu, \sigma^2)$ where μ is known as μ_0, but σ^2 is unknown. Then an U.M.P/U. test to test $H_0 : \sigma^2 = \sigma_0^2$ vs all alternatives at the level of significane 'α' has the C.R. given as :

(a) $W_0(\sigma_0^2) = \left\{ x \Big/ \sum^{n} \left(\frac{x_i - \mu_0}{\sigma_0} \right)^2 < \chi_1^2 \text{ or } \chi_2^2 \right\}$

(b) $W_0(\sigma_0^2) = \left\{ x \Big/ \sum^n \left(\frac{x_i - \mu_0}{\sigma_0} \right)^2 < \chi_1^2 \text{ or } > \chi_2^2 \right\}$

(c) $W_0(\sigma_0^2) = \left\{ x \Big/ \sum_i^n \left(\frac{x_i - \mu_0}{s} \right)^2 < \chi_1^2 \text{ or } > \chi_2^2 \right\}$

(d) $W_0(\sigma_0^2) = \left\{ x \Big/ \sum^n \left(\frac{x_i - \mu_0}{\sigma^2} \right)^2 < \chi_1^2 \text{ or } > \chi_2^2 \right\}$

where (χ_1^2, χ_2^2) are with n d.f.

Q.68 In Q.(67), the acceptance region is provided as :

(a) $A_0(\sigma_0^2) = \left\{ x \Big/ \chi_1^2 \le \sum^n \left(\frac{x_i - \mu_0}{\sigma_0} \right) \le \chi_2^2 \right\}$

(b) $A_0(\sigma_0^2) = \left\{ x \Big/ \chi_1^2 \le \sum^n \left(\frac{x_i - \mu}{\sigma_0} \right) \le \chi_2^2 \right\}$

(c) $A_0(\sigma_0^2) = \left\{ x \Big/ \chi_1^2 \le \sum^n \left(\frac{x_i - \mu_0}{\sigma} \right)^2 \le \chi_2^2 \right\}$

(d) $A_0(\sigma_0^2) = \left\{ x \Big/ \chi_1^2 \le \sum^n \left(\frac{x_i - \mu_0}{s} \right)^2 \le \chi_2^2 \right\}$

Q.69 In problem described in Q.(67), for given σ^2, $x \in A_0(\sigma^2)$ if :

(a) $\left(\sum_{i}^{n} \frac{(x_i - \mu_0)^2}{\chi_2^2} \leq \sigma^2 \leq \sum^{n} \frac{(x_i - \mu_0)^2}{\chi_1^2} \right)$

(b) $\left(\sum^{n} \frac{(x_i - \mu_0)^2}{\chi_1^2} \leq \sigma^2 \leq \sum^{n} \frac{(x_i - \mu_0)^2}{\chi_2^2} \right)$

(c) $\left(\sum^{n} \frac{(x_i - \overline{x})^2}{\chi_2^2} \leq \sigma^2 \leq \sum^{n} \frac{(x_i - \overline{x})^2}{\chi_1^2} \right)$

(d) $\left(\sum^{n} \frac{(x_i - \overline{x})^2}{\chi_1^2} \leq \sigma^2 \leq \sum^{n} \frac{(x_i - \overline{x})^2}{\chi_2^2} \right)$

Q.70 In case of problem discussed in Q.(67), an uniformally most accurate unbiased (U.M.A.U.) confidence set for σ^2 at the confidence level $(1 - \alpha)$ for given $\underline{x}$ is provied as :

(a) $S_0(\underline{x}) = \left\{ \sigma^2 \Big/ \sum^{n} \frac{(x_i - \mu_0)^2}{\chi_2^2} \leq \sigma^2 \leq \sum^{n} \frac{(x_i - \mu_0)^2}{\chi_1^2} \right\}$

(b) $S_0(\underline{x}) = \left\{ \sigma^2 \Big/ \sum^{n} \frac{(x_i - \mu_0)^2}{\chi_1^2} \leq \sigma^2 \leq \sum^{n} \frac{(x_i - \mu_0)^2}{\chi_2^2} \right\}$

(c) $S_0(\underline{x}) = \left\{ \sigma^2 \Big/ \sum^{n} \frac{(x_i - \overline{x})^2}{\chi_2^2} \leq \sigma^2 \leq \sum^{n} \frac{(x_i - \overline{x})^2}{\chi_1^2} \right\}$

(d) $S_0(\underline{x}) = \left\{ \sigma^2 \Big/ \sum^{n} \frac{(x_i - \overline{x})^2}{\chi_1^2} \leq \sigma^2 \leq \sum^{n} \frac{(x_i - \overline{x})^2}{\chi_2^2} \right\}$

Q.71 Suppose a family of sets $S(x)$, for varying $x \in X$ with space 6 at level $(1 - \alpha)$ is defined as :

$P\ 6', \eta\ [\ S_0\ (x) < 6\ ] < (1 - \alpha)\ \forall\ 6,\ 6', \eta.$

Then the sets $S(x)$ is said to provide a family of :

(a) Biased confidence sets

(b) Consistent confidence sets

(c) Unbiased confidence sets

(d) Efficient confidence sets

Q.72 One can state that there exists a relationship between the concepts of "Confidence sets" and "Similar sets" in the presence of :

(a) Sufficient statistic (b) Nuisance parameter

(c) Population parameter (d) Consistent estimator

Q.73 Let $A(6_0)$ be the acceptance region for each 6_0, at a level of significance "α" for a similar test for the hypothesis $H_0 : 6 = 6_0$ and for each $\underset{\sim}{x} \in X$, let $S(x)$ be the set of 6 defined as :

$S(x) = \{ 6 / \underset{\sim}{x} \in A(6) \}$

Then the set $S(x)$ is said to be the family of :

(a) Sufficient set for 6. (b) Consistent set for 6.

(c) Confidence set for 6. (d) Efficient set for 6.

Q.74 Let $A_0(6_0)$ be the acceptance region for each 6_0 at $\alpha\%$ level of significance for U.M.P. similar test for $H_0 : 6 = 6_0$ v.s. all alternatives. Then the set $S_0(x)$ defined as :

$S(x) = \{ 6 / x \in A_0(6) \}$

is a family of uniformally most accurate (..........) for 6 at confidence level $(1 - \alpha)$.

(a) Confidence set (b) Confidence level

(c) Confidence interval (d) Consistent set

Q.75 Let $A_0(6_0)$ for each 6_0 be the acceptance region of confidence level $(1 - \alpha)$ and a U.M.P.U. similar test for $H_0 : 6 = 6_0$ v.s. all alternatives. Then the set $S_0(x)$ defined as :

$S_0(x) = \{ 6 / x \in A_0(6) \}$

for varying $\underset{\sim}{x} \in X$, is the family of "Uniformally most accurate (..........) confidence sets for 6 at the confidence level $(1 - \alpha)$.

(a) Unbiased (b) Sufficient

(c) Biased (d) Consistent

Q.76 Let $x \sim N(\mu, \sigma^2)$ with (μ, σ^2) being unknown. Then an U.M.A.U. similar test to test $H_0\ \mu = \underset{0}{\mu}$ v.s. all alternatives at α level of significance, has the

C.R. given as $W_0(\mu_0)$:

(a) $\{ x / \dfrac{|\bar{x} - \mu_0| n}{s} > t_{\alpha/2,\, n-1} \}$

(b) $\{ x / \dfrac{|\bar{x} - \mu_0| \sqrt{n}}{s^2} > t_{\alpha/2,\, n-1} \}$

(c) $\{ x / \dfrac{|\bar{x} - \mu| \sqrt{n}}{s} > t_{\alpha/2,\, n-1} \}$

(d) $\{ x / \dfrac{|\bar{x} - \mu_0| \sqrt{n}}{s} > t_{\alpha/2,\, n-1} \}$

Q.77 In Q.(76), the acceptance region $A_0(\mu_0)$ is given as $A_0(\mu_0)$:

(a) $\{ x / \dfrac{|\bar{x} - \mu_0| \sqrt{n}}{s} \leq t_{\alpha/2,\, n-1} \}$

(b) $\{ x / \dfrac{|\bar{x} - \mu_0| n}{s} \leq t_{\alpha/2,\, n-1} \}$

(c) $\{ x / \dfrac{|\bar{x} - \mu_0| \sqrt{n}}{s^2} \leq t_{\alpha/2,\, n-1} \}$

(d) $\{ x / \dfrac{|\bar{x} - \mu_0| n}{s^2} \leq t_{\alpha/2,\, n-1} \}$

Q.78 In Q.(76), for given μ, the observation x belongs to the acceptance region $A_0(\mu)$ if :

(a) $\{ (\bar{x} - t_{\alpha/2,\, n-1} \cdot \dfrac{s^2}{n}) \leq \mu \leq (\bar{x} + t_{\alpha/2,\, n-1} \cdot \dfrac{s^2}{n})$

(b) $\{ (\bar{x} - t_{\alpha,\, n} , \dfrac{s}{\sqrt{n}}) \leq \mu \leq (\bar{x} + t_{\alpha,\, n} \cdot \dfrac{s}{\sqrt{n}})$

(c) $\{ (\bar{x} - t_{\alpha,\, n} , \dfrac{s^2}{n}) \leq \mu \leq (\bar{x} + t_{\alpha,\, n} \cdot \dfrac{s^2}{n})$

(d) $\{ (\bar{x} - t_{\alpha/2,\, n-1} \frac{s}{\sqrt{n}}) \leq \mu \leq (\bar{x} + t_{\alpha/2,\, n-1} \cdot \frac{s}{\sqrt{n}})$

Q.79 In Q.(76), for given x, an U.M.A.U. confidence set $S_0(x)$ for μ with confidence coefficient $(1-\alpha)$ is given $S_0(x)$ as :

(a) $\{ \mu / (\bar{x} - t_{\alpha/2,\, n-1} \cdot \frac{s^2}{n}) \leq \mu \leq (\bar{x} + t_{\alpha/2,\, n-1}, \frac{s^2}{n}) \}$

(b) $\{ \mu / (\bar{x} - t_{\alpha,\, n,} \frac{s}{\sqrt{n}}) \leq \mu \leq (\bar{x} + t_{\alpha,\, n} \cdot \frac{s}{\sqrt{n}}) \}$

(c) $\{ \mu / (\bar{x} - t_{\alpha/2,\, n-1} \cdot \frac{s}{\sqrt{n}}) \leq \mu \leq (\bar{x} + t_{\alpha/2,\, n-1} \cdot \frac{s}{\sqrt{n}}) \}$

(d) $\{ \mu / (\bar{x} - t_{\alpha,\, n,} \frac{s^2}{n}) \leq \mu \leq (\bar{x} + t_{\alpha,\, n} \cdot \frac{s^2}{n}) \}$

Q.80 In Q.(79), the confidence set $S_0(x)$ is a confidence interval with (..........) equal to $(1-\alpha)$:

(a) Confidence coefficient (b) Confidence level.

(c) Both (a) & (b) (d) Level of significance

Q.81 In Q.(76), to test $H_0 : \sigma^2 = \sigma_0^2$ v.s. all alternative, for an U.M.P.U. test of level 'α', the C.R. is given as :

$W_0\,(\sigma_0^2) = ...$

(a) $\left\{ x \Big/ \frac{(n-1)\, s^2}{\sigma_0^2} < \chi_1^2 \ or\ \frac{(n-1)\, s^2}{\sigma_0^2} > \chi_2^2 \right\}$

(b) $\left\{ x \Big/ \frac{(n-1)\, \sigma_0^2}{s^2} < \chi_1^2 \ or\ \frac{(n-1)\, \sigma_0^2}{s^2} > \chi_2^2 \right\}$

(c) $\left\{ x \Big/ \frac{n\, s^2}{\sigma_0^2} < \chi_1^2 \ or\ \frac{n\, s^2}{\sigma_0^2} > \chi_2^2 \right\}$

(d) $\left\{ x \Big/ \frac{n\,\sigma_0^2}{s^2} < \chi_1^2 \; or \; \frac{n\,\sigma_0^2}{s^2} > \chi_2^2 \right\}$

Q.82 In Q.(81), if $(\chi_1^2) f(\chi_1^2) = \chi_2^2 f(\chi_2^2)$, then we have :

(a) $P(\chi^2 < \chi_1^2) + P(\chi^2 > \chi_2^2) = \alpha/2$

(b) $P(\chi^2 < \chi_1^2) + P(\chi^2 > \chi_2^2) > \alpha$

(c) $P(\chi^2 < \chi_1^2) + P(\chi^2 > \chi_2^2) < \alpha$

(d) $P(\chi^2 < \chi_1^2) + P(\chi^2 > \chi_2^2) = \alpha$

Q.83 In Q.(81), the acceptance region is given as $A0(\sigma_0^2)$ equal to :

(a) $\{ x / \chi_1^2 \leq (n-1) \frac{\sigma_0^2}{s^2} \leq \chi_2^2 \}$

(b) $\{ x / \chi_1^2 \leq (n-1) \frac{s^2}{\sigma_0^2} \leq \chi_2^2 \}$

(c) $\{ x / \chi_1^2 \leq \frac{n\,s^2}{\sigma_0^2} \leq \chi_2^2 \}$

(d) $\{ x / \chi_1^2 \leq \frac{n\,\sigma_0^2}{s^2} \leq \chi_2^2 \}$

Q.84 In Q.(81), for given σ^2, $x \in A_0(\sigma^2)$ if :

(a) $\left\{ \frac{(n-1)\,s^2}{\chi_2^2} \leq \sigma^2 \leq \frac{(n-1)\,s^2}{\chi_1^2} \right\}$

(b) $\left\{ \frac{(n-1)\,\chi_2^2}{s^2} \leq \sigma^2 \leq \frac{(n-1)\,\chi_1^2}{s^2} \right\}$

(c) $\left\{ \frac{n\,s^2}{\chi_2^2} \leq \sigma^2 \leq \frac{n\,s^2}{\chi_1^2} \right\}$

(d) $\left\{ \frac{n\,\chi_2^2}{s^2} \le \sigma^2 \le \frac{n\,\chi_1^2}{s^2} \right\}$

Q.85 In Q.(81), for given x, an U.M.A.U. confidence set $S_0(x)$ for σ^2 with confidence coefficient $(1-\alpha)$ is given as $S_0(x)$ equal to :

(a) $\left\{ \sigma^2 / \frac{(n-1)\,\chi_2^2}{s^2} \le \sigma^2 \le \frac{(n-1)\,\chi_1^2}{s^2} \right\}$

(b) $\left\{ \sigma^2 / \frac{(n-1)\,s^2}{\chi_2^2} \le \sigma^2 \le \frac{(n-1)\,s^2}{\chi_1^2} \right\}$

(c) $\left\{ \sigma^2 / \frac{n\,s^2}{\chi_2^2} \le \sigma^2 \le \frac{n\,s^2}{\chi_1^2} \right\}$

(d) $\left\{ \sigma^2 / \frac{n\,\chi_2^2}{s^2} \le \sigma^2 \le \frac{n\,\chi_1^2}{s^2} \right\}$

Q.86 The confidence set $S_0(x)$ provided in Q.(85) is a confidence interval with confidence coefficient equal to $(1-\alpha)$ for :

(a) σ^2 (b) $\left(\sqrt{\sigma}\right)^2$

(c) Both (a) & (b) (d) σ

Q.87 In Q.(81), for given x, an U.M.A.U. confidence set for σ with confidence level $(1-\alpha)$, is given as S_0^* equal to :

(a) $\left\{ \sigma^2 / \frac{(n-1)\,s}{\chi_2} \le \sigma \le \frac{(n-1)\,s}{\chi_1} \right\}$

(b) $\left\{ \sigma^2 / (n-1)\sqrt{\frac{s^2}{\chi_2^2}} \le \sigma \le (n-1)\sqrt{\frac{s^2}{\chi_1^2}} \right\}$

(c) Both (a) & (b)

(d) $\left\{ \sigma / (n-1)\,\frac{s^2}{\chi_2^2} \le \sigma \le (n-1)\,\frac{s^2}{\chi_1^2} \right\}$

3.4 Key Answers

Write True/False

Q.	Ans.	Q.	Ans.	Q.	Ans.	Q.	Ans.
1.	True	31.	True	61.	True	91.	True
2.	True	32.	True	62.	True	92.	True
3.	True	33.	True	63.	True	93.	True
4.	True	34.	True	64.	True	94.	True
5.	True	35.	True	65.	True	95.	True
6.	True	36.	True	66.	True	96.	True
7.	True	37.	True	67.	True	97.	True
8.	True	38.	True	68.	True	98.	True
9.	True	39.	True	69.	True	99.	True
10.	True	40.	True	70.	True		
11.	True	41.	True	71.	True		
12.	True	42.	True	72.	True		
13.	True	43.	True	73.	True		
14.	True	44.	True	74.	True		
15.	True	45.	True	75.	True		
16.	True	46.	True	76.	True		
17.	True	47.	True	77.	True		
18.	True	48.	True	78.	True		
19.	True	49.	True	79.	True		
20.	True	50.	True	80.	True		
21.	True	51.	True	81.	True		
22.	True	52.	True	82.	True		
23.	True	53.	True	83.	True		
24.	True	54.	True	84.	True		
25.	True	55.	True	85.	True		
26.	True	56.	True	86.	True		
27.	True	57.	True	87.	True		
28.	True	58.	True	88.	True		
29.	True	59.	True	89.	True		
30.	True	60.	True	90.	True		

Fill in the Blanks

Q.	Ans.
1.	Parameter
2.	Interval, confidence
3.	(t_1, t_2)
4.	Confidence coefficient
5.	Probability
6.	Parameter
7.	$(1-\alpha)$
8.	Lower, upper
9.	$(1-\alpha)$
10.	Confidence belt, acceptance
11.	Coefficient
12.	$(1-\alpha)$
13.	Sufficient
14.	Central interval
15.	Non-central interval
16.	$\frac{\sigma}{\sqrt{n}}, t_{\alpha/2}$
17.	$\frac{s}{\sqrt{n}}, t_{\alpha/2}$
18.	$\chi^2_{(1-\alpha/2), n}$
19.	$\chi^2_{\alpha/2, n-1}, \chi^2_{(1-\alpha/2), n-1}$
20.	Sufficient statsitic
21.	Shortest confidence interval
22.	Minimum
23.	Expected length
24.	$(1-\alpha)$
25.	$\chi^2_{1-\alpha_1, n}, \chi^2_{\alpha_2, n}$
26.	Chi-square
27.	$(1-\alpha)$
28.	$(1-\alpha)$
29.	Randon interval
30.	Lower confidence, $(1-\alpha)$
31.	Lower one sided
32.	Upper confidence bound
33.	Uper one sided, Parameter
34.	Two
35.	Increasing, Decreasing
36.	$\psi(t), \infty$
37.	$-\infty, \psi(t_2)$
38.	$(1-\alpha)$, Interval
39.	Statistic
40.	$(t_2 - t_1)$, C.I.
41.	$(t_2 - t_1)$
42.	Expected length
43.	Parameter
44.	Minimum
45.	$(1-\alpha)$, $(1-\alpha)$
46.	$\leq, \theta$
47.	$\geq, \geq$
48.	Confidence
49.	Unique, θ
50.	$a^{-1}(x_1), b^{-1}(x_n)$
51.	$a^{-1}(x_1), b^{-1}(x_n)$
52.	n, n
53.	$\left(\frac{1}{x} + \frac{1.96}{\sqrt{n}\,\bar{x}}\right)$
54.	Parameter function
55.	Two tailed, sampling
56.	Confidence, $(1-\alpha)$

Q.	Ans.	Q.	Ans.
57.	Smallest	72.	Nuisance Parameter
58.	Unbiased confidence set	73.	Confidence set, $(1-\alpha)$
59.	Unbiased	74.	Confidence set, $(1-\alpha)$
60.	Unbiasedness	75.	U.M.A.U.
61.	Confidence	76.	$\mid \bar{x} - \mu_0 \mid / s$
62.	Uniformally most accurate unbiased (U.M.A.U.)	77.	$\mid \bar{x} - \mu_0 \mid / \sqrt{n}$
63.	σ, n		
64.	$(\bar{x} - \mu_0)$	78.	$(t_{\alpha/2, n-1} \cdot \frac{s}{\sqrt{n}}), (t_{\alpha/2, n-1}, \frac{s}{\sqrt{n}})$
65.	$(\bar{x}, + t_{\alpha/2} \frac{\sigma}{\sqrt{n}})$	79.	$(\bar{x} - t_{\alpha/2, n-1} \frac{s}{\sqrt{n}})$,
66.	$(\bar{x}, + t_{\alpha/2} \frac{\sigma}{\sqrt{n}})$		$(\bar{x} + t_{\alpha/2, n-1}, \frac{s}{\sqrt{n}})$
		80.	Confidence coefficient
67.	$\left(\frac{x_i - \mu}{\sigma_0}\right)$	81.	$(n-1)\, s^2/\sigma_0^2$
68.	$\left(\frac{x - \mu_0}{\sigma}\right)$	82.	$\chi^2 > \chi_2^2$
		83.	$(n-1)\, s^2/\sigma_0^2$
69.	$(x_i - \mu)$	84.	$(n-1)\, s^2$
70.	$(x_i - \mu)$	85.	$(n-1)\, s^2$, $(n-1)\, s^2$
71.	Unbiased confidence	86.	σ^2, $(1-\alpha)$
		87.	$(1-\alpha)$, $(n-1)$

Multiple Choice Questions and Answers

Q.	Ans.	Q.	Ans.	Q.	Ans.
1.	(d)	30.	(a)	59.	(a)
2.	(b)	31.	(a)	60.	(b)
3.	(c)	32.	(b)	61.	(a)
4.	(c)	33.	(b)	62.	(a)
5.	(a)	34.	(a)	63.	(d)
6.	(c)	35.	(a)	64.	(c)
7.	(b)	36.	(a)	65.	(a)
8.	(c)	37.	(a)	66.	(b)
9.	(c)	38.	(c)	67.	(a)
10.	(a)	39.	(b)	68.	(a)
11.	(d)	40.	(a)	69.	(a)
12.	(c)	41.	(c)	70.	(a)
13.	(b)	42.	(a)	71.	(c)
14.	(b)	43.	(b)	72.	(b)
15.	(c)	44.	(b)	73.	(c)
16.	(b)	45.	(c)	74.	(a)
17.	(a)	46.	(a)	75.	(a)
18.	(b)	47.	(b)	76.	(c)
19.	(d)	48.	(c)	77.	(a)
20.	(d)	49.	(a)	78.	(d)
21.	(a)	50.	(b)	79.	(c)
22.	(b)	51.	(a)	80.	(c)
23.	(d)	52.	(a)	81.	(a)
24.	(d)	53.	(a)	82.	(d)
25.	(a)	54.	(b)	83.	(b)
26.	(c)	55.	(b)	84.	(a)
27.	(a)	56.	(a)	85.	(b)
28.	(c)	57.	(c)	86.	(c)
29.	(a)	58.	(b)	87.	(c)

❑❑❑

CHAPTER - 4

Test of Hypothesis

4.1 Write True/False

4.1.1 Preliminaries

Q.1 In selection of a test, one should try to keep the probabilities of both types of error small.

Q.2 Under a non-sequential sampling scheme (i.e. fixed sample size n), it is not possible to minimise both the errors at the same time.

Q.3 The test which minimises the probabilities of error I, maximises the error II.

Q.4 In construction of a test, one just keeps the probabilities of error I fixed at a desired level say α, and try to minimise the probabilities of error II. i.e. β.

Q.5 The minimization of β means maximization of power of a test which is defined as $(1 - \beta)$.

Q.6 In construction of a test, we choose a critical region of size α which has atleast the same power as any other critical region of the same size.

Q.7 The C.R. Chosen in a fashion as described in Q.6, is known as the best critical region.

Q.8 Let Ω_θ be the parameter space and Ω_{θ_0} be some specified proper subset of $\Omega\theta$. Then we write the null hypothesis as :

$$H_0 : \theta \in \Omega_{\theta_0}.$$

Q.9 Let C be the critical region and A be the acceptance region such that $(C+A) = R_n$. Then we choose a test for H_0 such that :

$$P_\theta (c) = \alpha, \text{ for } \theta \in \Omega_{\theta_0}. \quad ...(1)$$

Q.10 Let β be the probability of error II, then β shouls be as small as possible.

Q.11 The probability 'α' defined in Q.9, is known as the 'Level of significance.'

Q.12 The complement of the level of significance i.e. $(1-\alpha)$ is known as the 'Level of confidence' or 'confidence coefficient.'

Q.13 The probability of error II i.e. β is the function of alternative hypothesis H_1.

Q.14 The probability of Ii kind of error i.e. β depends on C.R.

Q.15 The complement of error II i.e. $(1-\beta)$ is known as the 'power of the test.'

Q.16 The power of test criterion measures the sensitiveness of the test.

Q.17 The other things being identical, the comparison of two test is the comparison of their respective powers.

Q.18 By definition, the power function assumes the value 'α' at $H_0 : \theta = \theta_0$.

Q.19 Suppose that there are many C.R.'s of the some size 'α'. Then C.R. with minimum β i.e. maximum power, is known as "Best critical region."

Q.20 The test corresponding to best C.R., is known as most powerful (M.P.) test.

Q.21 Let $\Omega\,\theta_0$ be consists of single element θ_0 and its complement say $(\Omega_\theta - \Omega\theta_0)$ has also single element θ_1. Suppose we have to test :

$H_0 : \theta = \theta_0$ v.s. $H_1 : \theta = \theta_1$. ...(2)

Then C.R.(c) is known as M.P. critical region of size α for testing H_0 v.s. H_1. if :

$$\left.\begin{aligned} &P_{\theta_0}(c) = \alpha \\ &\text{and } P_{\theta_1}(c) \geq P_{\theta_1}(c'). \end{aligned}\right\} \quad \ldots(3)$$

for whatever is the other C.R.(c′) satisfying (3).

Q.22 Suppose we have to test $H_0 : \theta = \theta_0$ v.s. a alternative composite hypothesis $H_1 : \theta \neq \theta_0$. If a set 'C' is a best C.R. of size α for testing H0 against each simple hypothesis in H_1 then 'C' is known as uniformaly most powerful (U.M.P.) critical region of size α.

Q.23 A test defined by C.R. in Q.(22), is called a U.M.P. test with significance level α to test $H_0 : \theta = \theta_0$ v.s. alternative composite hypothesis $H_1 : \theta \neq \theta_0$.

Q.24 An UMP test does not always exist.

Q.25 If an UMP test exists then Neyman-Pearson Lemma, provides a technique to find it.

Q.26 Let C be the critical region of size α for testing $H_0 : \theta = \theta_0$ such that :

$$P_{\theta_0}(c) = \alpha$$

and $P_\theta(c) > P_\theta(c') \ \forall \ \theta \neq \theta_0$. ...(4)

for whatever the other region c′ satisfying (4). Then C is called UMP critical region.

Q.27 Let C be a C.R. of size α such that :

$P_\theta(c) > P_{\theta_0}(c) \ \forall \ \theta \neq \theta_0$.

Then critical region 'C' is said to be unbiased C.R.

Q.28 Suppose $6(\theta)$ be a test function for testing :

$H_0 : \theta \in \Omega \, \theta_0$ v.s. $H_1 : \theta \in (\Omega_\theta - \Omega_{\theta_0}) = \Omega_{\theta_1}$.

Then the test function $6(\theta)$ is said to be unbiased of size α if :

$$P_{\theta_0} [\ 6(\theta)\ ] < \alpha \text{ i.e. type one error}$$

and $$P_{\theta_1} (\ 6(\theta)\ ) > \alpha, \text{ i.e. power} > \alpha.$$

Q.29 A test is said to unbiased if its power does not fall below α.

Q.30 A critical region 'c' corresponding to unbiased test is known as "umbiased C.R."

Q.31 A test based on an unbiased C.R. is said to be an 'unbiased test.'

Q.32 If the power of a test is always greater than α then such test is known as 'Uniformly unbiased' or completely unbiased.

Q.33 A most powerful (MP) test is always an 'Unbiased test.'

Q.34 If we have to test a simple hypothesis $H_0 : \theta = \theta_0$. against a composite alternative H_1. $\theta \neq \theta_0$ then an. UMP test may or may not exist.

Q.35 In Q34, if an UMP does not exist then we require to search out a "Uniformaly most powerful unbiased test."

Q.36 The region C* is said to be "Uniformly most powerful among unbiased critical regions (U.M.P.U. region) of size α. for testing $H_0 : \theta = \theta_0$ v.s. $H_1 : \theta \neq \theta_0$ if :

$$P_{\theta_0}(c^*) = \alpha \quad ...(A)$$

$$P_\theta(c^*) \geq \alpha \ \forall \ \theta \neq \theta_0 \quad ...(B)$$

and $$P_\theta(c^*) > P_\theta(c) \ \forall \ \theta \neq \theta 0 \quad ...(C)$$

for whatever other region c may be and satisfies (A) and (B).

Q.37 The critical region c* is said to be a "type A" critical region of size α for testing $H_0 : \theta = \theta_0$ v.s. $H_1 : \theta \neq \theta_0$ if :

$$P_{\theta_0}(c^*) = \alpha \quad ...(l)$$

$$\left.\frac{d\, P_{\theta_0}(c^*)}{d_\theta}\right|_{\theta=\theta_0} = 0 \qquad \ldots(m)$$

and $$\left.\frac{d^2\, P_{\theta_0}(c^*)}{d_\theta}\right|_{\theta=\theta_0} \geq \left.\frac{d^2\, P_{\theta_0}(c)}{d_{\theta^2}}\right| \theta = \theta_0 \qquad \ldots(n)$$

for whatever the other region c may be which satisfies (l) and (m).

Q.38 In Q.(37), the second derivative of P_{θ_0}(c*) must be continuous in the neighbourhood of $\theta = \theta_0$.

Q.39 Let c* be a M.P. critical region for testing $H_0 : \theta_0$ v.s. $H_1 : \theta = \theta_1$ such that θ_1 is not too far from θ_0. Then in that situation 'type A' region of size α may also be called as a "Locally most powerful unbiased region" of size α.

Q.40 The region c* is said to be "type A_1 critical region" of size α for testing $H_0 : \theta_0$ v.s. $H_1 : \theta \neq \theta_0$ if :

$$P_{\theta_0}\,(c^*) = \alpha \qquad \ldots(p)$$

$$\frac{d}{d_\theta} P_{\theta.}\,(c^*)\Big|_{\theta=\theta_0} = 0 \qquad \ldots(q)$$

and $$P_\theta\,(c^*) \geq P_\theta\,(c) \;\forall\; \theta \neq \theta_0 \qquad \ldots(r)$$

for whatever other region c may be which satisfies (p) and (q).

Q.41 In most of the cases, a "type A_1 region" will also be a UMPU region.

Q.42 Although as a rule, a "type A region" may turn out to be a "type A_1" region.

Q.43 The 'type A_1 region' and 'type A region' are not identical.

Q.44 A simple null hypothesis H_0 can be tested against a simple alternative hypothesis H_1.

Q.45 A simple null hypothesis $H_0 : \theta = \theta_0$ can be tested against a composite hypothesis $H_1 : \theta \neq \theta_0$.

Q.46 A composite hypothesis $H_0 : \theta \neq \theta_0$ can be tested against a simple hypothesis $H_1 : \theta = \theta_1$.

Q.47 A composite hypothesis $H_0 : \theta \geq \theta_0$ can be tested against composite hypothesis $H_1 : \theta \leq \theta_0$.

Q.48 When both H_0 and H_1 are simple hypothesis then B.C.R. of size α exists if the parent population (distribution) is of continuous nature.

4.1.2 Neyman Pearson Lemma

Q.49 When both the hypothesis H_0 and H_1 are simple then BCR is obtainable through the use of Neyman-Pearson Lemma (NPL).

Q.50 Let $x_1, x_2, ..., x_n$ be a random sample draw from a population $f(x, \theta)$. The likelihood functions under H_0 and H_1 for testing.

$$H_0 : \theta = \theta_0 \text{ v.s. } H_1 : \theta = \theta_1$$

are given as :

$$P(x, H_0) = f(x_1, \theta_0)\, f(x_2, \theta_0)\,.........f(x_n, \theta_0)$$

and $$P(\underset{\sim}{x}, H_1) = f(x_1, \theta_1)\, f(x_2, \theta_1)\,.........f(x_n, \theta_1)$$

Then B.C.R. consists of points of the sample space R_n for which :

$$\frac{P(x, \theta_0)}{P(x, \theta_1)} \leq k \qquad ...(5)$$

where k is such that the size of C.R. is α.

Q.51 In Q.50, the k is the function of α and H_1.

Q.52 Neyman Pearson Lemma (N.P.L.) enables us to find a M.P.C.R. to test $H_0 : \theta = \theta_0$ v.s $H_1 : \theta = \theta_1$ in case of $(x = x_1, x_2, ... , x_n)$ being a continuous r.v.'s.

Q.53 N.P.L also enables us to get an U.M.P.C.R. for testing $H_0 : \theta = \theta_0$ vs. $H_1 : \theta \neq \theta_0 = \theta_1$ (composite hypothesis).

Q.54 Let the ratio $\lambda = f_{\theta_1}(x) / f_{\theta_0}(x)$ have a continuous distribution. There exists a constant $k > 0$ such that if W_0 be a region in the sample space R_n defined as :

$$W_0 = \{ X / (f_{\theta_1}(x) / f_{\theta_0}(x) > k \} \qquad ...(6)$$

Then we have the following result :

$$\int_{W_0} f_{\theta_0}(x)dx = \alpha.$$

Q.55 NPL states that any M.P.C.R. of size α for testing $H_0 : \theta = \theta_0$ vs $H_1 : \theta = \theta_1$, will be identical with W_0 (mentioned in (Q.54) except perhaps for a set of Lebesgue measure zero.

Q.56 The M.P.C.R. denoted as W_0 in Q.55, is called an unique M.P.C.R. for testing $H_0 : \theta = \theta_0$ v.s. $H_1 : \theta = \theta_1$.

Q.57 Let W_0 defined as in Q.54), be independent of $\theta_1 \in \Omega\, \theta_1$. Then W_0 will be the U.M.P.C.R. of size α, to test $H_0 : \theta = \theta_0$ vs composite alternative $H_1 : \theta \neq \theta_0$.

Q.58 The method of construction of an MP or UMPC will be similar to that of discrate as well as continous r.v.'s.

Q.59 In case of discrete random variable, the constant k will not be unique for testing $H_0 : \theta = \theta_0$ v.s. $H_1 : \theta = \theta_1 \neq \theta_0$.

Q.60 In fact in case of discrete distribution function, there may be two consecutive mass points y_0 and y_1. for y. such that :

$f(y_0) < \alpha < f(y_1)$

Q.61 For a given size a, there may not exist any MP or UMPCR of exact size α for testing $H_0 : \theta = \theta_0$ v.s. $H_1 : \theta = \theta_1 \neq \theta_0$.

Q.62 Consider a family of distribution $N(\theta, \sigma^2)$ where mean θ is unknown and variance σ^2 is known. Suppose we have to test $H_0 : \theta = \theta_0$ v.s. $H_1 : \theta = \theta_1$. If $\theta_1 > \theta_0$ then M.P.C.R. say W_0 is given as :

$$W_0 = \{ X \left| \frac{(\bar{x} - \theta_0)\sqrt{n}}{\sigma} > Z\alpha \right. \} \quad \ldots(7)$$

where $Z\alpha$ denotes the value of standard noramal variate at level of significance 'α'.

Q.63 In Q.62, let $\theta_1 < \theta_0$ then MPCR W_0^* is given as :

$$W_0^* = \{ X \left| \frac{(\bar{x} - \theta_0)\sqrt{n}}{\sigma} < Z\alpha \right. \} \quad \ldots(8)$$

Q.64 In Q.62, the MPCR say W0 does not explicitly dependent $\theta_1 > \theta_0$.

Q.65 In Q.63, the MPCR say W_0^* does not depend on $\theta_1 < \theta_0$.

Q.66 In Q.62, the MPCR say W_0 is also the UMPCR for testing $H_0 : \theta = \theta_0$ vs $H_1 : \theta = \theta_1 > \theta_0$.

Q.67 In Q.63, the MPCR say W_0^* is also U.M.P.C.R. for testing $H_0 : \theta = \theta_0$ vs $H_1 : \theta = \theta_1 < \theta_0$.

Q.68 In Q.62 if we have to test $H_0 : \theta = \theta_0$ vs $H_1 : \theta \neq \theta_0$ then there does not exist a CR of size α which is UMPCR for H_0 against a composite (two sided) alternative.

Q.69 Consider the family of distribution $N(\mu, \sigma^2)$ where μ is known and σ^2 is unknown. Let the simple hypothesis tested is $H_0 : \sigma^2 = \sigma_0^2$ against simple alternative $H_1 : \sigma^2 = \sigma_1^2$. Then MPCR say W_0 is given as :

$$W_0 = \{ X / \sum_{i}^{n} \frac{(x_i - \mu)^2}{\sigma_0^2} > \chi^2_{\alpha, n} \} \quad ...(9)$$

Q.70 In Q.69, suppose we have to test the hypothesis $H_0 : \sigma^2 = \sigma_0^2$ against the hypothesis $H_1 : \sigma^2 = \sigma_1^2 < \sigma_0^2$. The MPCR say W_0^* is given as :

$$W_0^* = \{ X / \sum_{i}^{n} \frac{(x_i - \mu)^2}{\sigma_0^2} < \chi^2_{\alpha, n} \} \quad ...(10)$$

Q.71 The critical regin (MPCR) provided in Q.69, does not explicitly depend on σ_1^2.

Q.72 The MPCR say W_0 provided in Q.69 is also UMPCR for $H_0 : \sigma^2 = \sigma_0^2$ vs composite one sided $H_1 : \sigma^2 > \sigma_0^2$.

Q.73 Suppose in Q.69, we have to test $H_0 : \sigma^2 = \sigma_0^2$ against one sided composite hypothes $H_1 : \sigma^2 = \sigma_1^2 < \sigma_0^2$. Then MPCR say W_0^* provided in Q (70), is also UMPCR.

Q.74 In Q.69, if we have to test the simple hypothesis $H_0 : \sigma^2 = \sigma_0^2$ vs composite hypothesis $H_1 : \sigma^2 \neq \sigma_0^2$, then there does not exist an U.M.P.C.R.

Q.75 Suppose there does not exist an. UMPCR for testing $H_0 : \sigma^2 = \sigma_0^2$ vs $H_1 : \sigma^2 \neq \sigma_0^2$. Then to overcome this problem, we require to construct an uniformaly most powerful unbiased critical region (UMPUCR).

Q.76 When a MPCR or UMPCR exist, then we need not to bother for finding an U.M.P.C.R. for testing such statistical hypothesis.

Q.77 If we have a M.P.C.R. say W_0 for testing $H_0 : \sigma^2 = \sigma_0^2$ v.s. $H_1 : \sigma^2 = \sigma_1^2$ then this W_0 is necessarily an unbiased.

Q.78 Consider a random sample from normal distribution $N(\mu, \sigma^2)$ where μ is unknown but σ^2 is known.

Then type "A CR region" say W_0 of size α, is given as :

$$W_0 = \{ X / \left| \frac{| x_i - \underset{0}{\mu} | \sqrt{n}}{\sigma} \right. > Z\alpha \} \qquad ...(11)$$

for testing $H_0 : \mu = \underset{0}{\mu}$ vs $H_1 : \mu \neq \underset{0}{\mu}$.

Q.79 Consider a random sample from $N(\mu, \sigma^2)$ where μ is known but σ^2 is unknown. Suppose we have to test $H_0 : \sigma^2 = \underset{0}{\sigma^2}$ vs $H_1 : \sigma^2 \neq \underset{0}{\sigma^2}$. Then type A region is provided as :

$$W_0 = \{ X / \sum_i^n \frac{(x_i - \mu)^2}{\sigma_0^2} < d_1 \text{ or } \sum_i^n \frac{(x_i - \mu)^2}{\sigma_0^2} > d_2 \} \qquad ...(12)$$

Q.80 In determining the critical values of d_1 & d_2 or say χ_1^2 & χ_2^2, we have to write the result that :

$$P [x^2 < \chi_1^2] + P [x^2 < \chi_2^2] = \alpha. \qquad ...(13)$$

Q.81 In solving the equation (13) in Q.80, some method of successive aproximation may be needed.

Q.82 Let a CR W_0 be defined as :

$$W_0 = \{ X / f_\theta (x) > k_1 \, f_{\theta_0} (x) + k_2 \, f'_{\theta_0} (x) \} \qquad ...(14)$$

and is same for all $\theta \neq \theta_0$ where k_1 & k_2 are constants such that :

$$\int_{w_0} f_{\theta_0} (x) \, dx = \alpha \text{ and } \int_{w_0} f'_{\theta_0} (x) \, dx = 0 \qquad ...(15)$$

Then W_0 is known as 'Type A_1 critical region of size α for testing $H_0 : \theta = \theta_0$ v.s. composite $H_1 : \theta \neq \theta_0$.

Q.83 If we presume that for a testing problem, a type A region and a type A_1 region both exist then both will be identical.

Q.84 Let $\{ x_i, i = 1, 2, \ldots , n \}$ be a random sample from $N (\mu, \sigma^2)$. Suppose we have to test the hypothesis $H_0 : \sigma^2 = \underset{0}{\sigma^2}$ vs $H_1 : \sigma^2 \neq \underset{0}{\sigma^2}$. Then type A_1 critical region if size α is identical with the type A critical region.

Q.85 Let T be a minimal sufficient statistic for μ. Let $t = T(x)$ is ithe value of sufficient statistic T for the set of observations $x = x_1, x_2, \ldots\ldots , x_n$. Then for testing $H_0 : \mu = \mu_0$ vs a compsoite alternative $H_1 : \mu \neq \mu_0$, the MPCR as UMPCR may be defined in terms of marginal distribution of T alone instead of being defined in terms of joint distribution of $x_1, x_2, \ldots, X_n$.

Q.86 Suppose a set of observations x on $x_1, x_2, \ldots$, xn is obtained and one should perform a random experiment which has two possible outcomes say O_1 leading to the rejection of H_0 and O_2 leading to its acceptance. Let $\phi(x)$ be the probability for O_1 to occur. Then a test of this kind is known as "Randomised Test."

Q.87 Let f_0 be the joint p.d.f. or pmf of $x_1, x_2, \ldots, x_n$. Then to test $H_0 : \mu = \mu_0$ vs $H_1 : \mu \neq \mu_1$, there exist a test $\phi(x)$ and a constant $k > 0$, such that :

$$E_{\mu_0} [\ \phi(x_1, x_2, \ldots, x_n\ ] = \alpha \qquad \ldots(16)$$

$$\text{and } \phi(x) = \begin{cases} 1, \text{ when } f_{\mu_1}(x) > k\ f_{\mu_0}(x) \\ 0, \text{ when } f_{\mu_1}(x) < k . f_{\mu_0}(x) \end{cases} \qquad \ldots(17)$$

Q.88 If a test satisfies (16) & (17) in Q.87, then it is M.P. for testing $H_0 : \mu = \mu_0$ vs $H_1 : \mu = \mu_1$ at the level α.

Q.89 If $\phi(x)$ is MP at the level α to test H_0 : vs H_1. Then it satisfies (16) and for some k, it satisfies (17) almost everywhere.

Q.90 When we have to test H_0 against H_1, then MP test at the level α will be given by critical function $\phi(x)$ such that :

$$\phi(x) = \begin{cases} 1, \text{ when } f_{\theta_1}(x) > k\ f_{\theta_0}(x) \\ \alpha, \text{ when } f_{\theta_1}(x) = k . f_{\theta_0}(x) \\ 0, \text{ when } f_{\theta_1}(x) < k . f_{\theta_0}(x) \end{cases}$$

Q.91 In case if $x_1, x_2, \ldots, x_n$ are continuous r.v.'s, then MP randomized test to test H_0 : against H_1 will generally be identical with the MP non-randomised test.

Q.92 In case $x_1, x_2, \ldots, x_n$ being discreate r.v.'s and MPCR of size α to test H_0 vs H_1, exists, then MP randomised test will be equivalent to the non-randomised test.

Q.93 The M.P. randomised test will lead to an increase in power while it still affects the desired level of control on the probability of type I error.

Q.94 If a sufficient statsitc T exists for the family of distribution say $\{f(\theta), \theta \in \Omega\ \theta\}$. then Neyman Pearson "MP test" will be a function of T.

4.2 Fill in the Blanks

4.2.1 Preliminaries

Q.1 In selection of a test, one should try to keep the probabilities of both type of (..........) small.

Q.2 Under a non sequential sampling schem, it is not possible to minimised both (..........) say (α & β) at the same time.

Q.3 The test which minimises the error I, it (..........) the error II.

Q.4 In construction of a test, one just keeps the probability of error I fixed at a fixed level say 'α' and try to (..........) the error II.

Q.5 The minimsation of error β means (..........) of power of the test.

Q.6 In construction of a test, we choose a C.R. of size 'α' which has atleast the same power as any other (..........) of the same size α.

Q.7 The C.R. choosen in a fashion as described in Q(6) is known as (..........).

Q.8 Let Ω_θ be the parameter space and $\Omega\theta_0$ be some specified proper subset of Ω_θ. Then we write the null hypothesis as : $H_0 : \theta \in$ (..........).

Q.9 Let C be the C.R. and A be the acceptance region such that $(C+A) = R_n$. Then we choose a test H_0 such that :

P_θ (C) = α, for (..........).

Q.10 Let β be the probability of error II, then β should beas (..........) as possibile.

Q.11 The probability 'α' defined in Q(9) is known as the (..........).

Q.12 The complement of the level of significance i.e. $(1-\alpha)$ is known as the level of confidence or (..........).

Q.13 The probability of error II i.e. β is the function of (..........).

Q.14 The probability of type (..........). i.e. β depends on C.R.

Q.15 The complement of the error II i.e. $(1-\beta)$ is known as (..........) of the test.

Q.16 The power of a test criterion measures the sensitiveness of (..........).

Q.17 The other things being identical, the comparison of two test is the comparision of their respective (..........).

Q.18 Let θ_0 be the true value of θ. Then by definition, the power function assumes the value 'α' at H_0 : ....

Q.19 Suppose that there are many C.R.'s of same size 'α'. Then the C.R. with minimum β i.e. maximum (..........), is knownm as best (..........).

Q.20 The statistical test corresponding to best C.R., is known as most (..........).

Q.21 Let Ω_{θ_0} and its complement say $(\Omega_\theta - \Omega_{\theta_0})$ have single element $(\theta_0 \& \theta_1)$ vs $H_1 : \theta = \theta_1$.

Then C.R. (c) choosen is known as (..........) of size α if :

$$P_{\theta_0}\ (c) = \alpha \quad \}$$

and $$P_{\theta_1}\ (c) \geq P_{\theta_1}\ (c') \quad \} \qquad ...(1)$$

for any other C.R. say (c′).

Q.22 Suppose we have to test $H_0 : \theta = \theta_0$ v.s. a alternative composite hypothesis $H_1 : \theta \neq \theta_0$. If a set 'C' is a best C.R. of size α, for testing H_0 against each simple hypothesis in H_1 then 'C' is known as (..........) critical region of size α.

Q.23 The test defined in Q.(22) is known as an (..........) test with significance level (α) to test $H_0 : \theta = \theta_0$ vs alternative composite hypothesis $H_1 : \theta \neq \theta_0$.

Q.24 If an U.M.P. test exists then (..........) Lemma provides a technique to find it.

Q.25 Let 'C' be the C.R. of size α to test $H_0 : \theta = \theta_0$ v.s. $H_1 : \theta \neq \theta_0$ such that :

$$P_{\theta_0}\ (c) = \alpha \quad \} \qquad ...(2)$$

and $P_\theta\ (c) \geq P_\theta(c') \ \forall\ \theta \neq \theta_0$ }

for an arbitrary C.R. (c) satisfying (2). Then 'c' is known as (..........) critical region.

Q.26 Let 'c' be a C.R. of size 'α' to test $H_0 : \theta = \theta_0$ v.s. $H_1 : \theta \neq \theta_0$ such that :

$$P_\theta\ (c) \geq P_{\theta_0}(c) \ \forall\ \theta \neq \theta_0$$

Then C.R. 'c' is said to be (..........) C.R.

Q.27 Let $6(\theta)$ be a test function to test :

$$H_0 : \theta \in \Omega_{\theta_0}.\ \text{vs}\ H_1 : \theta \in \Omega_{\theta_1}.$$

If we define the test as if :

$$P_{\theta_0}\ [\ 6(\theta)\ ] < \alpha \text{ i.e. type I error} < \text{C.R.}$$

and $P_{\theta_1}\ [\ 6(\theta)\ ] > \alpha$ i.e. Power $> \alpha$.

then the test function $6(\theta)$ is said to be (..........).

Q.28 A test is said to be (..........) if its power does not fall below α.

Q.29 A C.R. 'c' corresponding to unbiased test is known as (..........). C.R.

Q.30 A test based on an unbiased C.R. is said to be an (..........).

Q.31 If the power of a test is always greater than 'α' then such test is known as (..........).

Q.32 A most power (M.P.) test is always an (..........).

Q.33 If we have to test a simple hypothesis $H_0 : \theta = \theta_0$ vs a composite alternative $H_1 : \theta \neq \theta_0$ then an (..........). test may or may not exist.

Q.34 If in Q.(33), an U.M.P. test does not exist then we require to search out an (..........) test.

Q.35 Let 'c*' be the C.R. of size α to test $H_0 : \theta = \theta_0$ vs $H_1 : \theta \neq \theta_0$ and if :

$$P_{\theta_0}(c^*) = \alpha \qquad \text{...(A)}$$

$$P_{\theta}\cdot(C^*) \geq \alpha \ \forall \ \theta \neq \theta_0 \qquad \text{...(B)}$$

and $P_\theta(c\bullet) \geq P_\theta(c) \ \forall \ \theta \neq \theta_0$ ...(C)

where 'C' is an arbitrary C.R. Then 'C*' is said to be (..........) of size a and satisfies A and B.

Q.36 Let C* be a C.R. of size 'α' to test $H_0 : \theta = \theta_0$ vs $H_1 : \theta \neq \theta_0$ and if :

$$P_{\theta_0}(c^*) = \alpha \qquad \text{...}(l)$$

$$\frac{d}{d\theta} P_{\theta_0}(C^*) \,|_{\,\theta = \theta 0} = 0 \qquad \text{...}(m)$$

and $\dfrac{d^2}{d\theta^2} P_{\theta_0}(c^*) \Big/ _{\theta = \theta_0} \geq \dfrac{d^2}{d\theta^2} P_{\theta_0}(c) / \theta = \theta_0$ ...(n)

for an arbitrary region 'c' which satisfiess (l) & (m). Then c* is said to be a (..........).

Q.37 Let c* be a M.P. critical region to test $H_0 : \theta = \theta_0$ vs $H_1 : \theta \neq \theta_0$ such that θ_1 is not too far from θ_0. Then "type A" C.R. of size α may also be called as a (..........) C.R. of size α.

Q.38 Let 'c*' be a C.R. to test $H_0 : \theta = \theta_0$ vs $H_1 : \theta \neq \theta_0$ and if

$$P_{\theta_0}(c^*) = \alpha \qquad \text{...}(p)$$

$$\frac{d}{d\theta} P_\theta(c^*) \Big/ _{\theta = \theta_0} = 0 \qquad \text{...}(q)$$

and $P_\theta(c^*) \geq P_\theta(c) \ \forall \ \theta \neq \theta_0$. ...(r)

for an arbitrary region 'c' which statisfies (p) & (q). Then 'c*' is said to be (..........).

Q.39 In most of the cases, a "type A_1 critical region" will also be a (.........)

Q.40 The "type A_1 region" and "type A region" are not (..........).

Q.41 A simple null hypothesis H_0 can be tested against a simple (..........). hypothesis H_1.

Q.42 A simple null hypothesis $H_0 : \theta = \theta_0$ can be tested against a (..........) hypothesis vs $H_1 : \theta \neq \theta_0$.

Q.43 A composite hypothesis vs $H_0 : \theta \geq \theta_0$ can be tested against (..........) $H_1 : \theta \leq \theta_0$.

4.2.2 Neyman-Pearson Lemma

Q.44 When both H_0 and H_1 are simple hypothesis then B.C.R. of size 'a' exists if the parent distribution is of (..........) nature.

Q.45 When both the hypothesis H_0 & H_1 are simple then B.C.R. of size a is obtainable through the use of (..........).

Q.46 Let x be a random sample from the distribution $f(x, \theta)$.

The likelihood functions under H_0 & H_1 to test :

$H_0 : \theta = \theta_0$ vs $H_1 : \theta = \theta_1$.

are given as :

$$P(x, H_0) = \sum_{i=1}^{n} f(x_i, \theta_0)$$

and
$$P(x, H_1) = \sum^{n} f(x_i, \theta_1)$$

Then B.C.R. consists of points of the (..........) R_n for which :

$$\frac{P(x, \theta_0)}{P(x, \theta_1)} < k \text{ (According to N.P. Lemma).}$$

where k is such that the size of C.R. is 'a'.

Q.47 In Q.(46), the 'k' is the function of (..........).

Q.48 Neyman Pearson Lemma (N.P.L.) enable us to find a MPCR to test $H_0 : \theta = \theta_0$ vs $H_1 : \theta = \theta_1$ if the sample $x = (x_1, x_2, ..., x_n)$ is of (..........). nature.

Q.49 N.P. Lemma also enable us to get an UMP critical region to test $H_0 : \theta = \theta_0$ vs H_1 (..........) the composite hypothesis.

Q.50 The method of construction of (MP or U.M.P.) critical region of size 'α' will be similar to that of discreate as well as (..........). r.v.'s.

Q.51 In case of discrete distributions, the constant k used in N.P. Lemma will not be unique to test simple $H_0 : \theta = \theta_0$ vs composite H_1 : ....

Q.52 In fact in case of discrete distribution function, there may be two consecutive mass points y_0 & y_1 for the variable y such that :
$f(...)$ .... $< \alpha < f(..........)$.

Q.53 For a given size α, there may not exist any (M.P. or U.M.P.) critical region of exact size a, to test a simple $H_0 : \theta = \theta_0$ vs (..........) $H_1 : \theta = \theta_1$.

Q.54 Let $X \sim N(\mu, \sigma^2)$ where μ is unknown and σ^2 is known.

Then to test $H_0 : \mu = \mu_0$ vs $H_1 : \mu = \mu_1 > \mu_0$, the M.P.C.R. say W_0 is given as :

$W_0 = \{ X / |$ .... $|$ .... $> Z\alpha \}$

where $Z\alpha$ has the usual meaning.

Q.55 In Q.(54), let $\mu_1 < \mu_0$, then M.P.C.R. say W_0^* is given as :

$$W_0^* = \{ X / \frac{|\bar{x} - \mu_0| \sqrt{n}}{\sigma} \text{ .... } \}$$

Q.56 In Q.(54 & 55), the M.P.C.R. say W_0 does not explicitely depend on (.......... and ....).

Q.57 In Q.(54 & 55), the M.P.C.R. say (W_0 & W_0^*) are also the U.M.P. critical region to test :

$H_0 : \mu = \mu_0$ vs H_1 : ....

and $H_0 : \mu = \mu_0$ vs H_1 : ....

respectively.

Q.58 If in Q.(54), we have to test $H_0 : \mu = \mu_0$ vs $H_1 : \mu \neq \mu_0$, then there does not exist a critical region of size α which is (..........) critical region for H_0 vs H_1.

Q.59 Let $X \sim N(\mu, \sigma^2)$ where μ is known but σ^2 is unknown. Suppose we have to test $H_0 : \sigma^2 = \sigma_0^0$ vs $H_1 : \sigma^2 = \sigma_1^2 > \sigma_0^2$. Then M.P.C.R say W_0 is given as :

$$W_0 = \{ X / \sum^{n} (\ldots\ldots\ldots\ldots)^2 > \chi^2_{\alpha, n} \}$$

Q.60 Suppose in Q.(59), we have to test $H_0 : \sigma^2 = \sigma_0^2$ v.s. $H_1 : \sigma^2 = \sigma_1^2 < \sigma_0^2$. Then M.D.C.R. say W_0^* is given as :

$$W_0^* = \{ X / \sum_{i}^{n} \left(\frac{x_1 - \mu}{\sigma_0} \right)^2 < \ldots\ldots\ldots \}.$$

Q.61 The critical region (M.P.C.R.) provided in Q.(54) does not explicitely depend on. (.....................).

Q.62 The M.P.C.R. say W_0 provided in Q.(59) is also (..........). for $H_0 : \sigma^2 = \sigma_0^2$ v.s. one sided composite $H_1 : \sigma^2 > \sigma_0^2$.

Q.63 Suppose in Q.(54), we have to test $H_0 : \sigma^2 = \sigma_0^2$ v.s. one sided composite hypothesis $H_1 : \sigma^2 = \sigma_1^2 < \sigma_0^2$. Then M.P.C.R. say W_0^* provided in Q.(60) is also (..........) critical region.

Q.64 In Q.(54), if we have to test the simple hypothesis $H_0 : \sigma^2 = \sigma_0^2$ v.s. composite hypothesis $H_1 : \sigma^2 \neq \sigma_0^2$, then there does not exist an (..........) critical region.

Q.65 Suppose there does not exist an U.M.P.C.R. to test $H_0 : \sigma^2 = \sigma_0^2$ v.s. $H_1 : \sigma^2 \neq \sigma_0^2$. Then to overcome this difficulty, we require to construct an "Uniformally most powerful (..........). i.e. (U.M.P.U.C.R.).

Q.66 If we have a MPCR say W_0 for testing $H_0 : \sigma^2 = \sigma_0^2$ v.s. $H_1 : \sigma^2 \neq \sigma_0^2$ then W_0 is necessarily an (...........) critical region.

Q.67 If we have an U.M.P.C.R. say W_0^* to test $H_0 : \sigma^2 = \sigma_0^2$ v.s. $H_1 : \sigma^2 = \sigma_1^2$, then this W_0^* is necessarily (..........) critical region.

Q.68 Consider a random sample drawn from $N(\mu, \sigma^2)$ where μ is unknown but σ^2 is known. Then to test $H_0 : \mu = \mu_0$ vs $H_1 : \mu \neq \mu_0$, then "type A region" say W_0 of size α, is given as :

$$W_0 = \{ X / \frac{|.........|}{(..........)} > Z\alpha \}$$

Q.69 Let $\underset{\sim}{x}$ be a random sample of size n drawn from $N(\mu, \sigma^2)$ where μ is known and σ^2 is unknown. Suppose we have to test $H_0 = \sigma^2 = \sigma_0^2$ v.s. $H_1 : \sigma^2 \neq \sigma_0^2$. Then "Type A" region W_0 is provided as :

$$W_0 = \{ X / \sum_{i}^{n}\left(\frac{x_1 - \mu}{\sigma_0}\right)^2 < d_1 \text{ or } \sum^{n}(..........)^2 > d_2 \}$$

Q.70 To determine the critical values $d_1 = \chi_1^2$ and $d_2 = \chi_2^2$, we have to write the result that :

$$P[\chi^2 < \chi_1^2] + P[\chi^2 > \chi_2^2] = ...$$

Q.71 In solving the equation $P[\chi^2 < \chi_1^2] + P[\chi^2 > \chi_2^2] = \alpha$, in Q.(70), some method of successive (..........) may be needed.

Q.72 Suppose we have to test the hypothesis $H_0 : \theta = \theta_0$ vs composite $H_1 : \theta \neq \theta_0$. Let a crical region W_0 be defined as :

$$W_0 = \{ X / f\theta(x) > k_1 f\theta_0(x) + k_2 f'\theta_0(x) \}$$

and is same $\forall\ \theta \neq \theta_0$ where (k_1, k_2) are constants such that :

$$\int_{w_0} f\theta_0(x)\, dx = \alpha, \text{ and } \int_{w_0} f'\theta_0(x)\, dx = 0$$

Then W_0 is known as (..........) of size α.

Q.73 If we presume that for testing a hypothesis, "type A region" and "type A_1 region" both exist then both will be (..........).

Q.74 Let x be a random sample from $N(\mu, \sigma^2)$. Suppose we have to test the hypothesis $H_0 : \sigma^2 = \sigma_0^2$ v.s. $H_1 : \sigma^2 \neq \sigma_0^2$. Then "Type A_1 C.R." of size α will be identical with the ".........." of size α.

Q.75 Let $x_1, x_2, ... x_n$ be n observations obtained and one has to perform a random experiment which has two possible outcomes say $(\theta_1 \& \theta_2)$

leading to rejection and acceptance of H_0 respectively. Then a test of this kind is known as "..........".

Q.76 Let f_θ be the joint p.d.f. of a random sample $x_1, x_2, ..., x_n$. Then to test $H_0 : \mu = \underset{0}{\mu}$ v.s. $H_1 : \mu = \underset{1}{\mu}$, there exists a text $\phi(x)$ and a constant $k > 0$, such that :

$$E_{\underset{0}{\mu}} [\ \phi(x)\ ] = \alpha$$

$$\text{and } \phi(x) = \begin{cases} 1 \ when \ \dfrac{f_{\underset{1}{\mu}}(x)}{f_{\underset{0}{\mu}}(x)} .... \\ 0 \ where \ \dfrac{f_{\underset{1}{\mu}}(x)}{f_{\underset{0}{\mu}}(x)} .... \end{cases}$$

Q.77 If a test satisfies the conditions mentioned in Q.(76) then to test $H_0 : \mu = \underset{0}{\mu}$ v.s. $H_1 : \mu = \underset{1}{\mu}$ at the level α, then the test is (..........).

Q.78 Suppose we have to test H_0 against H_1, then M.P. test at the level α will be given by the critical function $\phi(x)$ such that :

$$f(x) = \begin{cases} 1, \ when \ f_{\theta_1}(x) / f_{\theta_0}(x) ... \\ \alpha, \ when \ f_{\theta_1}(x) / f_{\theta_0}(x) ... \\ 0, \ when \ f_{\theta_1}(x) / f_{\theta_0}(x) ... \end{cases}$$

Q.79 In case if $x_1, x_2, ... x_n$ are continuous r.v.'s, then M.P. randomised test to test H_0 : against H_1 will generally be identical with the M.P. (..........) test.

Q.80 If $x_1, x_2, ..., x_n$ are discrete r.v.'s and MPCR of size α exists to test H_0 v.s. H_1, then M.P. randomised test will be equivalent to (..........) test.

Q.81 The M.P. randomised test will lead to an increase in power of the test. while it still affects the desired level of control on the probability of "..........".

Q.82 If a sufficient statistic T(x) exists for the family of distribution say $\{f(\theta), \theta \in \Omega\ \theta\}$, then Neymam-Pearson (..........) will be a function of T(x).

4.3 Multiple Choice Questions and Answers

4.3.1 Preliminaries

Q.1 In selection of a test, one should try to keep the probabilities of both type of error (..........).

(a) Minimum (b) Least
(c) none (d) Both (a) & (b

Q.2 Under a non-sequential sampling plan, it is not possible to (..........) both the errors at the same time.

(a) Minimize (b) Maximise
(c) Average (d) None

Q.3 The test which minimise the error I, it (..........) the error II.

(a) Minimise (b) Maximise
(c) Decreased (d) Increased

Q.4 On construction of a test, one just keep the probabiliy of type-I error fixed at a fixed level 'α' and try to (.........) the error II.

(a) Minimise (b) Decrease
(c) Maximise (d) Both (a) & (b)

Q.5 The minimisations of error β means (..........) of power of the text.

(a) Maximisation (b) Increasing
(c) Both (a) & (b) (d) Decreasing

Q.6 In construction of a test. we choose a C.R. of size 'α' which has atleast the same power as any other C.R. of the same size 'α'. That C.R. is known as :

(a) Best critical region (b) Average critical region
(c) Central critical region (d) Confidence interval

Q.7 Let Ω_θ be the parametric space and $\Omega\theta_0$ be some specified proper subset of Ω_θ. Then we write the null hypothesis as :

(a) $\theta \in \Omega\theta$. (b) $\theta \in \Omega\theta_0$.
(c) $\theta \in R_n$. (d) $\theta \notin \Omega\theta_0$.

Q.8 Let c be the C.R. of size α and A be the acceptance region such that $(C+A) = R_n$. Then we choose a test H_0 such that :

(a) $P_\theta(c) = \alpha, \forall\ \theta \in \Omega\ \theta_0$.
(b) $P_\theta(c) = \alpha, \forall\ \theta \notin \Omega\ \theta_0$.
(c) $P_\theta(c) = \alpha, \forall\ \theta \in \Omega\ \theta$.
(d) $P_\theta(c) = \alpha, \forall\ \theta \in R_n$.

Q.9 Let β be the probability of error II, then β should be as (.........) as possible.

(a) Maximum (b) Minimum
(c) Average value (d) Median value

Q.10 The probability 'α' defined in Q.(8) is known as :

(a) Confidence level (b) Level of significance

(c) Probability of error I, (d) Both (b) & (c)

Q.11 The complement of level of significance i.e. $(1 - \alpha)$ is known as :

(a) Confidence level (b) Power of the test

(c) Coefficient of confidence (d) Both (a) & (c)

Q.12 Probability of error II i.e. β is the function of :

(a) Alternative hypothesis H_1 :

(b) Null hypothesis H_0 :

(c) Complementary hypothesis H_1 :

(d) Both (a) & (c).

Q.13 The probability of error II i.e. β depends on :

(a) Probability of error I (b) Level of significance 'α'

(c) Size of critical region (d) All the above three.

Q.14 The complement of error II i.e. $(1-\beta)$ is known as (...........).

(a) Power of the test (b) Size of C.R.

(c) Confidence coefficient (d) Confidence level.

Q.15 The power of a test criterion measures the sensitiveness of the :

(a) Confidence coefficient (b) Erroro β

(c) Test (d) Error α

Q.16 The other things being identical, the comparison of two test is the comparision of their respective :

(a) Level of confidence (b) Size of C.R.

(c) Power (d) Level of significance

Q.17 Let θ_0 be the true value of θ. Then by definition, the power function assumes the value 'α' at the :

(a) Hypothesis $H_0 : \theta = \theta 0$ (b) Hypothesis $H_1 : \theta = \theta,$

(c) Size of CR (d) Probability of error II

Q.18 Suppose there are many C.R.'s of the same size 'α'. Then the C.R. with minimum β is known as :

(a) Maximum power (b) Best critical region

(c) Confidence coefficient (d) Confidence level

Q.19 The statistical test corresponding to best C.R. is known as :

(a) Most confident test (b) Most sufficient test

(c) Most powerful test (d) Most unbiased test

Q.20 Let Ω_{θ_0} and its complement say $(\Omega_\theta - \Omega_\theta) = \Omega_{\theta_1}$ have single element $(\theta_0$ & $\theta_1)$ respectively. Let a C.R. "C" is chosen such that if :

P_{θ_0} (c) = α

and P_{θ_1} (c) ≥ P_{θ_1}(c')

for any arbitrary C.R. say (c'). Then the critical 'c' is known as :

(a) Confident C.R. (b) Most efficient C.R.

(c) Central confident C.R. (d) Most powerful C.R.

Q.21 Suppose we have to test $H_0 : \theta = \theta_0$ v.s. alternative composite hypothesis $H_1 : \theta \neq \theta_0$. If a set 'c' is a best C.R. of size α then 'C' is known as :

(a) Most power critical region

(b) Uniformly most powerful critical region

(c) Most powerful unbiased critical region

(d) Most powerful sufficient critical region

Q.22 The test defined in Q. (21) is known as (..........) with confidence level (1–α) to test $H_0 : \theta = \theta_0$ v.s. $H_0 : \theta \neq \theta_0$:

(a) M.P. unbiased test (b) M.P. test

(c) Unbiased test (d) U.M.P. test

Q.23 If an U.M.P. test exists then (..........) Lemma provides a technique to find it.

(a) Rao-Blackwell (b) Neyman-Pearson

(c) Prof. Mahalanobi's (d) Cochram & Cox.

Q.24 Let 'C' be the C.R. of size α to test $H_0 : \theta = \theta_0$ v.s. $H_1 : \theta \neq \theta_0$ such that :

P_{θ_0} (c) = a

and P_θ (c) > P_θ (c') $\forall\ \theta = \theta_0$. ...(A)

where c' is an arbitrary C.R. satisfying (A). Then 'C' is known as :

(a) M.P. critical region (b) U.M.P. critical region

(c) M.P.U. critical region (d) M.P. biased critical region

Q.25 Let 'C' be a C.R. of size α to test $H_0 : \theta = \theta_0$ v.s. $H_0 : \theta \neq \theta_0$ such that :

P_θ (c) ≥ P_{θ_0} (C) $\forall\ \theta \neq \theta_0$

Then 'C' is said to be :

(a) Sufficient C.R. (b) Unbiased C.R.

(c) Biased C.R. (d) Efficient C.R.

Q.26 Let $6(\theta)$ be a test function to test :

$H_0 : \theta \in \Omega\, \theta_0$ v.s. $H_0 : \theta \in \Omega\, \theta_1$.

Let us define the test function as if :

P_{θ_0} [$6(\theta)$] $\leq \alpha$ i.e. type I error $\leq$ C.R.

and P_{θ_1} [$6(\theta)$] $> \alpha$ i.e. if power $> \alpha$.

then the test function $6(\theta)$ is said to :

(a) Biased test function (b) Sufficient test function

(c) Unbiased test function (d) Efficient test function

Q.27 A test of size α is said to be unbiased test if its :

(a) Power $> \alpha$ (b) Power $< \alpha$

(c) Power $< 2\alpha$ (d) Power $> 2\alpha$

Q.28 A critical region corresponding to an unbiased test is known as :

(a) Biased C.R. (b) Sufficient C.R.

(c) Unbiased C.R. (d) Efficient C.R.

Q.29 A test based on an unbiased critical region is said to be :

(a) A biased test (b) A sufficient rest

(c) An unbiased test (d) A consistent test

Q.30 If the power of a test is always greater than the size α of C.R. then the test is known as :

(a) Biased test (b) Unbiased test

(c) Consistent test (d) M.P. test

Q.31 A most powerful (M.P.) test is always :

(a) An unbiased test (b) A biased test

(c) Consistent test (d) Efficient test

Q.32 If we have to test a simple hypothesis $H_0 : \theta = \theta_0$ v.s. a composite alternative hypothesis $H_1 : \theta \neq \theta_0$, then an uniformally most powerful test :

(a) May exist (b) May not exist

(c) May or may not exist (d) Always exist

Q.33 If in Q.(32), an U.M.P. does not exists, then we require to search out :

(a) A.M. P. test (b) An U.M.P. unbiased test

(c) A consistent test (d) A biased test.

Q.34 Let 'C' be the C.R. of size α to test $H_0 : \theta = \theta_0$ v.s. $H_1 : \theta \neq \theta_0$ and if :

$P_{\theta_0}(C^*) = \alpha$ ....(A)

$P_\theta(C^*) \geq \alpha \ \forall \ \theta \neq \theta_0$ ...(B)

and $P_\theta(C^*) > P_\theta(C) \ \forall \ \theta \neq \theta_0$ where'C' is an arbitrary C.R. Then 'C*' is said to be (..........) C.R. of size a and satisfies A & b.

(a) U.M.P.U. critical region (b) M.P. critical region

(c) M.P.U. critical region (d) M.P. confidence coefficient

Q.35 Let 'C*' be the C.R. of size a to test $H_0 : \theta = \theta_0$ v.s. $H_1 : \theta \neq \theta_0$ and if :

$P_{\theta_0}(C^*) = \alpha$ ....(*l*)

$\frac{d}{d_\theta} P_{\theta_0}(C^*) / \theta = \theta_0$ ...(*m*)

and $\frac{d^2}{d_{\theta^2}} P_{\theta_0}(C^*) \Big/ _{\theta = \theta_0} \geq \frac{d^2}{d_{\theta^2}} P_{\theta_0}(C) \Big/ _{\theta = \theta_0}$ ...(*n*)

for an arbitrary region 'C' which satisfies (*l*) & (*m*). Then 'C*' is said to be :

(a) Type A_1 critical region

(b) Type A critical region

(c) Type A confidence interval

(d) Type A_1 confidence interval

Q.36 Let 'C*' be a M.P. critical region to test $H_0 : \theta = \theta_0$ v.s. $H_1 : \theta = \theta_1$ such that θ_1 is not too far away from θ_0.

Then "Type A" C.R. of size α may also be called as (..........) C.R. of size α.

(a) Locally M.P. unbiased (b) Centrally M.P. unbiased

(c) U.M.P. (d) U.M.P.U.

Q.37 Let 'C*' be a C.R. to test $H_0 : \theta = \theta_0$ v.s. $\theta \neq \theta_0$ and if :

$P_{\theta_0}(C^*) = \alpha$ ...(*p*)

$\frac{d}{d_\theta} P_{\theta_0}(C^*) \Big/ _{\theta = \theta_0} = 0$ ...(*q*)

and $P_\theta(C^*) \geq P_\theta(C) \ \forall \ \theta \neq \theta_0$ ...(r)

for an arbitrary region 'C' which satisfies (p) & (q). Then 'C*' is said to be :

(a) Type A critical region (b) Type A_1 critical region

(c) Locally M.P. critical region (d) U.M.P. critical region

Q.38 In most of the cases, a "type A_1" C.R. will also be :

(a) An U.M.P.U. critical region

(b) Locally M.P. critical region

(c) Centrally M.P. critical region

(d) None

Q.39 The "Type A_1" C.R. and "Type A" C.R. are :

(a) Not identical (b) Identical

(c) Different (d) Both (a) & (c).

Q.40 A simple null hypothesis H_0 can be tested against a :

(a) Simple alternative $H_1 : \theta = \theta_1$.

(b) Composite alternative $H_1 : \theta \neq \theta_0$.

(c) Both (a) & (b)

(d) None.

Q.41 If both the hypotheses H_0 & H_1 are simple hypotheses, then B.C.R. of size 'α' exists if the parent distribution is :

(a) Continuous distribution (b) Discrete distribution

(c) Normal distribution (d) Both (a) & (c)

4.3.2 Neyman-Pearson Lemma

Q.42 When both the hypotheses H_0 & H_1 are simple then B.C.R. of size 'α' is obtainable through the use of (..........) Lemma.

(a) Cramer-Rao (b) Neyman-Pearson

(c) Prof. Mahalanobis (d) Cochram & Cox

Q.43 Let x be a random sample drawn from the distribution $f(x, \theta)$. The likelihood function under H_0 & H_1 to test $H_0 : \theta = \theta_0$ v.s. $H_1 : \theta \neq \theta_0$, are given as :

$$P(x, \theta_0) = \prod_{i=1}^{n} f(x_i, \theta_0) \text{ and } P(x, \theta_1) = \prod_{i=1}^{n} f(x_i, \theta_1)$$

Then B.C.R. consists of the points of sample space say R_n for which according to N.P. Lemma, we have $\dfrac{P(x, \theta_0)}{P(x, \theta_1)} \cdots$

(a) $\leq k$ (b) $= k$

(c) $> k$ (d) $< 2k$

where k is such that the size of C.R. is equal to 'α'.

Q.44 In Q.(43), the "k" is the function of :

(a) (β, H_1) (b) (β, H_0)

(c) (α, H_1) (d) (α^{α}, H_0)

Q.45 Neyman-Pearson Lemma enables us to find a MPCR to test $H_0 : \theta = \theta_0$ v.s. $H_1 : \theta = \theta_1$ if the sample (x) is drawn from :

(a) Continuous distribution (b) Normal distribution

(c) Both (a) & (b) (d) Binomial distribution

Q.46 Neyman-Pearson Lemma also enables us to get an U.M.P. critical region to test $H_0 : \theta = \theta_0$ v.s. composite hypothesis :

(a) $H_1 : \theta > \theta_0$ (b) $H_1 : \theta < \theta_0$

(c) $H_1 : \theta \neq \theta_0$ (d) All the above

Q.47 The method of construction of (M.P. or U.M.P.) critical region of size 'α' will be similar to that of :

(a) Discreate and continuous distribution

(b) Discreate and normal distribution

(c) Binomial and normal distribution

(d) All the above.

Q.48 In case of discrete distributions, to test $H_0 : \theta = \theta_0$ v.s. $H_1 : \theta \neq \theta_0$, the constant '$k$' used in N.P. lemma. will :

(a) Not be unique (b) Be Unique

(c) Be different (d) Both (a) & (c)

Q.49 In fact for discrete distribution function, there may be two consecutive mass points (y_0, y_1) of the variable y such that :

(a) $f(y_0) < \alpha < f(y_1)$ (b) $f(y_0) \leq \alpha \leq f(y_1)$

(c) $f(y_0) \leq \alpha < f(y_1)$ (d) $f(y_0) < \alpha \leq f(y_1)$

Q.50 Let x~ $N(\mu, \sigma^2)$ where μ is unknown and σ^2 is known. Suppose we have to test $H_o : \mu = \mu_0$ v.s. $H_1 : \mu = \mu_1 > \mu_0$.

Then M.P.C.R. of size 'α' is given as :

(a) $W = \{ X / \dfrac{\left| \bar{x} - \mu_0 \right| \sqrt{n}}{\sigma} \geq Z\alpha \}$

(b) $W = \{ X / \frac{\left| \bar{x} - \mu_0 \right| \sqrt{n}}{\sigma} \leq Z\alpha \}$

(c) $W = \{ X / \frac{\left| \bar{x} - \mu_0 \right| \sqrt{n}}{\sigma} \geq Z\alpha \}$

(d) $W = \{ X / \frac{\left| \bar{x} - \mu_0 \right| \sqrt{n}}{\sigma} \leq Z\alpha \}$

Q.51 In Q.(50), let $\mu_1 < \mu_0$, then M.P.G.R. say W* is given as :

(a) $W^* = \{ X / \frac{\left| \bar{x} - \mu_0 \right| \sqrt{n}}{\sigma} \geq Z\alpha \}$

(b) $W^* = \{ X / \frac{\left| \bar{x} - \mu_0 \right| \sqrt{n}}{\sigma} \leq Z\alpha \}$

(c) $W^* = \{ X / \frac{\left| \bar{x} - \mu_0 \right| \sqrt{n}}{\sigma} \geq Z\alpha \}$

(d) $W^* = \{ X / \frac{\left| \bar{x} - \mu_0 \right| \sqrt{n}}{\sigma} \leq Z\alpha \}$

Q.52 In Q.(50), we have to test $H_0 : \mu = \mu_0$ v.s. $H_1 : \mu > \mu_0$. Then existing M.P.C.R. of size α is :

(a) Also U.M.P.C.R. (b) Not U.M.P.C.R.

(c) Unbiased C.R. (d) Biased C.R.

Q.53 In Q.(51) we have to test $H_0 : \mu = \mu_0$ v.s. $H_1 : \mu < \mu_0$. Then the existing M.P.C.R. of size α is :

(a) Not U.M.P.C.R. (b) Unbiased C.R.

(c) Also U.M.P.C.R. (d) Biased C.R.

Q.54 Let X~ N(μ, σ^2) where μ is known but σ^2 is unknown. Supose we have to test $H_0 : \sigma^2 = \sigma_0^2$ v.s. $H_1 : \sigma^2 > \sigma_0^2$.

Then M.P.C.R. of size α say W* is given as :

(a) $W = \{ X / \sum_i^n \left(\frac{x_i - \mu}{\sigma_0}\right)^2 > \chi^2_{\alpha}, n \}$

(b) $W = \{ X / \sum_i^n \left(\frac{x_i - \mu}{\sigma_0}\right)^2 < \chi^2_{\alpha}, n \}$

(c) $W = \{ X / \sum_i^n \left(\frac{x_i - \mu}{\sigma_0}\right)^2 \geq \chi^2_{\alpha}, n \}$

(d) None.

Q.55 In Q.(54), to test $H_0 : \sigma^2 = \sigma_0^2$ v.s. $H_1 : \sigma^2 < \sigma_0^2$, then M.P.C.R. of size α say W* is :

(a) $\sum_i^n \left(\frac{x_i - \mu}{\sigma_0}\right)^2 > \chi^2_{1-\alpha, n}$

(b) $\sum_i^n \left(\frac{x_i - \mu}{\sigma_0}\right)^2 < \chi^2_{1-\alpha, n}$

(c) $\sum_i^n \left(\frac{x_i - \mu}{\sigma_0}\right)^2 \geq \chi^2_{\alpha/2, n}$

(d) $\sum_i^n \left(\frac{x_i - \mu}{\sigma_0}\right)^2 < \chi^2_{1-\alpha, n}$

Q.56 In Q.(54), to test $H_0 : \sigma^2 = \sigma_0^2$ v.s. $H_1 : \sigma^2 > \sigma_0^2$, then M.P.C.R. provided as W is :

(a) Also U.M.P.C.R. (b) Not U.M.R.P.C.R.

(c) Unbiased C.R. (d) Biased C.R.

Q.57 In Q.(55), we have to test $H_0 : \sigma^2 = \sigma_0^2$ v.s. $H_1 : \sigma^2 < \sigma_0^2$. Then the M.P.C.R. provide as W* is :

(a) Not U.M.P.C.R. (b) Also U.M.P.C.R.

(c) Unbiased C.R. (d) Biased C.R.

Q.58 Suppose there does not exist an U.M.P.C.R. to test $H_0 : \sigma^2 = \sigma_0^2$ v.s. $H_1 : \sigma^2 < \sigma_0^2$. Then to overcome this problem, we require to construct :

(a) An U.M.P.C.R. (b) M.P.C.R.

(c) M.P. Unbiased C.R. (d) M.P. biased C.R.

Q.59 If there exists a MPCR say W_0 to test $H_0 : \sigma^2 = \sigma_0^2$ v.s. $H_1 : \sigma^2 = \sigma_1^2$. then this W* is necessarily :

(a) An unbiased CR (b) Biased CR

(c) Uniformally Unbiased CR (d) Uniformally biased CR

Q.60 If there exists an U.M.P.C.R. say W* to test $H_0 : \sigma^2 = \sigma_0^2$ v.s. $H_0 : \sigma^2 = \sigma_1^2$, then this W* is necessarily :

(a) Biased C.R. (b) Unbiased C.R.

(c) Consistent C.R. (d) Sufficient C.R.

Q.61 Let X~ N(μ, σ^2) where μ is unknown but σ^2 s known. Suppose we have to test $H_0 : \mu = \mu_0$ v.s. $H_1 : \mu \neq \mu_0$. Then "Type A region" say W of size α, is given as :

(a) $W = \{ X / \dfrac{\left| \bar{x} - \mu_0 \right| \sigma}{\sqrt{n}} < Z\alpha \}$

(b) $W = \{ X / \dfrac{\left| \bar{x} - \mu_0 \right| \sigma}{\sqrt{n}} < Z\alpha/2 \}$

(c) $W = \{ X / \dfrac{\left| \bar{x} - \mu_0 \right| \sigma}{\sqrt{n}} > Z\alpha \}$

(d) $W = \{ X / \frac{\left| \bar{x} - \mu_0 \right| \sigma}{\sqrt{n}} > Z\alpha/2 \}$

Q.62 Let $X \sim N(\mu, \sigma^2)$ where μ is known but σ^2 is unknown. Suppose we have to test $H_0 : \sigma^2 = \sigma_0^2$ v.s. $H_1 : \sigma^2 \neq \sigma_0^2$. Then "Type A region" say W_0 is provided as :

(a) $W_0 = \{ X / \sum_{i}^{n} \left(\frac{x_i - \mu}{\sigma_0} \right)^2 < \chi_1^2 \}$

(b) $W_0 = \{ X / \sum_{i=1}^{n} \left(\frac{x_i - \mu}{\sigma_0} \right)^2 > \chi_2^2 \}$

(c) Both (a) & (b)

(d) None

Q.63 To determine the critical values (χ_1^2 and χ_2^2) in Q.(62), we have to consider the equation that $P(\chi^2 < \chi_2^2) + P(\chi^2 > \chi_2^2)$ is equal to :

(a) α, (b) 2α,

(c) $\alpha / 2$ (d) None

Q.64 Suppose we have to test $H_0 : \theta = \theta_0$ v.s. $H_1 : \theta \neq \theta_0$. Let a C.R. say W_0 be defined as :

$W_0 = \{ X / f_\theta(x) > k_1 f_{\theta_0}(x) + k_2 f'_{\theta_0}(x) \}$

is same H_1 $\theta \neq \theta_0$ where (k_1 & k_2) are constants such that :

$\int_{W_0} f_{\theta_0}(x)\, dx = \alpha$ and $\int_{W_0} f'_{\theta_0}(x)\, dx = 0$

Then W_0 of size α, is known as :

(a) U.M.P.C.R. (b) Type A_1 C.R.

(c) M.P.C.R. (d) 'Type A' C.R.

Q.65 Suppose we have $X \sim N(\mu, \sigma^2)$ and we have to test $H_0 : \sigma^2 = \sigma_0^2$ v.s. $H_1 : \sigma^2 \neq \sigma_0^2$. Then the 'Type A' and 'Type A_1' critical regions both of size

α will be :

(a) Identical (b) Not identical

(c) Different (d) None

Q.66 Let x be a random sample of size n and one has to perform a random experiment which has two possibile outcomes say (θ_1 & θ_2) leading to acceptance and rejection of H_0 respectively. Then a test of this kind is known as :

(a) Biased test (b) Consistent test

(c) Randomized test (d) M.P. test

Q.67 Let f_θ be the joint p.d.f. of a random sample of size n. Then to test $H_0 : \mu = \mu_0$ v.s. $H_1 : \mu = \mu_1$, there exist a test $f(x)$ and a constant ($k > \theta$) such that :

$$E_{\mu_0} [f(x)] = \alpha$$

$$\text{and } \phi(x) = \begin{cases} 1, \text{ when } \dfrac{f_{\mu_1}(x)}{f_{\mu_0}(x)} > k \\ 0, \text{ when } \dfrac{f_{\mu_1}(x)}{f_{\mu_0}(x)} < k \end{cases}$$

Then such test is known as :

(a) U.M.P. test (b) M.P. test

(c) U.M.P.U. test (d) M.P. consistent test

Q.68 Suppose we have to test H_0 against H_1, then M.P. test of size α will be given by the critical function $\phi(x)$ such that :

(a) $\phi(x) = 1$ when $f_{\theta_1}(x) / f_{\theta_0}(x) > k$.

(b) $\phi(x) = \alpha$ when $f_{\theta_1}(x) / f_{\theta_0}(x) = k$.

(c) $\phi(x) = 0$ when $f_{\theta_1}(x) / f_{\theta_0}(x) < k$.

(d) All the three are correct.

Q.69 Let x be a continuous random variable then M.P. randomised test to test H_0 against H_1 will generally be identical with the :

(a) MP non-randomized test

(b) MP consistent test

(c) U.M.P consistent test

(d) None.

Q.70 If x_1, x_2, ..., x_n are discrete r.v.'s and MPCR of size α exists to test H_0 : v.s. H_1, then M.P. randomised test will be equivalent to :

(a) M.P. non-randomised test.

(b) M.P. consistent test.

(c) U.M.P. consistent test.

(d) None

Q.71 The M.P. randomised test will lead to an increase in the power of test while it still affects the desired level of control on :

(a) Probability of type I error.

(b) Level of significance.

(c) Size of critical reason.

(d) All the three.

4.4 Key Answers

Write True/False

Q.	A.	Q.	A.	Q.	A.	Q.	A.
1.	True	30.	True	59.	True	88.	True
2.	True	31.	True	60.	True	89.	True
3.	True	32.	True	61.	True	90.	True
4.	True	33.	True	62.	True	91.	True
5.	True	34.	True	63.	True	92.	True
6.	True	35.	True	64.	True	93.	True
7.	True	36.	True	65.	True	94.	True
8.	True	37.	True	66.	True		
9.	True	38.	True	67.	True		
10.	True	39.	True	68.	True		
11.	True	40.	True	69.	True		
12.	True	41.	True	70.	True		
13.	True	42.	True	71.	True		
14.	True	43.	True	72.	True		
15.	True	44.	True	73.	True		
16.	True	45.	True	74.	True		
17.	True	46.	True	75.	True		
18.	True	47.	True	76.	True		
19.	True	48.	True	77.	True		
20.	True	49.	True	78.	True		
21.	True	50.	True	79.	True		
22.	True	51.	True	80.	True		
23.	True	52.	True	81.	True		
24.	True	53.	True	82.	True		
25.	True	54.	True	83.	True		
26.	True	55.	True	84.	True		
27.	True	56.	True	85.	True		
28.	True	57.	True	86.	True		
29.	True	58.	True	87.	True		

Fill in the Blanks

1. Error
2. Type of errors
3. Maximises
4. Minimise
5. Maximization
6. Critical Region
7. Best C.R.
8. $\Omega\theta_0$
9. $\theta \in \Omega\ \theta_0$
10. Minimum or small
11. Level of significance
12. Confidence coefficient
13. Alternative hypothesis H_1
14. II error
15. The power
16. Test
17. Power
18. $\theta = \theta_0$
19. Power, C.R.
20. Powerful Test
21. Most powerful critical region (M.P.C.R.)
22. Uniformly most powerful (U.M.P.)
23. U.M.P. Test
24. Neyman-Pearson
25. U.M.P.
26. Unbiased
27. Unbiased
28. Unbiased
29. Unbiased
30. Unbiased Test
31. Uniformly Unbiased
32. Unbiased Test
33. U.M.P.
34. U.M.P.U.
35. U.M.P.U. Region
36. Type A critical region
37. Locally most powerful unbiased
38. Type A1 Critical region
39. U.M.P.U. region
40. Not identical
41. Alternative
42. Composite alternative
43. Composite hypothesis
44. Continuous
45. N.P. Lemma
46. Sample space
47. α, H_1
48. Continuous r.v.
49. $\theta \neq \theta_0$
50. Continuous
51. $\theta \neq \theta_0$
52. y0, y1
53. Composite
54. $\dfrac{|\bar{x} - \mu_0|\sqrt{n}}{\sigma}$
55. $\leq Z\alpha$
56. $\mu_1 > \mu_0$, $\mu_1 < \mu_0$
57. $\mu_1 > \mu_0$, $\mu_1 < \mu_0$
58. U.M.P.

59. $\left(\dfrac{x_i - \mu}{\sigma_0}\right)$

60. $\chi^2_{1-\alpha,\, n}$

61. σ_1^2

62. U.M.P.C.R.

63. U.M.P.

64. U.M.P.

65. Unbiased critical region

66. Unbiased

67. Unbiased

68. $\dfrac{|\bar{x} - \mu_0|\sqrt{n}}{\sigma}$

69. $\left(\dfrac{x_i - \mu}{\sigma_0}\right)$

70. α.

71. Approximation

72. Type A critical region

73. Identical

74. Typa A critical region

75. Randomised test

76. $> k$, $< k$

77. M.P.

78. $> k$, $= k$, $< k$

79. Non-randomized

80. Non-randomized

81. Type I error

82. M.P. test.

Multiple Choice Questions and Answers

Q.	A.	Q.	A.	Q.	A.
1.	(d)	25.	(b)	49.	(a)
2.	(a)	26.	(c)	50.	(a)
3.	(b)	27.	(b)	51.	(b)
4.	(d)	28.	(c)	52.	(a)
5.	(c)	29.	(c)	53.	(c)
6.	(a)	30.	(b)	54.	(a)
7.	(b)	31.	(a)	55.	(b)
8.	(a)	32.	(c)	56.	(a)
9.	(b)	33.	(b)	57.	(b)
10.	(d)	34.	(a)	58.	(a)
11.	(d)	35.	(b)	59.	(a)
12.	(d)	36.	(a)	60.	(b)
13.	(d)	37.	(b)	61.	(c)
14.	(a)	38.	(a)	62.	(c)
15.	(c)	39.	(d)	63.	(a)
16.	(c)	40.	(c)	64.	(d)
17.	(a)	41.	(d)	65.	(a)
18.	(b)	42.	(b)	66.	(c)
19.	(c)	43.	(a)	67.	(b)
20.	(d)	44.	(c)	68.	(d)
21.	(b)	45.	(c)	69.	(a)
22.	(d)	46.	(d)	70.	(a)
23.	(b)	47.	(d)	71.	(d)
24.	(b)	48.	(d)		

❑❑❑

CHAPTER - 5

More on Testing of Hypothesis

5.1 Write True/False

Q.1 The null hypothesis H_0 is said to be composite if it does not specify the joint probability of $x_1, x_2, ..., x_n$ completely.

Q.2 We may write null hypothesis as :

$$H_0 : \theta \in \Omega_{\theta_0}. \qquad ...(1)$$

Q.3 Let W be the C.R. of size α then to test $H_0 : \theta = \theta_0$ vs. $H_1 : \theta \neq \theta_0$ we require that :

$$P_\theta(w) = \alpha \qquad \forall \ \theta \in \Omega_{\theta_0}. \qquad ...(2)$$

Q.4 We should note that the C.R.'s 'W' must be such that the size of W should be same for all $\theta \in \Omega\theta_0$.

Q.5 The size of all such C.R.'s say 'W' of the sample space R^n may not be independent of $\theta \in \Omega \ \theta_0$.

Q.6 Those C.R.'s say 'W' which has same size α in the sample space R^n, are called "Similar regions."

Q.7 The sample space R^n has the property that $P_\theta (R^n) = 1$.

Q.8 In case of composite null hypothesis, the selection of suitable test must be such that all the similar region are of the same size α.

Q.9 One should find a similar region of size α that is the best from the point of power.

Q.10 Let T be the sufficient statistic for $H_0 : \theta \in \Omega \ \theta_0$. Also let I_w be an indicator function of the CR say W then we have :

$$Iw = \begin{cases} 1, \text{if } x \in w \\ 0, \text{otherwise} \end{cases} \qquad ...(3)$$

Q.11 If W is the critical region such that the conditional expectation.

$E\ (I_W /\ T = t\ ) = \alpha$, a.e. $\theta \in \Omega\ \theta_0$. ...(4)

then W is the similar critical region of size α.

Q.12 A test with critical region say W is said to be of "Neyman Structure" w.r. to sufficient statistic T if :

$E(I_w / T = t)$

is same almost every where $\forall\ \theta \in \Omega_{\theta_0}$.

Q.13 Every test based on a simillllar region will have a "Neyman Structure" w.r. to a sufficient statisitic T if the family $F(T_\theta)$ of distribution of T_0 for $\theta \in \Omega\ \theta_0$ i.e.

$F(T_0) = \{P_\theta\ (T_0) / \theta \in \Omega_{\theta_0})$ ...(5)

is boundedly complete.

Q.14 Suppose $\theta = (\theta_1, \theta_2, \ldots.., \theta_n)$ and that H_0 : specifies the values of first k < m, parameters. i.e.

$H_0 : \theta_1 = \theta_{10}, \theta_2 = \theta_{20}, \ldots.., \theta k = \Omega_{k_0}$ ...(6)

Then H_0 is said to be composite hypothesis with $(m - k)$ degree of freedom.

Q.15 Suppose 6 is the whole set of parameters whose values are specified and η be the set of parameters with unspecified values. Then H_0 may be expressed as :

$H_0 : \theta \in 6$. ...(7)

Q.16 Those parameters whose values are not specified, are also called as "nuisance parameters."

Q.17 Consider the normal distribution $N(\mu, \sigma^2)$ where μ & σ^2, both are unknown. Suppose we have to test the following simple hypothesis against simple alternative written as :

$H_0 : \mu = \mu_0\ \&\ \sigma^2 = \sigma_0^2$

$H_1 : \mu = \mu_1\ \&\ \sigma^2 = \sigma_1^2$

Then M.P. similar region W_0 of size a for testing H_0 against H_1 is given as :

$$W_0 = \{ X / \frac{f\ (x, \mu_1, \sigma_1^2)}{f\ (x, \mu_0, \sigma_0^2)} > k_1\ (v)\ \} \quad \ldots(8)$$

where $k_1(v)$ is such that the conditional size of W_0 given V = v is equal to α.

Q.18 In Q.17, let $\mu_1 > \mu_0$ then C.R. say W_0 may be written as :

$$W_0 = \{ X / (\bar{x} - \mu_0) \sqrt{\frac{n}{v}} > k_1(v) \text{ (say) } \} \qquad \ldots(9)$$

where $k_1(v)$ is to be determined as :

$P_{\theta_0}(W_0/v) = \alpha.$

Q.19 In Q.17, we finally write that :

$$W_0 = \{ X / \frac{(\bar{x} - \mu_0)\sqrt{n}}{s} > t_{\alpha, (n-1)} \} \qquad \ldots(10)$$

Q.20 The critical region provided in Q.19, is independent of σ_0^2 & σ_1^2 and hence it is the M.P. similar region of size α for testing H_0 : against H_1.

Q.21 In Q.17, let $\mu < \mu_0$ then the MP similar region of size α for testing H_0 against H_1 is given as :

$$W_0' = \{ X / (\bar{x} - \mu_0) \frac{\sqrt{n}}{s} < t_{\alpha, n-1} \} \qquad \ldots(11)$$

Q.22 In Q.19 and Q.21, the C.R.'s W_0 & W_0' are both independent of μ_1. Hence W_0 and W_0' are both U.M.P. similar regions for testing $H_0 : \mu = \mu_0$ against $H_1 : \mu > \mu_0$ and $\mu < \mu_0$.

Q.23 Let $x_1, x_2, \ldots\ldots, x_n$ be a random sample from $N(\mu, \sigma^2)$ where μ & σ^2 both are unknown. Now consider the problem of testing a composite hypothesis :

$H_0 : \sigma^2 = \sigma_0^2$ vs composite $H_1 : \sigma^2 \neq \sigma_1^2$

Let $\sigma_1^2 > \sigma_0^2$ then the MP similar region say W_0 is given as :

$$W_0 = \{ X / \sum_i^n \frac{(x_i - \bar{x})^2}{\sigma_0^2} > \chi^2_{\alpha, n-1} \} \qquad \ldots 12$$

Q.24 In Q.23, let $\sigma_1^2 < \sigma_0^2$ then MP similar region for testing H_0 against H_1 say W_0^* is given as :

$$W_0^* = \left\{ X \Big/ \sum_{i-1}^{n} \frac{(x_i - \overline{x})^2}{\sigma_0^2} < \chi^2_{\alpha}, n-1 \right\} \qquad \ldots(13)$$

Q.25 In Q.23, the W_0 is same for all $\sigma_1^2 > \sigma_0^2$, it is fact that W_0 is the U.M.P similar region of size α for testing $H_0 : \sigma^2 = \sigma_0^2$ against one sided alternative $H_1 : \sigma_1^2 > \sigma_0^2$.

Q.26 In Q.24, the region W_0^* is same for all $\sigma_1^2 < \sigma_0^2$, it is the U.MP similar region of size α for testing $H_0 : \sigma^2 = \sigma_0^2$ against one sided alternative $H_1 : \sigma_1^2 < \sigma_0^2$.

Q.27 Consider a simple hypothesis $h_0 \in H_0$ and an alternative simple hypothesis $H_1 \in H_1$. Such that

$H_0 : \mu = \mu_0, \sigma^2 = \sigma_0^2$

$H_1 : \mu = \mu_1, \sigma^2 = \sigma_1^2$

Then MP similar region of size α for testing H_0 against H_1 say W_0 is given as :

$$W_0 = \left\{ X \Big/ \frac{f(x, \mu_1, \sigma_1^2)}{f(x, \mu_0, \sigma_0^2)} > k(\overline{x}) \right\} \qquad \ldots(14)$$

where $k(\overline{x})$ is such that the conditional size of W_0 given $\overline{X} = \overline{x}$ is equal to α.

Q.28 When a null hypothesis (either simple or composite) is to be tested against a composite alternative, the power of the test should increase with the increase in the divergence of the true parametric value from the null hypothesis.

Q.29 Suppose we have to test the hypothesis $H_0 : \theta = \theta_0$ v.s. $H_1 : \theta \neq \theta_0$ and W is the chosen C.R. Then the power P_0 (W) of the test should increase monotonically as θ moves farther and farther away from θ_0.

Q.30 All the UMP and UMPU tests possess the monotonicity property for their power function.

Q.31 Consider the normal distribution $N(\mu, \sigma^2)$ where σ^2 is known. Suppose we have to test the hypothesis $H_0 : \mu = \underset{0}{\mu}$ v.s. $H_1 : \mu > \underset{0}{\mu}$. Then U.MP test of size α has the CR say W. written as :

$$W = \{ X / \bar{x} > \underset{0}{\mu} + Z\alpha. \frac{\sigma}{\sqrt{n}} \} \qquad ...(15)$$

Q.32 The power function of the test cited in Q.31, is written as :

$$P_\mu(W) = P_\mu \left[\frac{(\bar{x} - \mu)\sqrt{n}}{\sigma} > Z\alpha - \frac{(\mu - \underset{0}{\mu})\sqrt{n}}{\sigma} \right] \qquad ...(16)$$

Q.33 The power function $P_\mu(W)$ stated in Q.32, is a monotonically increasing function of $(\mu - \underset{0}{\mu})$.

Q.34 The power function $P_\mu(W)$ of the U.M.P. test of the level α to test the hypothesis $H_0 : \mu = \underset{0}{\mu}$ v.s. $H_1 : \mu < \underset{0}{\mu}$ is monotonically increasing function of $(\underset{0}{\mu} - \mu)$.

Q.35 The power function $P\mu(W)$ of the UMPU test of level α for testing the hypothesis $H_0 : \mu = \underset{0}{\mu}$ v.s. composite alternative $H_1 : \mu \neq \underset{0}{\mu}$, is monotonically increasing function in $| \mu - \underset{0}{\mu} |$.

Q.36 The power of a test should increase wit hthe increase in sample size n drawn from a population.

Q.37 There must be high probability of rejecting a false hypothesis (The power of the test) based on a large sample say $x_1, x_2, ..., x_n$ from the $N(\mu, \sigma^2)$.

Q.38 The power function of a test criterion must possess the concept of "consistency".

Q.39 A sequence of the tests corresponding to the C.R.'s say (W_n) is said to be "consistent" if for every value of $\mu \in (\Omega\mu - \Omega\underset{0}{\mu})$, the power $P_\mu (W_n)$ tends to 1.0 as $n \to \infty$.

Q.40 A consistent estimator of a parameter θ provides a consistent sequence of the test for the hypothesis $H_0 : \theta = \theta_0$.

Q.41 A simple method of test construction related to M.L.E. method, has been suggested by "Neyman-Pearson".

Q.42 Let $L(\theta / x_1, x_2, \ldots\ldots, x_n)$ be the likelihood function of θ corresponding to a set of values $(x_1, x_2, \ldots\ldots, x_n)$ of $(X_1, X_2, \ldots\ldots, X_n)$. Suppose we have to test the hypothesis $H_0 : \theta = \theta_0$ against a composite hypothesis $H_1 : \theta \neq \theta_0$. Then one should reject H0 : if :

$L(\theta_0 / x_1, x_2, \ldots\ldots, x_n)$

is too small > 0.

Q.43 In Q.42 cited above, one should accept H_0 if $L(\theta_0 / x_1, x_2, \ldots\ldots, x_n)$ is too large < 1.0.

Q.44 The test illustrated in Q.42 should be based on the C.R. written as :

$$W = \{ X / \lambda(x_1, x_2, \ldots\ldots, x_n) < \lambda_0 \} \qquad \ldots(17)$$

where λ is the likelihood ratio.

Q.45 The likelihood ratio λ is defined as :

$$\lambda = \frac{L(\theta_0 / x_1, x_2, \ldots\ldots\ldots, x_n)}{Sup_\theta \rightarrow \Omega_\theta \; L(\theta / x_1, x_2, \ldots\ldots\ldots, x_n)} \qquad \ldots(18)$$

Q.46 The content λ_0 cited ub Q.44 is chosen such that the probability of error I, associated with the test must be equal to 'α'.

Q.47 Let $(x_1, x_2, \ldots\ldots, x_n)$ be a random sample from $N(\mu, \sigma^2)$ where μ is unknown but σ^2 is known. Suppose we have to test the hypothesis as :

$H_0 : \mu = \mu_0$ v.s. $H_1 : \mu \neq \mu_0$.

Then LR. test has the C.R. say W written as :

$$W = \left\{ X / \frac{|\bar{x} - \mu_0| \sqrt{n}}{\sigma} > \lambda_0 \right\} \qquad \ldots(19)$$

where λ_0 is such that :

$$P_{\mu_0}\left[\left| \frac{|\bar{x} - \mu_0| \sqrt{n}}{\sigma} \right| > \lambda_0 \right] = \alpha \qquad \ldots(20)$$

Q.48 In Q.47, $\frac{|\bar{x} - \mu_0| \sqrt{n}}{\sigma} \sim N(0, 1)$, λ_0 must be the upper $\alpha/2$ point of the distribution and is denoted by $Z_{\alpha/2}$.

Q.49 If in 47, illustrated above, we have to test the hypothesis $H_0 : \mu = \mu_0$ v.s. $H_1 : \mu > \mu_0$ then MLE of μ for $\mu \in \Omega_\mu$ is given as :

$$\mu = \begin{cases} \bar{x} & \text{if } \bar{x} \geq \mu_0 \\ \mu_0 & \text{if } \bar{x} < \mu_0 \end{cases}$$

Q.50 As stated earlier in Q.49, to test the hypothesis $H_0 : \mu = \mu_0$ v.s. $H_1 : \mu > \mu_0$. the L.R. test has the critical region say W_1 and it is given as :

$$W_1 = \left\{ X / \frac{(\bar{x} - \mu_0)\sqrt{n}}{\sigma} > \lambda_1 \right\} \quad \ldots(21)$$

where $\lambda 1$ is such that :

$$P_{\mu_0}\left[\frac{|\bar{x} - \mu_0|\sqrt{n}}{\sigma} > \lambda_1\right] = \alpha$$

Q.51 In Q.50, we have $\frac{|\bar{x} - \mu_0|\sqrt{n}}{\sigma} \sim N(0, 1)$, the λ_1 must be equal to $Z\alpha$.

Q.52 As stated in Q.47, if we have to test the hypothesis as :

$H_0 : \mu = \mu_0$ v.s. $H_1 : \mu < \mu_0$.

Then we have the M.L.E. of μ for $\mu \in \Omega\, \mu$ written as :

$$\mu = \begin{cases} \bar{x} & \text{if } \bar{x} \leq \mu_0 \\ \mu_0 & \text{if } \bar{x} > \mu_0 \end{cases}$$

Q.53 In Q.52, the L.R. test corresponds to the C.R. say W_2 written as :

$$W_2 = \left\{ X \Big/ \frac{(\bar{x} - \mu_0)\sqrt{n}}{\sigma} < -Z_\alpha \right\} \qquad \ldots(22)$$

Q.54 When $H_0 : \mu = \mu_0$ is to be tested again one sided alternative, the L.R. tests are same as the corresponding U.M.P. tests.

Q.55 To test the hypothesis $H_0 : \mu = \mu_0$ against two sided alternative $(H_0 : \mu \neq \mu_0)$, the L.R. test is identical with the U.M.P.U. test.

Q.56 Let $x_1, x_2, \ldots\ldots, x_n$ be a random sample from $N(\mu, \sigma^2)$ where both μ & σ^2 are unknown. Suppose one has to test the hypothesis :

$H_0 : \mu = \mu_0$ v.s. $H_1 : \mu \neq \mu_0$.

We have $\Omega_{\theta_0} = \{ (\mu, \sigma^2), \mu = \mu_0, 0 < \sigma^2 < \infty \}$

and $\Omega_\theta = \{ (\mu, \sigma^2) / -\infty < \mu < \infty, 0 < \sigma^2 < \infty \}$

Q.57 In Q.56, to test the hypothesis :

$H_0 : \mu = \mu_0$ v.s. $H_1 : \mu \neq \mu_0$.

the C.R. of the L.R. test of size α is given as :

$$W = \{ X / \frac{|\bar{x} - \mu_0|\sqrt{n}}{s} > t_{\alpha/2, n-1} \} \qquad \ldots(23)$$

Q.58 For the problem cited in Q.56, if we have to test the hypothesis :

$H_0 : \mu = \mu_0$ v.s. $H_1 : \mu > \mu_0$.

the C.R. of L.R. test of size α is given as :

$$W_1 = \{ X / \frac{|\bar{x} - \mu_0|\sqrt{n}}{s} > t_{\alpha/2, n-1} \} \qquad \ldots(24)$$

Q.59 If for the problem cited in Q.56, we have to test the hypothesis :

$H_0 : \mu = \mu_0$ against $H_1 : \mu < \mu_0$.

the C.R. say W2 of L.R. test of size α, is given as :

$$W_2 = \{ X / \frac{|\bar{x} - \mu_0| \sqrt{n}}{s} < -t_{\alpha/2,\, n-1} \}$$

Q.60 The student's t tests defined as :

$$t = \frac{|\bar{x} - \mu_0| \sqrt{n}}{s} \sim t_{\alpha,\, n-1},$$

are the L.R. tests.

Q.61 Let x_{ij} $\{ i = 1, 2, \ldots.. k, j = n_i \}$ be k. independent random samples from normal distributions $N(\mu_i, \sigma^2, i = 1, 2, \ldots.., k)$. We may require to test the hypothesis :

$$H_0 : \mu_1 = \mu_2 = \ldots\ldots\ldots = \mu_k$$

$$H_1 : \mu_1 \neq \mu_2 \neq \ldots\ldots\ldots \neq \mu_k$$

We assume that variances of all k normal distribution are same σ^2 but unknown.

The M.L.E of μ & σ^2 are for $\theta \in \Omega\theta$.

$$\mu = \sum^{k} n_i \, \bar{x}_i / n = \bar{x} \text{ (say)}$$

$$\text{and } \sigma^2 = \frac{1}{n} \sum_i \sum_j (x_{ij} - \bar{x})^2 = \frac{Sw}{n} \text{ (Say)}$$

Q.62 For the problem illustrated in Q.61 and for $\theta \in \Omega_{\theta 0}$, the MLE of the parameters μ & σ^2 are given as :

$$\mu = \sum_i^{k} n_i \, \bar{x}_i / n = \bar{x} \text{ (Say)}$$

$$\text{and } \sigma^2 = \sum_i \sum_j (x_{ij} - \bar{x})^2 / n = S_T / n \text{ (say)}$$

Here $S_T = S_B + S_W$ where S_B is MS between samples, & S_w is the MS within samples.

Q.63 The likelihood ratio F for testing H0 : $\mu_1 = \mu_2 = \ldots.. = \mu_k$ is defined as :

$$F = \frac{S_B / (k-1)}{S_w / (n-k)} = \frac{M.S.B}{M.S.W} \sim F_{k-1,\, n-1}$$

Q.64 In Q.63, the M.S.B. is known as "Mean Square Between" samples and M.S.W. is known as M.S. within sample.

Q.65 Denoting the upper α point by $F_{\alpha,\, k-1,\, n-k}$, we have the CR say W. for testing the null hypothesis as stated in Q.61 is given as :

$$W = \{ X / \frac{\text{MSB}}{\text{MSW}} > F_{\alpha,\, k-1,\, n-k} \} \qquad \ldots(25)$$

Q.66 In case of applying L.R. test, the control is effective over the probability of "Type one Error" by choosing suitably the cut point say λ_0.

Q.67 In regard of power of a L R test, it is generally UMP if an UMP test at all exists at all.

Q.68 In some cases, L R. critical region works as a short of highest common factor among the M.P. regions for individual simple alternatives.

Q.69 In many cases, the L R tests are unbiased but this is not universally true.

Q.70 When null hypothesis H_0 is composite then in that case the L R. critical region will generally be similar but may not be universally true.

Q.71 Let θ be a MLE. for θ then θ has assymptotic variance equal to C.R. lower bound.

Q.72 Let us write $\theta' = (\theta_k, \theta'_{m-k})$ where θ'_{m-k} represents nuisence parameters. When null hypothesis H_0 holds then every component of $(\theta - \theta)$ is zero.

Q.73 If θ is the MLE of θ and every component of $(\theta - \theta)$ is equal to zero then.

$L(\theta_k, \theta_{m-k}) \propto 1.0$

Q.74 Let $\lambda = \dfrac{L(\theta_{k_0} - \hat{\theta}_{m-k})}{L(\hat{\theta}_k - \hat{\theta}_{m-k})}$

Then $-2 \log \lambda$ for a hypothesis imposing k constraints on θ, is asymptotically distributed as a "Non-Central" χ_k^2 distribution.

Q.75 The result given in Q.74 is due to Wald.

Q.76 When H_0 holds good, then $-2 \log \lambda$ is distributed asymptotically as "Central χ^2" with k.d.f.

Q.77 Under a very general condition (assumption), the LR test is found to be "Consistent" test.

5.2 Fill in the Blanks

Q.1 The null hypothesis H_0 is said to be (.........) if it does not specifly the joint probability of a sample x of size n completely.

Q.2 We may write (..........) hypothesis as $H_0 : \theta \in \Omega_{\theta 0}$.

Q.3 Let W be the C.R. of size α, then to test $H_0 : \mu = \underset{0}{\mu}$ v.s. $H_1 : \mu \neq \underset{0}{\mu}$, we require that :

$P_\theta(w) =$ .... $\forall \in \Omega_{\theta 0}$.

Q.4 We should note that the C.R. say 'w' must be such that the size of w should be same for all $\theta \in$ .....

Q.5 The C.R.'s say 'W' having the same size 'α' in the sample space 'R^n' are called (..........).

Q.6 The sample space R^n has the property that :

$P_\theta(R^n) =$ ....

Q.7 Incase of composite null hypothesis, the selection of suitable test must be such that all the (.........) regions are of the same size 'α'.

Q.8 One should find that (..........) out of many similar regions of size α which is the best w.r. to the power of the test.

Q.9 Let I_W be an indicator function of the C.R. say w then we have :

$I_W =$ .... of $X \in W$.

Q.10 Let w be the C.R. of size α such that the conditional distribution :

$E(I_W / T = t) = \alpha$, a.e. $\theta \in \Omega_{\theta_0}$.

Then 'W' is called (..........) of size α.

Q.11 A test with C.R. say 'W' is said to be of (..........) structure w.r. to a sufficient statistic 'T' if:

$E (I_W / T = t)$ is same almost everywhere (a.e.) $\forall\ \theta \in \Omega_\theta$.

Q.12 Every test based on a similar region will have a 'Neyman structure' w.r. to a (..........) T_0 if the family $F(T_0)$ of distribution of T_0 for $\theta \in \Omega_{\theta_0}$ is boundedly complete.

Q.13 Let θ be a vector as $\theta = (\theta_1, \theta_2, ..., \theta m)$ and the H_0 specifies the values of first $(k < m)$ parameters i.e. :

$H_0 : \theta_1 = \theta_{10}, \theta_2 = \theta_{20}, ..., \theta_k = \theta_{ko}$.

Then H_0 is said to be (..........) hypothesis with $(m–k)$..........

Q.14 The population parameters whose values are not specified are also called as (..........).

Q.15 Let $X \sim N(\mu, \sigma^2)$ where (μ & σ^2) both are unknown. We have to test the hypothesis H_0 as :

$$H_0 : \mu = \mu_0 \ \& \ \sigma^2 = \sigma_0^2 \text{ VS } H_1 : \mu = \mu_1, \sigma^2 = \sigma_1^2$$

Then M.P. similar regions W_0 of size α is given as :

$$W_0 = \left\{ X \,/\, \left| \frac{f(x, \mu_1, \sigma_1^2)}{\cdots\cdots\cdots} \right| > k(v) \right\}$$

where $k(v)$ is such that the conditional size of W_0 given that $V = v$ is equal to α.

Q.16 In Q.(15) let $\mu_1 > \mu_0$ then the (..........) say W_0 is written as :

$$W_0 = \left\{ X \,/\, (\bar{x} - \mu_0) \sqrt{\frac{n}{v}} > k_1(v) \right\}$$

where $k_1(v)$ is to be determined such that :

$$P_{\mu_0} (W_0/v) = \cdots\cdots\cdots$$

Q.17 In Q.(16), we finally write that :

$$W_0 = \left\{ X \,/\, \frac{|\bar{x} - \mu_0| \sqrt{n}}{s} > \cdots\cdots\cdots \right\}$$

Q.18 The C.R. say W_0 provided in Q.17 is independent of σ_0^2 & σ_1^2. Hence it is the M.P. (..........) of size α to test H_0 vs H_1.

Q.19 In Q.(15), let $\mu < \mu_0$, then M.P. similar region of size α to test H_0 : vs H_1 is given as :

$$W_0' = \left\{ X \,/\, | \bar{x} - \mu_0 | \frac{\sqrt{n}}{s} < \cdots\cdots\cdots \right.$$

Q.20 In Q.(17 & 19), the C.R.'s W_0 & W_0' are both independent of μ_1, Hence (W_0 & W_0') are both U.M.P. (..........) to test $H_0 : \mu = \mu_0$ against $H_1 : \mu_1 \neq \mu_0$ and $\mu_1 < \mu_0$.

Q.21 Let X~ N(μ, σ^2) where both (μ, σ^2) are unknown. Suppose we have to test $H_0 : \sigma^2 = \sigma_0^2$ v.s. composite $H_1 : \sigma^2 \neq \sigma_1^2 > \sigma_0^2$. Then M.P. similar region say W_0 is given as :

$$W_0 = \{ X / \sum_i^n \left(\frac{x_i - \overline{x}}{\sigma_0} \right)^2 > \text{..........} \}$$

Q.22 Let in Q.(21), we have to test $H_0 : \sigma^2 = \sigma_0^2$ v.s. $H_1 : \sigma^2 = \sigma_1^2 < \sigma_0^2$. Then M.P. similar region say W_0^* is given as :

$$W_0^* = \{ X / \sum_i^n \left(\frac{x_i - \overline{x}}{\sigma_0} \right)^2 < \text{..........} \}$$

Q.23 In Q.(21), the W_0 is same for all $\sigma_1^2 > \sigma_0^2$. Hence infact W_0 is the (..........) similar region of size α to test $H_0 : \sigma^2 = \sigma_0^2$ v.s. $H_1 : \sigma^2 = \sigma_1^2 > \sigma_0^2$.

Q.24 In Q.(22), the W_0^* is same for all $\sigma_1^2 < \sigma_0^2$. Hence infact W_0^* is the (..........) similar region of size α to test $H_0 : \sigma^2 = \sigma_0^2$ v.s. $H_1 : \sigma^2 = \sigma_1^2 < \sigma_0^2$.

Q.25 Consider a simple hypothesis $h_0 \in H_0$ and an alternative simple hypothesis $h_1 \in H_1$, such that :

$h_0 : \mu = \mu_0, \sigma^2 = \sigma_0^2$ vs $h_1 : \mu = \mu_1, \sigma^2 = \sigma_1^2$.

Then MP similar region of size α to test H_0 : against H_1, say W_0 is given as :

$$W_0 = \{ X / \frac{f\,(x, \mu_1, \sigma_1^2)}{\text{..........}} > k\,(\overline{x}) \}$$

where $k(\overline{x})$ is such that the conditional size of W_0 given $\overline{X} = \overline{x}$ is equal to α.

Q.26 When a null hypothesis (either simple or composite) is to be tested against a composite alternative, the power of the test should (..........) with the (..........) in the divergence of true parametric value from the null hypothesis.

Q.27 Suppose we have to test the hypothesis $H_0 : \theta = \theta_0$ vs $H_1 : \theta \neq \theta_0$ and 'W' is the chosen C.R. Then the power $P_\theta(w)$ of the test should (..........). monotonically as θ moves farther and farther away from θ_0.

Q.28 All the U.M.P. and U.MPU tests possess the monotonically property for their (..........) function.

Q.29 Let $X \sim N(\mu, \sigma^2)$ where σ^2 is known. We have to test $H_0 : \mu = \mu_0$ v.s. $H_1 : \mu > \mu_0$. Then U.M.P. test of size α has the C.R. say W written as :

$$W = \{ X / \bar{x} > \mu_0 + \text{..........} \}$$

Q.30 The power function of the test cited in Q.(29) is written as :

$$P_\mu(W) = P_\mu \left[\frac{(\bar{x} - \mu)\sqrt{n}}{\sigma} > Z_\alpha - (\mu - \mu_0)\sqrt{\text{..........}} \right]$$

Q.31 The power function $P_\mu(w)$ stated in Q.(30) is a monotonically (..........) function of $(\mu - \mu_0)$.

Q.32 The power function $P_\mu(w)$ of UMP testof the confidence level $(1-\alpha)$ to test the $H_0 : \mu = \mu_0$ v.s. $H_1 : \mu < \mu_0$ is monotonically (..........) function of $(\mu_0 - \mu)$.

Q.33 The power function $P_\mu(w)$ of the U.M.P.U test of the level $(1-\alpha)$ to test $H_0 : \mu = \mu_0$ v.s. $H_1 : \mu \neq \mu_0$ is (..........) increasing function of (| ....|).

Q.34 The power of a test should (.........) with the increase in sample size n drawn from population.

Q.35 There must be high probability of (...........). a false hypothesis based on a large sample of size n from $N(\mu, \sigma^2)$.

Q.36 The power function of a test criterion must possess the property of (..........).

Q.37 A sequence of the tests corresponding to the C.R.'s say (W_n) is said to be (..........) if for every value of $\mu \in (\Omega\mu - \Omega\mu_0)$, the power $P_\mu(W_n)$ tends to 1.0 as $n \to \infty$.

Q.38 A consistent estimator of the parameter θ provides a (..........) sequence of the tests to test the hypothesis $H_0 : \theta = \theta_0$.

Q.39 A simple method of test construction related to M.L.E. method, has been suggested by (..........).

Q.40 Let $L(\theta / x_1, x_2, ..., x_n)$ be the likelihood function of θ. Corresponding to a set of values $(x_1, x_2, ... x_n)$. Suppose we have to test the $H_0 : \theta = \theta_0$ v.s. $H_1 : \theta \neq \theta_0$. Then one should reject H0 if :

$L(\theta_0 / x_1, x_2, ..., x_n)$ is too (.....) > 0.

Q.41 In Q.(40),j one should accept H0 if :

$L (\theta_0 / x_1, x_2, ..., x_n)$ is too (.....) < 1.0.

Q.42 The test illustrated in Q.(40) should be based on the C.R. say 'w' written as :

$W = \{ X / \lambda (x_1, x_2, ..., x_n) < \lambda_0 \}$

where λ is the (..........) ratio.

Q.43 The likelihood ratio λ is defined as :

$$\lambda = \frac{L\ (......................)}{Sup_{\ \theta \in \Omega\theta}\ L\ (\theta / x_1, x_2, ..., x_n)}$$

Q.44 The constant λ_0 cited in Q(42) is chosen such that the probability of error I associated with the test must be equal to (.........).

Q.45 Let $X \sim N(\mu, \sigma^2)$ where μ is unknown but σ^2 is known. Suppose we have to test $H_0 : \mu = \underset{0}{\mu}$ v.s. $H_1 : \mu \neq \underset{0}{\mu}$.

Then L.R. test has the C.R. say 'w' written as :

$$W = \{ X / \frac{|..........|\sqrt{n}}{\sigma} > \lambda_0 \}$$

Q.46 In Q.(45), $\frac{(\bar{x} - \underset{0}{\mu})\sqrt{n}}{\sigma}$ ~ ...(..........), and λ_0 must be the upper $\alpha/2$ point of the distribution and is denoted by (.........).

Q.47 Suppose in Q.(45), we have to test $H_0 : \mu = \underset{0}{\mu}$ v.s. $H_1 : \mu \neq \underset{0}{\mu}$.

Then LR test has the C.R. say W_1 and it is given as :

$$W_1 = \{ X / \frac{(..........)\sqrt{n}}{\sigma} > \lambda_1 \}$$

Q.48 Suppose in Q.(45), we have to test $H_0 : \mu = \underset{0}{\mu}$ v.s. $H_1 : \mu > \underset{0}{\mu}$. Then L.R. test corresponds to the C.R. say W_2 which is given as :

$$W_2 = \{ X / \frac{(\bar{x} - \mu_0)\sqrt{n}}{\sigma} < -...... \}$$

Q.49 When $H_0 : \mu = \mu_0$ is to be tested against one sided alternative, the L.R. tests are same as the corresponding (.........).

Q.50 If we have to test $H_0 : \mu = \mu_0$ v.s. $H_1 : \mu \neq \mu_0$, then L.R. test is identical with (..........) test

Q.51 Let $X \sim N(\mu, \sigma^2)$ where both (μ & σ^2) are unknown. Suppose we have to test $H_0 : \mu = \mu_0$ v.s. $H_1 : \mu \neq \mu_0$. Then the C.R. of L.R. test of size α say 'W' is given as :

$$W = \{ X / \frac{(\bar{x} - \mu_0)\sqrt{n}}{s} > .... \}$$

Q.52 Suppose in Q.(51), we have to test $H_0 : \mu = \mu_0$ v.s. $H_1 : \mu > \mu_0$. Then CR. of L.R test of size a say 'W_1' is given as :

$$W_1 = \{ X / \frac{(\bar{x} - \mu_0)\sqrt{n}}{s} > .... \}$$

Q.53 Suppose in Q.(51), we have to test $H_0 : \mu = \mu_0$ v.s. $H_1 : \mu < \mu_0$. The C.R. of L.R. test of size α say 'W_2' is given as :

$$W_2 = \{ X / \frac{(\bar{x} - \mu_0)\sqrt{n}}{s} < .... \}$$

Q.54 The student's t tests which are the L.R. tests are defined as :

$t = ... \sim t_{\alpha/2, (n-1)}$

Q.55 Let x_{ij} ($i = 1, 2, ..., k, j = n_i$) be k independent random samples drawn from k normal population N (μ_i, σ^2, $i = 1, 2, ..., k$). We have to test :

$$H_0 : \mu_1 = \mu_2 = \mu_3 = ... = \mu_k$$

$$H_1 : \mu_1 \neq \mu_2 \neq \mu_3 \neq ..... \neq \mu_k$$

We assume that the variances of all normal populations are same but unknown. Then L.R. test to test H_0 is given by F statistic defined as :

$$F = \frac{S_B^2/(k-1)}{S_W^2/(n-k)} = \frac{MSB}{MSW} \sim \text{..........}$$

Q.56 In Q.(55), the MSB is known as "Mean Square ...." samples and MSW is known as "Mean Square ...." sample.

Q.57 Denoting the upper α point by $F_{\alpha, k-1, n-k}$, we have the C.R. say 'W' to the hypothesis as stated in Q.(55), given as :

$$W = \{ X / \frac{MSB}{MSW} > \text{..........} \}$$

Q.58 In regard of power of L.R. test, the LR. test is generally (..........) if an UMP test exists at all.

Q.59 In many cases, the L.R. tests are (..........) but it is not universally true.

Q.60 When Null hypothesis H_0 is composite then in that case, the L.R. critical region will generally be (..........) but may not be universally true.

Q.61 Let θ be a M.L.E. of parameter θ then θ has (..........) variance equal to C.R. lower bound.

Q.62 Let $\lambda = L(\theta_k - \hat{\theta}_{m-k}) / L(\hat{\theta}_k - \hat{\theta}_{m-k})$

Then $-2 \log \lambda$ for a hypothesis with k constraint on θ, is asymptotically distributed as a (..........) distribution.

Q.63 The result given in Q.(62) is poineered by (..........).

Q.64 When H_O hold good in Q.(62), then $-2 \log \lambda$ is distributed asymptotically as (..........) with k.d.f.

Q.65 Under a very general assumption, the L.R. test is found to be (..........) test.

5.3 Multiple Choice Questions and Answers

Q.1 If the null hypothesis H_0 does not specify the joint probability of a sample x of size n completely then it is known as :

(a) Composite hypothesis (b) Simple hypothesis

(c) Complex hypothesis (d) Both (a) & (c).

Q.2 If we write a statistical hypothesis as $H_0 : \theta \in \Omega\theta_0$. then this hypothesis is known as :

(a) Simple hypothesis (b) Null hypothesis

(c) Composite hypothesis (d) Both (b) & (c)

Q.3 Let W be the C.R. of size α, then to test $H_0 : \theta = \theta_0$ v.s. $H_1 : \theta \neq \theta_0$, we require that :

$P_\theta(W) =$ .... $\forall\ \theta \in \Omega\theta_0$.

(a) α (b) $\alpha/2$

(c) 2 α (d) None

Q.4 The C.R.'s say 'W' must be such that the size of W should be same for all $\theta \in$ ....

(a) $\Omega\theta_0$ (b) $\Omega\theta$

(c) Null hypothesis (d) None

Q.5 The C.R.'s say 'W' having the same size α, in the sample space 'Rn' are called ....

(a) Similar region (b) Central region

(c) M.P.C.R. (d) Alternative region

Q.6 The space R^n has the property that $P_\theta(R^n)$ is equal to :

(a) 1 (b) $\frac{1}{2}$

(c) $> \frac{1}{2}$ (d) < 1

Q.7 In case of composite null hypothesis, the selection of suitable test must be such that all the (..........) regions must be of the same size α.

(a) Similar (b) MP Critial

(c) Alternative (d) None

Q.8 One should find that 'similar region' out of many similar regions of size a which is the best w.r. to the .....

(a) Size of the test (b) Null hypothesis

(c) Power of the test (d) Simple hypothesis

Q.9 Let I_W be an indicator function of C.R. say 'W' then we have for $x \in w$, the value of I_W as :

(a) $I_W = \frac{1}{2}$ (b) $I_W = 1$

(c) $I_W = \alpha$ (d) $I_W = \frac{\alpha}{2}$

Q.10 Let W be the C.R. of size α such that the conditional distribution $E(I_W / T = t) = \alpha$, a.e. $\theta \in \Omega\theta$, then 'W' is called :

(a) Similar region of size α. (b) M.P.C.R. of size α.

(c) U.M.P.C. of size α. (d) None

Q.11 Let 'W' be a C.R. of size α and T be a sufficient statistic. If an indicator function 'I_W' is such that :

$E(I_W / T = t)$ is same almost every where (a.e).

$\forall\ \theta \in \Omega\ \theta$. Then 'W' is said to be of ....

(a) Neyman-Pearson structure

(b) Neyman structure

(c) Cochran structure

(d) Fisher structure

Q.12 Every test based on a similar region will have a "Neyman structure" w.r. to a (..........). T if the family F(T) of distribution for $\forall\ \theta \in \Omega\ \theta_0$ is boundedly complete.

(a) Unbiased statistic (b) Efficient statistic

(c) Consistent estimator (d) Sufficient statistic

Q.13 Let θ be a vector as $\theta = (\theta_1, \theta_2, \ldots\ldots, \theta_m)$ and the hypothesis H_0 : specifies the values of first $(k < m)$ parameters. The H_0 is said to be (.........). with $(m - k)$ d.f.

(a) Composite hypothesis (b) Simple hypothesis

(c) Complex hypothesis (d) Both (a) & (c)

Q.14 The population parameters whose values are not specified, are called :

(a) Consistent estimators (b) Unbiased estimators

(c) Sufficient estimators (d) Nuisance parameters

Q.15 Let X~ N(μ, σ^2) where (μ, σ^2) both are unknown. We have to test the hypothesis as :

$$H_0 : \mu = \mu_0\ \&\ \sigma^2 = \sigma_0^2 \text{ v.s. } H_1 : \mu = \mu_1, \sigma^2 = \sigma_1^2$$

The M.P. similar region W0 of size α is given as :

$$W_0 = \left\{ X \middle/ \left| \frac{f(x, \mu_1, \sigma_1^2)}{f(x, \mu_0, \sigma_0^2)} \right| \ldots\ldots\ldots\ldots \right\}$$

where k(v) is such that the conditional size of W_0 given that V = v is :

(a) $< k(v)$ (b) $= k(v)$

(c) $= \alpha$ (d) $> k(v)$

Q.16 In Q.(15), let $H_1 : \mu_1 > \mu_0$ then the C.R. say W_0 is written as :

(a) $W_0 = \{ X / (\bar{x} - \mu_0) \sqrt{\frac{n}{v}} > k_1(v)$

(b) $W_0 = \{ X / (\bar{x} - \mu_0) \frac{n}{v} > k_1(v)$

(c) $W_0 = \{ X / (\bar{x} - \mu_0) \frac{n}{v} < k_1(v)$

(d) $W_0 = \{ X / (\bar{x} - \mu_0) \sqrt{\frac{n}{v}} = k_1(v)$

where $k_1(v)$ is to be determined such that $P_{\mu_0}(W_0/v) = \alpha$.

Q.17 In Q.(16), finally, we have to write that :

$$W0 = \{ X / | \bar{x} - \mu_0 | \sqrt{\frac{n}{s}} \text{ ....} \}$$

(a) $> t_{\alpha, n,}$ (b) $> t_{\alpha, (n-1)}$

(c) $< t_{\alpha, (n-1)}$ (d) $= t_{\alpha, (n-1)}$

Q.18 The C.R. say W_0 provided in Q.(17) is independent of σ_0^2 & σ_1^2. Hence it is the (..........) of size α to test H_0 : vs H_1 :

(a) MPCR (b) UMPCR

(c) M.P. Similar region (d) U.M.P.U.C.R.

Q.19 In Q.(15) let $\mu < \mu_0$, then M.P. similar region of size α to test H_0 vs H_1 is given as :

(a) $W_0' = \{ X / | \bar{x} - \mu_0 | \sqrt{\frac{n}{s}} < -t_{\alpha(n-1)} \}$

(b) $W_0' = \{ X / | \bar{x} - \mu_0 | \sqrt{\frac{n}{s}} < -t_{\alpha(n)} \}$

(c) $W_0' = \{ X / | \bar{x} - \mu_0 | \sqrt{\frac{n}{s}} > t_{\alpha(n-1)} \}$

(d) $W_0 = \{ X / | \overline{x} - \mu | \sqrt{\frac{n}{s}} > -t_{\alpha(n)} \}$

Q.20 In Q.(17, 19), the C.R.'s (W_0 & W_0') are both independent of μ_1, hence both (W_0 & W_0') are (..........) to test :

$H_0 : \mu = \mu_0$ v.s. $H_1 : \mu > \mu_0$ and $\mu_1 < \mu_0$.

(a) M.P. similar region (b) UMP similar region

(c) M.P. critical region (d) M.P. consistent region

Q.21 Let X~ N(μ, σ^2) where both (μ, σ^2) are unknown. Suppose we have to test $H_0 : \sigma^2 = \sigma_0^2$ v.s. composite $H_1 : \sigma^2 = \sigma_1^2 > \sigma_0^2$. Then M.P. similar region say W_0 is given as :

(a) $W_0' = \{ X / \sum_i^n \left(\frac{x_i - \overline{x}}{\sigma_0} \right)^2 < \chi^2_{\alpha,(n-1)} \}$

(b) $W_0' = \{ X / \sum^n \left(\frac{x_i - \overline{x}}{\sigma_0} \right)^2 > \chi^2_{\alpha, n} \}$

(c) $W_0 = \{ X / \sum^n \left(\frac{x_i - \overline{x}}{\sigma_0} \right)^2 < \chi^2_{\alpha, n} \}$

(d) $W_0' = \{ X / \sum^n \left(\frac{x_i - \overline{x}}{\sigma_0} \right)^2 < \chi^2_{\alpha, n-1)} \}$

Q.22 Let in Q.(21), we have to test $H_0 : \sigma^2 = \sigma_0^2$ v.s. $H_1 : \sigma^2 < \sigma_0^2$. Then M.P. similar region say :

$W_0^* = \sum_i^n \left(\frac{x_i - \overline{x}}{\sigma_0} \right)^2$ is given as :

(a) $< \chi^2_{\alpha, (n-1)}$ (b) $> \chi^2_{\alpha, n}$,

(c) $< \chi^2_{\alpha, n}$, (d) $> \chi^2_{\alpha, (n-1)}$

Q.23 In Q.(21), the W_0 is same for all $\sigma_1^2 > \sigma_0^2$. Hence infact, W_0 of size α to test $H_0 : \sigma^2 = \sigma_0^2$ v.s. $H_1 : \sigma^2 = \sigma_1^2 > \sigma_0^2$ is the :

(a) M.P. similar region (b) UMP similar region

(c) M.P. critical region (d) M.P. consistent region

Q.24 In Q.(22) the W_0^* is same for all $\sigma_1^2 < \sigma_0^2$. Hence the W_0^* of size α to test $H_0 : \sigma^2 = \sigma_0^2$ v.s. $H_1 : \sigma^2 < \sigma_0^2$ is the :

(a) M.P. similar region (b) MP critical region

(c) U.M.P. similar region (d) M.P. consistent region

Q.25 Consider a hypothesis $h_0 \in H_0$ vs : Simple $H_1 : h_1 \in H_1$ such that :

$h_0 : \mu = \mu_0 : \sigma^2 = \sigma_0^2$ vs. $h_1 : \mu = \mu_1, \sigma^2 = \sigma_1^2$

The M.P. similar region say W_0 of size α to test H_0 vs. H_1 is written as :

$$W_0 = \left\{ X \Big/ \frac{f(x, \mu_1, \sigma_1^2)}{f(x, \mu_0, \sigma_0^2)} : .... \right\}$$

(a) $> k(\bar{x})$ (b) $< k(\bar{x})$

(c) $> \alpha$, (d) $< \alpha$,

where $k(\bar{x})$ is such that the conditional size of W_0 given $\bar{X} = \bar{x}$ is equal to α.

Q.26 When a null hypothesis (either simple or composite) is to be tested against a composite alternative, the power of the test should (..........). in the divergence of true parametric value from the null hypothesis.

(a) Increase with increase

(b) Increase with decrease

(c) Decrease with decrease

(d) Both (a) & (c)

Q.27 Suppose we have to test $H_0 : \theta = \theta_0$ v.s. $H_1 : \theta \neq \theta_0$ and 'W' is the chosen C.R. Then the Power $P_\theta(w)$ of the test should (.........) monotonically as θ moves farther and farther away from θ_0.

(a) Increase (b) Decrease

(c) Be constant (d) None

Q.28 Let $X \sim N(\mu, \sigma^2)$ where σ^2 is known. We have to test $H_0 : \mu = \mu_0$ v.s. $H_1 : \mu \neq \mu_0$. Then U.M.P. test of size α has the C.R. say W is written as :

(a) $W = \{ X / \bar{x} > \mu_0 + Z_\alpha \frac{\sigma}{\sqrt{n}} \}$

(b) $W = \{ X / \bar{x} > \mu_0 + Z_\alpha . \frac{\sigma^2}{\sqrt{n}} \}$

(c) $W = \{ X / \bar{x} < \mu_0 + Z_\alpha . \frac{\sigma}{\sqrt{n}} \}$

(d) $W = \{ X / \bar{x} < \mu_0 + Z_\alpha . \frac{\sigma^2}{\sqrt{n}} \}$

Q.29 The power function of the test cited in Q.(28) is written as :

$$P_\mu(w) = P_\mu \left[\frac{(\bar{x} - \mu)\sqrt{n}}{\sigma} > Z\alpha - .... \right]$$

(a) $(\mu, \mu_0) \sqrt{\frac{n}{\sigma^2}}$ (b) $(\mu, \mu_0) \frac{\sqrt{n}}{\sigma}$

(c) $(\mu, \mu_0) \frac{n}{\sigma^2}$ (d) Both (a) & (b)

Q.30 The power function $P\mu(w)$ stated in Q.(29) is the monotonically (..........). function of $(\mu_0 - \mu)$.

(a) Increasing (b) Decreasing

(c) Consistent (d) None

Q.31 The power function $P\mu(w)$ of U.M.P. test to test the $H_0 : \mu = \mu_0$ v.s. $H_1 : \mu < \mu_0$ is monotonically (..........) function of $(\mu_0 - \mu)$.

(a) Decreasing (b) Increasing

(c) Consistent (d) None

Q.32 The power function Pμ(w) of U.M.P.U. test to test $H_0 : \mu = \mu_0$ v.s. $H_1 : \mu \neq \mu_0$ is monotonically increasing function of (..........).

(a) $|\mu - \mu_0|$ (b) $|\mu_0 - \mu|$

(c) Both (a) & (b) (d) None

Q.33 The power of test function should (..........) with the increase in sample size n drawn from population.

(a) Increase (b) Decrease

(c) Be consistent (d) None

Q.34 There must be high probability of (..........) a false hypothesis based on a large sample of size n from $N(\mu, \sigma^2)$.

(a) Rejecting (b) Accepting

(c) Not accepting (d) Both (a) & (c)

Q.35 The power function of a test criterion must posses the property of :

(a) Consistency (b) Sufficiency

(c) Unbiasedness (d) None

Q.36 A sequence of tests corresponding to the C.R.'s say (Wn) is said to be (..........) if for every value of $\mu \in (\Omega\theta - \Omega\theta_0)$, the power $P_\theta(W_n)$ tends to 1.0 as $n \to \infty$.

(a) Consistent (b) Unbiased

(c) Efficient (d) Most powerful

Q.37 A consistent estimator of parameter θ provides a (..........) sequence (W_n) of the tests to test $H_0 : \theta = \theta_0$.

(a) Unbiased (b) M P, .

(c) Consistent (d) Sufficient

Q.38 A simple method of test construction related to M.L.E. method, has been suggested by (..........).

(a) Neyman-Pearson (b) Cochram-Cox

(c) Prof. C.R. Rao (d) Prof. Cramer-Rao

Q.39 Let $L(\theta / x)$ be the likelihood function of θ corresponding to a set of values $(x_1, x_2, \ldots, x_n)$. Suppose we have to test the hypothesis $H_0 : \theta = \theta_0$ v.s. $H_1 : \theta \neq \theta_0$. Then one reject H_0 if :

$L(\theta_0 / x)$ is :

(a) Too small > 0 (b) $= \alpha$

(c) $= \alpha/2$ (d) $\leq \alpha$

Q.40 In Q.(39), one should accept H0 if $L(\theta_0 / x)$ is :

(a) Too small (b) Too large

(c) Equal to α (d) $\geq \alpha$

Q.41 The test illustrated in Q.(39) should be based on C.R. say 'W' written as $W = \{ X / \lambda(x_1, x_2, ..., xn)$ where λ is the :

(a) Probability (b) Power function

(c) Likelihood ratio (d) Level of significance

Q.42 The constent λ_0 cited in Q.(41) is chosen such that the probability of error I, associated with the test must be :

(a) Equal to α (b) Greater than α

(c) Less than α (d) Equal to $\alpha/2$

Q.43 Let $X \sim N(\mu, \sigma^2)$ where μ is unknown but σ^2 is known. Suppose we have to test $H_0 : \mu = \underset{0}{\mu}$ v.s. $H_1 : \mu \neq \underset{0}{\mu}$. Then L.R. test has the C.R. say 'W' written as :

$W = \{ X / .... > \lambda_0 \}$

(a) $| \bar{x} - \underset{0}{\mu} | \dfrac{\sigma}{n} > \lambda_0$ (b) $| \bar{x} - \underset{0}{\mu} | \dfrac{\sigma}{\sqrt{n}} > \lambda_0$

(c) $| \bar{x} - \underset{0}{\mu} | \dfrac{\sigma^2}{\sqrt{n}} > \lambda_0$ (d) $| \bar{x} - \underset{0}{\mu} | \sqrt{\dfrac{\sigma}{\sqrt{n}}} > \lambda_0$

Q.44 In Q.(43), the $| \bar{x} - \underset{0}{\mu} | \dfrac{\sigma}{\sqrt{n}} \sim N(0, 1)$ and λ_0 must be chosen as :

(a) $\lambda_0 > Z_{\alpha/2}$ (b) $\lambda_0 > Z_\alpha$

(c) $\lambda_0 < Z_{\alpha/2}$ (d) None

at α% level of significance.

Q.45 Suppose in Q.(43), we have to test $H_0 : \mu = \underset{0}{\mu}$ v.s. $H_1 : \mu > \underset{0}{\mu}$. Then L.R. test has the C.R. say W_1 of size a. Then $W_1 = \{ X / (\bar{x} - \underset{0}{\mu}) \dfrac{\sigma}{\sqrt{n}} > \lambda_1 \}$ such that :

(a) $\lambda_1 = Z_{\alpha/2}$ (b) $\lambda_1 = Z_\alpha$

(c) $\lambda_1 = -Z_{\alpha/2}$ (d) None

Q.46 Suppose in Q.(43), we have to test $H_0 : \mu = \mu_0$ v.s. $H_1 : \mu < \mu_0$. Then L.R. test has the C.R. say W_2 of size α. Then, we have

$$W_2 = \{ X / \frac{|\bar{x} - \mu_0|\sqrt{n}}{\sigma} < \lambda_2 \} \text{ such that } \lambda_2 \text{ is :}$$

(a) $< -Z_\alpha$, (b) $< -Z_{\alpha/2}$

(c) $< -Z_{\alpha/2}$, (d) None

Q.47 If we have to test $H_0 : \mu = \mu_0$ against one sided alternative hypothesis then the L.R. is same as corresponding to ....

(a) M. P. test (b) U.M.P. test

(c) Randomized test (d) Consistent test

Q.48 Let $X \sim N(\mu, \sigma^2)$ where both (μ, σ^2) are unknown. Suppose we have to test $H_0 : \mu = \mu_0$ v.s. $H_1 : \mu \neq \mu_0$. Then C.R. of L R test of sized α say 'W' is given as :

$$W = \{ X / |\bar{x} - \mu_0| \frac{\sqrt{n}}{s} \text{ ....} \}$$

(a) $> t_{\alpha/2, n-1}$ (b) $< t_{\alpha, n-1}$

(c) $> t_{\alpha/2, n-1}$ (d) $< t_{\alpha, n-1}$

Q.49 Suppose in Q.(48), we have to test $H_0 : \mu = \mu_0$ v.s. $H_1 : \mu > \mu_0$. Then C.R. of LR test of size α say W_1 is given as :

$$W_1 = \{ X / (\bar{x} - \mu_0) \frac{\sqrt{n}}{s} \text{ ....} \}$$

(a) $> t_{\alpha, n-1}$ (b) $< t_{\alpha, n-1}$

(c) $> t_{\alpha/2, n-1}$ (d) $< t_{\alpha/2, n-1}$

Q.50 Suppose in Q.(48), we have to test $H_0 : \mu = \mu_0$ v.s. $H_1 : \mu < \mu_0$. Then C.R. of L. R. test of size α 'W_2' is given as :

$$W_2 = \{ X / (\bar{x} - \mu_0) \frac{\sqrt{n}}{s} \text{ ....} \}$$

(a) $> t_{\alpha, n-1}$ (b) $< t_{\alpha/2, n-1}$

(c) $< -t_{\alpha, n-1}$ (d) $> t_{\alpha/2, n-1}$

Q.51 The student's t test which is LR. test to test $H_0 : \mu = \mu_0$ v.s. $H_1 : \mu \neq \mu_0$ is defined as :

(a) $t = | \bar{x} - \mu | \frac{n}{s^2} \sim t_{\alpha/2, n-1}$.

(b) $t = (\bar{x} - \mu) \frac{n}{s^2} \sim t_{\alpha, n-1}$.

(c) $t = | \bar{x} - \mu_0 | \frac{\sqrt{n}}{s} \sim t_{\alpha/2, n-1}$.

(d) $t = (\bar{x} - \mu_0) \frac{s}{n} \sim t_{\alpha, n-1}$.

Q.52 Let x_{ij} ($i = 1, 2, ..., k, j = ni$) be k independent random samples drawn from k normal populations (μ_i, σ^2, $i = 1, 2, ..., k$) we have to test :

$H_0 : \mu_1 = \mu_2 = \mu_3 = ... = \mu_k$

$H_1 : \mu_1 \neq \mu_2 \neq \mu_3 = ... \neq \mu_k$

We assume that the variances of all nomal populations are same but unknown. Then L.R. test to test H_0 is given by F statistic defined as :

$$F = \frac{S_B^2 / (k-1)}{S_w^2 / (n-k)} = \frac{M\,S.B}{M\,S.W} \text{ which follows :}$$

(a) $F_{k, n-1}$ (b) $F_{k, n-k-1}$

(c) $F_{k-1, n-k}$ (d) $F_{k-2, n-k}$

Q.53 In Q.(52), the M.S.B. is known as 'Mean square' (..........).

(a) Between samples (b) Within samples

(c) Between populations (d) Within populations

Q.54 In Q.(52), the M.S.W. is known as "Mean square" (..........).

(a) Between samples (b) Within samples

(c) Within populations (d) Between populations

Q.55 Denoting the upper α point by $F_{\alpha, k-1, n-k}$ we have the C.R. say 'W' of L.R. test to test the hypothesis stated as Q.(52) is given as :

(a) $\frac{\text{MSB}}{\text{MSW}} > F_{\alpha/2, k-1, n-k}$ (b) $\frac{\text{MSB}}{\text{MSW}} < F_{\alpha/2, k-1, n-k}$

(c) $\frac{MSB}{MSW} > F_{\alpha, k-1, n-k}$ (d) $\frac{MSB}{MSW} < F_{\alpha, k-1, n-k}$

Q.56 In regard of power of the L.R. test, the L.R. tests are generally (..........) if an U.M.P. test exists at all.

(a) Randomised test (b) U.M.P. test

(c) Consistent test (d) none

Q.57 When null hypothesis H_0 is composite, then in that situation, the L.R. critical region will generally be (..........) but may not be universally true.

(a) Similar (b) M.P.

(c) M.P. unbiased (d) None

Q.58 Let q_θ be a MLE. of the parameter θ then $_\theta$ has (..........) variance equal to C.R. lower bound.

(a) Consistent (b) Asymptotic

(c) Minimum (d) None

Q.59 Let $\lambda = \frac{L(\theta_k - \theta_{m-k})}{L(\theta_k - \theta_{m-k})}$

Then $-2 \log \lambda$ for a hypothesis with k constraints on θ is asymptotically distributed as (..........) distribution.

(a) Non-central χ^2_{k-1}, (b) Non-central χ^2_k

(c) Central χ^2_k (d) Central χ^2_{k-1}

Q.60 The result given in Q.(59) is pioneered by :

(a) Cramer-Rao (b) J. Neyman

(c) Wald (d) Prof. Mahalanobias

Q.61 When H_0 is true in Q.(59), then $-2\log \lambda$ is distributed assymptotically as :

(a) Non-central χ^2_k (b) Central χ^2_k

(c) Non-central χ^2_{k-1}, (d) Central χ^2_{k-1}

Q.62 Under a very general assumption, the L.R. test is found to be (..........).

(a) Randomised test (b) Inconsistent test

(c) Consistent test (d) Efficient test

5.4 Key Answers

Write True/False

Q.	A.	Q.	A.	Q.	A.
1.	True	30.	True	59.	True
2.	True	31.	True	60.	True
3.	True	32.	True	61.	True
4.	True	33.	True	62.	True
5.	True	34.	True	63.	True
6.	True	35.	True	64.	True
7.	True	36.	True	65.	True
8.	True	37.	True	66.	True
9.	True	38.	True	67.	True
10.	True	39.	True	68.	True
11.	True	40.	True	69.	True
12.	True	41.	True	70.	True
13.	True	42.	True	71.	True
14.	True	43.	True	72.	True
15.	True	44.	True	73.	True
16.	True	45.	True	74.	True
17.	True	46.	True	75.	True
18.	True	47.	True	76.	True
19.	True	48.	True	77.	True
20.	True	49.	True		
21.	True	50.	True		
22.	True	51.	True		
23.	True	52.	True		
24.	True	53.	True		
25.	True	54.	True		
26.	True	55.	True		
27.	True	56.	True		
28.	True	57.	True		
29.	True	58.	True		

Fill in the Blanks

Q.	A.	Q.	A.
1.	Composite	29.	$Z_\alpha \dfrac{\sigma}{\sqrt{n}}$
2.	Null	30.	$\dfrac{n}{\sigma^2}$
3.	α	31.	Increasing
4.	$\Omega\theta_0$	32.	Increasing
5.	Similar region	33.	Monotonically, $\lvert \mu - \mu_0 \rvert$
6.	1	34.	Increase
7.	Similar	35.	Rejecting
8.	Similar region	36.	Consistency
9.	1	37.	Consistent
10.	Similar C.R.	38.	Consistent
11.	Neyman	39.	Neyman-Pearson
12.	Sufficient Statistic	40.	Small
13.	Composite, d.f.	41.	Large
14.	Nuisense Parameter	42.	Likelihood
15.	$f(x, \mu_0, \sigma_0^2)$	43.	$L(\theta_0 / x_1, x_2, ..., x_n)$
16.	C.R., α	44.	α
17.	$t_{\alpha, (n-1)}$	45.	$\lvert x - \mu_0 \rvert$
18.	Similar region	46.	N (0, 1), $Z_{\alpha/2}$
19.	$-t_{\alpha, (n-1)}$	47.	$(\bar{x} - \mu_0)$
20.	Similar region	48.	$< -Z_\alpha$
21.	$\chi^2_{k}, (n-1)$	49.	UMP test
22.	$\chi^2_{k}, (n-1)$	50.	UMPU
23.	U.M.P.	51.	$t_{\alpha/2, (n-1)}$
24.	U.M.P.	52.	$t_{\alpha, (n-1)}$
25.	$f(x, \mu_0, \sigma_0^2)$	53.	$-t_{\alpha, (n-1)}$
26.	Increase, Increase	54.	$\lvert \bar{x} - \mu \rvert \dfrac{\sqrt{n}}{s}$
27.	Increase		
28.	Power		

55. $\sim F_{(k-1),(n-k)}$
56. Between, Within
57. $F_{\alpha,(k-1),(n-k)}$
58. U.MP test
59. Unbiased
60. Similar
61. Assymptotic
62. Non-central χ_k^2
63. Wald
64. Central χ_k^2
65. Consistent

Multiple Choice Questions and Answers

Q.	A.	Q.	A.	Q.	A.
1.	(d)	22.	(a)	43.	(b)
2.	(c)	23.	(b)	44.	(a)
3.	(a)	24.	(c)	45.	(b)
4.	(a)	25.	(a)	46.	(a)
5.	(a)	26.	(d)	47.	(b)
6.	(a)	27.	(a)	48.	(a)
7.	(a)	28.	(a)	49.	(a)
8.	(c)	29.	(d)	50.	(c)
9.	(b)	30.	(a)	51.	(c)
10.	(a)	31.	(b)	52.	(c)
11.	(b)	32.	(c)	53.	(a)
12.	(d)	33.	(a)	54.	(b)
13.	(d)	34.	(d)	55.	(c)
14.	(d)	35.	(a)	56.	(b)
15.	(d)	36.	(a)	57.	(a)
16.	(a)	37.	(c)	58.	(b)
17.	(b)	38.	(a)	59.	(b)
18.	(c)	39.	(a)	60.	(c)
19.	(a)	40.	(b)	61.	(b)
20.	(b)	41.	(c)	62.	(c)
21.	(d)	42.	(a)		

❑❑❑

CHAPTER - 6

Sequential Analysis

6.1 Write True/False

Q.1 A procedure to make an inference about the unknown values of parameters of one or more than one distribution in which the sample size say n is a ramdom variable, is known as "Sequential procedure."

Q.2 The principal feature of sequential procedure is a 'Sampling Scheme' which provides a rule under which one decides at each stage of sampling whether to stop or continue the sampling work.

Q.3 The decision about stopping or continuing the sampling work, is taken in the light of observations already obtained.

Q.4 A rule which tells us when to stop the sampling work in a sequential procedure is known as "stopping rule."

Q.5 A rule which tells us about the type of action (decision) to make after the sampling work is stopped, is known a "stopping rule".

Q.6 Let here be a sequence of rv's say $X_1, X_2, ..., X_m$. The mth value x_m is an observation on X_m. Let $X_m = (x_1, x_2, ..., x_m)$ and let χ_m be the set of all values of X_m. Then stopping rule is equivalent to a set $A_m \in \chi_m$ for $m = 1, 2, ...$ such that the sampling is to be stopped as soon as $X_m \in A_m$.

Q.7 The action rule specifies that for each m, and for each possible value x_m of X_m which decision has to be taken if sampling is stopped after m observations taken.

Q.8 Suppose we have to test a hypothesis H_0 about the parameter θ then action rule will dictate whwich of two actions a_0 (acceptance of H_0) and a_1 (rejection of H_0) is to be taken after stopping of sampling work.

Q.9 Let us consider two sets A_{0m} and A_{1m} ($m = 1, 2, ...$) where A_{0m} and A_{1m} are disjoint sets of χ_m. Then take action a_0 if :

$x_m \in A_{0m}$ ...(1)

Q.10 In Q.9, take action a_1 if $x_m \in A_{1m}$.

Q.11 To judge the relative merits of two or more sequential tests, the two criteria say (i) O.C. function and (ii) A.S.N. function are considered in study.

Q.12 The O.C. function is defined as "The probability of accepting H_0 when θ_0 is true value of the parameter "θ" and is the function of θ.

Q.13 The OC function is generally denoted by $L(\theta)$ and is defined as :

$L(\theta) = P_\theta (A_0)$ ...(2)

Q.14 The OC function is closely related to the notion of the power function in the traditional theory of test of hypothesis.

Q.15 For any θ not consistent with H_0, the power is defined as :

Power of the test $= 1 - L(\theta)$ ...(3)

Q.16 If θ_0 is consistent with H_0 then $L(\theta_0)$ denotes the "Probability of taking correct decision."

Q.17 If θ_0 is consistent with H_0 then :

$1 - L(\theta_0)$ = Probability of type one error. ...(4)

Q.18 If θ_0 is consistent with H_0 then OC function is considered more favourable for higher value of $L(\theta_0)$.

Q.19 If θ is not consistent with H0 then OC function is considered less favourable for higher value of $L(\theta_0)$.

Q.20 Let N be the number of observations actually taken according to a sequential procedure. Then we call a sequential procedure favourable if it requires small value of N on the average.

Q.21 The average value of N (the number of observations taken in a sequential provedure) is known as the average sample number (A.S.N.) of the sequential procedure.

Q.22 The A.S.N. say 'N' depend on parameter θ and is denoted by $E_\theta(N)$.

Q.23 Every sequential procedure is associated with an operating character (O.C.) function.

Q.24 Every sequential procedure is associated with an "Average sample number" function.

Q.25 The O.C. function in a sequential procedure describes the extent of taking correct decision about the objective of the test.

Q.26 The ASN function represents the price to be paid in reaching a decision in terms of number of observations required by a sequential test procedure.

Q.27 The sequential probability ratio test (S.P.R.T.) developed by 'Wald' is the "Best sequential test procedure."

Q.28 If the null hypothesis $H_0 : \theta = \theta_0$ is true then for a positive integer m, the "Probability" or "Probability density" that the observations $x_1, x_2, \ldots\ldots, x_m$ are required to take a decision in SPR test is given as :

$$f_{0m} = \prod_{i=1}^{m} f_{\theta_0}(x_i) \qquad \ldots(5)$$

Q.29 In a situation described in Q.(28), if the null hypothesis H_0 is false and alternative H_1 is true then the probability for the sample $(x_1, x_2, \ldots\ldots, x_m)$ required to take a decision in 'SPPT' is given as :

$$f_{1,m} = \prod_{i=1}^{m} f_{\theta_1}(x_i) \qquad \ldots(6)$$

Q.30 The SPRT for testing H_0 against H_1, is defined in terms of the ratio λ written as :

$$\lambda = \frac{f_{1m}}{f_{0m}} \qquad \ldots(7)$$

Q.31 Let there two constents A and B such that $0 < B < 1$ and $A > 1$. Then in applying SPRT, to continue by taking next observations as long as :

$$B < \frac{f_{1m}}{f_{0m}} < A \qquad \ldots(8)$$

Q.32 In problem described in Q.(31) stop to take further observations as soon and take the decision as.

$$\text{Accept } H_0 \text{ if } \frac{f_{1m}}{f_{0m}} \leq B \qquad \ldots(9)$$

Q.33 In Q.(32), take the decision as below :

$$\text{Reject } H_0 \text{ if } \frac{f_{1m}}{f_{0m}} \geq A \qquad \ldots(10)$$

Q.34 In Q.(30), taking logrithm, we have

$$\log \lambda = \log \frac{f_{1m}}{f_{0m}} = \sum_{i}^{m} \log \frac{f_{\theta_1}(x_i)}{f_{\theta_0}(x_i)} = \sum_{i}^{m} Z_i \qquad \ldots(11)$$

where $Z_i = \log (f_{\theta_1} (x) / f_{\theta_0}(x))$

Q.35 Using the quantity Z_i as defined in equation (11), we take the decision as :

continue taking observations as long as :

$$\log B < \sum_{i}^{m} Z_i < \log A \qquad \ldots(12)$$

Q.36 In problem discussed in Q.(35) stop taking further obervation and accept

H_0 if $\sum^{m} Z_i < \log B$

Q.37 Stop taking further observations and accept H_0 if :

$$\sum_{i}^{m} Z_i \leq \log B \qquad \ldots(14)$$

Q.38 Stop taking further observations and accept H_1 (i.e. reject H_0) if :

$$\sum_{i=1}^{m} Z_i \geq \log A \qquad \ldots(15)$$

Q.39 Let us denote the event A_0 that ultimately H_0 is accepted then we have :

$$P_{\theta_0} (A_\theta) \geq \frac{1}{B} P_{\theta_1} (A_\theta) \qquad \ldots(16)$$

Q.40 Let A_1 denotes the event that ultimately A_1 is accepted then we have :

$$P_{\theta_0} (A_1) \leq \frac{1}{A} P_{\theta_1} (A_1) \qquad \ldots(17)$$

Q.41 Let α' be the probability of error of "Rejecting H_0 when it is true and β' be the probability of error of accepting H_0 when H_1 is true then we can write :

$$P_{\theta_0} (A_1) = \alpha' \qquad \ldots(18)$$

Q.42 We can write :

$$P_{\theta_1} (A_0) = \beta' \qquad \ldots(19)$$

Q.43 In sequential test procedure, we can write :

$$B \geq \frac{\beta'}{(1-\alpha')} \quad \text{...(20)}$$

Q.44 In sequential test procedure we can write that :

$$A \leq \frac{(1-\beta')}{\alpha'} \quad \text{...(21)}$$

Q.45 In sequential test procedure, we can write the relation among (B, β and α) as :

$$B = \beta / (1-\alpha) \quad \text{...(22)}$$

Q.46 In sequential test procedure, we can write the relation among (A, α, and β) as :

$$A = (1-\beta) / \alpha \quad \text{...(23)}$$

Q.47 The equations (22 & 23) jointly determine the constants (A & B) in terms of errors (α & β).

Q.48 In sequential test procedure, we can write the relation among error probabilities (α, β) & ($\alpha 1$, β_1) as :

$$(1-\alpha)\beta' \leq \beta(1-\alpha') \quad \text{...(24)}$$

Q.49 In sequential test procedure, we have the following relation between (α, β) and (α_1, β_1) as :

$$(1-\beta)\alpha' < \alpha(1-\beta') \quad \text{...(25)}$$

Q.50 We can write the relation in case of SPRT as :

$$\alpha' \leq \frac{\alpha}{(1-\beta)} \quad \text{...(26)}$$

Q.51 In SPRT, the following relation exists :

$$\beta' < \frac{\beta}{(1-\alpha)}$$

Q.52 When the prescribed level of probabilities of wrong decision are (α, β) then for practical purpose, we may write :

$$A = \frac{\alpha}{(1-\beta)} \quad \text{...(27)}$$

Q.53 When prescribed level of probabilities of wrong decision are (α, β), then for practical purpose, we may write :

$$B = \frac{\beta}{(1-\alpha)} \quad \text{...(28)}$$

Q.54 The SPRT procedure is a test for testing a simple null hypothesis $H_0 : \theta = \theta_0$. against the simple alternative hypothesis $H_1 : \theta = \theta_1$.

Q.55 In SPRT, the statistician generally make study about the nature of operating characteristic (O.C.) function $L(\theta)$ of the test over the entire parametric space.

Q.56 Suppose there exists a function $h(\theta) \neq 0$, of the parameter θ then we have :

$$\int_{-\infty}^{\infty} \left[\frac{f_{\theta_1}(x)}{f_{\theta_0}(x)} \right]^{h(\theta)} f_\theta(x)\, dx = 1 \qquad \ldots(29)$$

Q.57 In situation (problem) discussed in Q.(56), we write :

$$g_\theta(x) = \int_{-\infty}^{\infty} \left[\frac{f_{\theta_1}(x)}{f_{\theta_0}(x)} \right]^{h(\theta)} f_\theta(x) \qquad \ldots(30)$$

then $g_\theta(x)$ may be regarded as "Probability density function."

Q.58 Let us consider the following hypothesis.

H : f_θ is the true p.d.f of X against.

H* : g_θ is the true p.d.f of X.

Then under the SPRT for testing H_0 against H*, we continue to take observations as long as :

$$B^{h(\theta)} < \prod_{i=1}^{m} \frac{g_\theta(x_i)}{f_\theta(x_i)} < A^{h(\theta)} \qquad \ldots(31)$$

in case if $h(\theta) > 0$.

Q.59 For the problem discussed in Q.58, we accept H_0 if :

$$\prod_{i}^{m} \frac{g_\theta(x_i)}{f_\theta(x_i)} \leq B^{h(\theta)} \qquad \ldots(32)$$

Q.60 For the problem discussed in Q.58, we reject H_0 and accept H* if :

$$\prod_{i=1}^{m} \frac{g_\theta(x_i)}{f_\theta(x_i)} \geq A^{h(\theta)} \qquad \ldots(33)$$

Q.61 Considering $(A^{h(\theta)}) > 0$, (α' & β') as defined earlier, we can write :

$$A^{h(\theta)} \leq \frac{(1-\beta')}{\alpha'} \quad \ldots(34)$$

Q.62 Considering ($B^{h(\theta)}$, α' & β') as reported earlier, one can write :

$$B^{h(\theta)} \geq \frac{\beta'}{(1-\alpha')} \quad \ldots(35)$$

Q.63 When the excess of $\prod_{i=1}^{\infty} \frac{g_\theta(x_i)}{f_\theta(x_i)}$ over the boundary A & B at the termination of the experiment is negligible; the equality sign holds in an approximate sense and we can write :

$$A^{h(\theta)} \cong \frac{(1-\beta')}{\alpha'} \quad \ldots(36)$$

Q.64 In situation discussed as in Q.63, we can write :

$$B^{h(\theta)} \cong \frac{\beta'}{(1-\alpha')} \quad \ldots(37)$$

Q.65 In SPRT procedure, we can write the relation :

$$\alpha' \cong \frac{(1 - B^{h(\theta)})}{(A^{h(\theta)} - B^{h(\theta)})} \quad \ldots(38)$$

Q.66 In applying SPRT procedure, since $\alpha' = a - L(\theta)$, we can write :

$$L(\theta) \cong \frac{A^{h(\theta)} - 1}{A^{h(\theta)} - B^{h(\theta)}} \quad \ldots(39)$$

Q.67 Let X be a discrete r.v. with pmf written as :

$$f_\theta(X) = \begin{cases} \theta^x (1-\theta)^{1-x}, \text{ for } x = 0, 1 \\ 0, \text{ otherwise} \end{cases}$$

Let us consider $h(\theta) \cong h$ as a parameter, we can write :

$$\theta = \frac{1-\left(\frac{1-\theta_1}{1-\theta_0}\right)^h}{\left(\frac{\theta_1}{\theta_0}\right)^h - \left(\frac{1-\theta_1}{1-\theta_0}\right)^h} \quad \ldots(40)$$

Q.68 Considering A ≅ (1 – β) / α and B = β / (1 – α), we can write OC. function L(θ) in applying SPRT procedure, as :

$$L(\theta) \cong \frac{\left(\frac{1-\beta}{\alpha}\right)^h - 1}{\left(\frac{1-\beta}{\alpha}\right)^h - \left(\frac{\beta}{1-\alpha}\right)^h} \qquad \ldots(41)$$

Q.69 For any chosen arbitrary value of h, the points with co-ordinate (θ, L(θ)) provided in equations (40, 41) respectivelly, will be a point on the OC function.

Q.70 In SPRT procedure, one can draw the OC function easily by plotting a sufficient large number of co-ordinate (θ, L(θ)) on graph paper.

Q.71 Let Z1, Z2, ... are identically distributed r.v.'s. Then an approximate formula for the A.S.N. function is given as :

$$E_\theta(N) = \frac{L(\theta)\log B + [1 - L(\theta)]\log A}{E_\theta(Z_1)} \qquad \ldots(42)$$

Q.72 Consider a class of sequential tests of a given strength (α, β). Then a test is said to be more efficient than the other test of the class if the former needs a smaller number of observations on the average than that of later.

Q.73 If there exists a test in SPRT procedure for which E_{θ_0}(N) & E_{θ_1}(N) are smaller than that of any other test in the class, then the former test may be called as "An optimum Test."

Q.74 Let us denote the minimum value of E_{θ_0}(N) by N_0 (α, β) in the class and minimum value of E_{θ_1}(N) by N_1 (α, β). Then the efficiency of a given test in the class under H_0 is defined by the ratio λ_0 written as :

$$\lambda_0 = \frac{N_0(\alpha, \beta)}{E_{\theta_0}(N)} \qquad \ldots(43)$$

Q.75 As in case of Q.74, the efficiency of any test in the calss under H_1 is defined by the ratio λ_1 written as :

$$\lambda_1 = \frac{N_1(\alpha, \beta)}{E_{\theta_1}(N)} \qquad \ldots(44)$$

Q.76 The efficiences λ_0 and λ_1 defined under H_0 and H_1 respectively are in the limit (0, 1) ie.

$0 < \lambda_0 < 1$ and $0 < \lambda_1 < 1$ ...(45)

Q.77 An optimum test in SPRT procedure has the efficiency say, e, is equal to 1.0 under the situations H_0 and H_1 both. In notation, we can write it as :

$e = 1.0$ ...(46)

Q.78 The statistician Wald has proved that the efficiency e under SPRT procedure is very nearly equal to 1.0 under H_0 and H_1 both i.e.

$e \to 1.0$ under H_0 & H_1 ...(47)

Q.79 If the value of θ lies as $(\theta_0 < \theta < \theta_1)$, the ASN value $E_\theta(N)$ of the SPRT may exceed the sample size n of an equivalent non-sequential plan with the same error I and Error II probabilities. i.e.

$E_\theta(N) \geq n$ ...(48)

Q.80 The theoretical and empirical study made indicates that the distribution of 'N' for the SPRT is negatively skewed.

Q.81 Suppose that the simple null hypothesis $H_0 : \theta = \theta_0$ is tested against one sided composite hypothesis $H_1 : \theta > \theta_0$. Then the zone for acceptance of H_0 may be said to constitute the single value θ_0.

Q.82 The degree of preference to reject H0 generally increases with the increase of θ in the domain $\theta > \theta_0$.

Q.83 We can find a value $\theta_1 > \theta_0$ such that the acceptace of H_0 is an error of practical importance whenever $\theta > \theta_1$.

Q.84 When θ lies as $(\theta_0 < \theta < \theta_1)$, the acceptance of H_0 is an error of no particular importance.

Q.85 The $\theta > \theta_1$ may be said to constitute the zone of preference of rejecting H_0.

Q.86 The limit $(\theta_0 < \theta < \theta_1)$ constitute the zone of indifference.

Q.87 If the random variable X has a normal, binomial or Poisson distribution, the SPRT test of strength (α, β) to test $H_0 : \theta = \theta_0$ vs $H_1 : \theta = \theta_1$ satisfies the required assumptions.

Q.88 The $L(\theta)$ monotonically decreases in θ for $\theta \geq \theta_1$.

Q.89 In all cases, the SPRT for testing $H_0 : \theta = \theta_0$ against $H_1 : \theta = \theta_1$, provides a satisfactory solution to the problem.

Q.90 Suppose we have to test simple $H_0 : \theta = \theta_0$ against composite alternative $H_1 : \theta \neq \theta_0$ then the zone for the acceptance of H_0 may be said to constitute the single value θ_0.

Q.91 The degree of preference to reject H_0 generally will increase with the increase of θ in the domain $\theta < \theta_0$.

Q.92 We can find a value $\theta_1 < \theta_0$ such that the acceptance of H_0 is an error of practical importance whenever :

$\theta < \theta_0$. ...(50)

Q.93 The $\theta < \theta_0$ may be said to constitute the zone of preference of rejecting H_0.

Q.94 For $\theta < \theta_0$, the $L(\theta)$ monotonically decreases in θ as θ decreases.

Q.95 Suppose we have to test composite $H_0 : \theta \leq \theta_0$ against composite $H_1 : \theta_1 \geq \theta_0$ where $(\theta_0 < \theta_1)$ then we impose the following restriction :

$1 - L(\theta) \leq \alpha$ if $\theta < \theta_0$ ...(51)

Q.96 In Q.(95), the following restriction is imposed :

$L(\theta) \leq \beta$ if $\theta \geq \theta_1$ ...(52)

Q.97 Under the restrictions (51 & 52), the SPRT for testing $H_0 : \theta = \theta_0$ v.s. $H_1 : \theta = \theta_1$, will generally provide a satisfactory solution to the problem.

Q.98 To the problem of testing composite null hypothesis H_0 against composite alternative H_1, we consider an approach suggested by "Wald".

Q.99 Let $X \sim N(\mu, \sigma^2)$ where (μ, σ^2) are both unknown. Suppose we have to test the hypothesis as :

$H_0 : \mu = \mu_0$ v.s. $H_1 : \mu \neq \mu_0$.

and $|\mu - \mu_0| / \sigma$. is small then acceptance of H_0 will not be a "serious Error".

Q.100 The importance of error committed in accepting H_0 will increase with the increase of $|\mu - \mu_0| / \sigma$.

Q.101 In Q.99, we can find a value $\delta > 0$ such that the acceptance of H_0 is considered as "an error of practical importance" only when :

$$\frac{|\mu - \mu_0|}{\sigma} \geq \delta \quad ...(53)$$

Q.102 Suppose we have to test the hypothesis as :

$H_0 : \theta \in \Omega\, \theta_0$ v.s. $H_1 : \forall\, \theta \in \Omega\, \theta_1$.

then we have to report Ω_{θ_0} as :

$$\Omega_{\theta_0} = \{ (\mu, \sigma^2) / \mu = \mu_0 \} \quad ...(54)$$

Q.103 In Q.102, the parametric space Ω_{θ_1} will be reported to be as :

$$\Omega_{\theta_1} = \{ (\mu, \sigma^2) / \mid \mu = \underset{0}{\mu} \mid = \delta\sigma \} \qquad \ldots(55)$$

Q.104 For the problem stated in Q.99, Wald choose the weight function 6_0 under H_0 of σ such that :

$$6_0(\sigma) = \begin{cases} \frac{1}{C} \; if \; \sigma \leq C \\ 0, \text{ otherwise, where } c \to \infty \end{cases} \qquad \ldots(56)$$

Q.105 As in Q.104, Wald chooses the weight function 6_1 under H_1 of σ such that :

$$6_1(\sigma) = \begin{cases} \frac{1}{2C} \text{ for } \sigma < C \\ 0, \text{ otherwise} \end{cases} \qquad \ldots(57)$$

Q.106 For fixed 'c' let (g_{0m}, g_{1m}) be the probability under (H_0, H_1) respectively. Wald has shown that the probability ratio :

$$\lambda = \frac{g_{1m}}{g_{0m}} \qquad \ldots(58)$$

for $c \to \infty$, is strictly increasing function of :

$$\frac{\mid \bar{x} - \underset{0}{\mu} \mid \sqrt{n}}{s}.$$

Q.107 The test procedure stated in Q.106, is called the sequential "t-test" and may be illustrated as follows.

Continue to take observation as long as :

$$B < \frac{g_{1n}}{g_{0n}} < A \qquad \ldots(59)$$

Q.108 As illustrated in Q.107, stop to take observation as soon as the condition mentioned in (59) fails and accept H_0 if :

$$\frac{g_{1n}}{g_{0n}} \leq B \qquad \ldots(60)$$

Q.109 In situation illustrated in Q.108, we reject H_0 if :

$$\frac{g_{1n}}{g_{0n}} \geq A. \qquad \ldots(61)$$

Q.110 In order to make $\alpha(\theta) < \alpha$ in Ω_{θ_0} and $\beta(\theta) < \beta$ in Ω_{θ_1}, we take (A & B) as follows :

$$A = \frac{(1-\beta)}{\alpha}, B = \frac{\beta}{(1-\alpha)} \qquad ...(62)$$

Q.111 Let us consider the sequence of statistics T_1, T_2, ... such that :
$T_n = T_n (X_1, X_2, ..., X_n)$.
For each n, T_n has a distribution that depends on θ through λ(θ) alone. Then the hypothesis $H_0 : \lambda(\theta) = \lambda_0$ against $H_1 : \lambda(\theta) = \lambda_1$ is tested through SPRT corresponding to the sequence $\{T_n\}$.

Q.112 Let X~ N(μ, σ²) where both μ & σ² are unknown. Then we can find a sequential test to test the :

$$H_0 : \frac{\mu}{\sigma} = \lambda_0 \text{ v.s. } H_1 : \frac{\mu}{\sigma} = \lambda_1.$$

Q.113 Consider the random variable y_i defined as :

$$y_i = \frac{X_i}{|X_i|}, i = q, 2, ... \qquad ...(63)$$

Q.114 The SPR test is advised to apply onthe sequence $\{y_i\}$.

Q.115 In Q.113, the SPR test may be expressed in terms of "Student's t statistic".

Q.116 As illustrated in Q.112, the SPR test procedure will be based on the probability ratio η defined as :

$$\eta = \frac{f_{\lambda_1 n}(y_1, y_2, ..., y_n)}{f_{\lambda_0 n}(y_1, y_2, ..., y_n)} \qquad ...(64)$$

Q.117 In case if $\lambda_0 < \lambda_1$, the ratio η defined in Q.116, is monotonically increasing function of t_n.

Q.118 If $\lambda_0 < \lambda_1$, the SPR test procedure may be stated as :
Continue sampling as long as :

$$b_n < t_n < a_n \qquad ...(65)$$

Q.119 In following S P R test procedure, the sampling is stopped as soon as either inequality fails to hold and accept H_0 if :

$$t_n \leq b_n \qquad ...(66)$$

Q.120 In following S.P.R. test procedure, we reject H0 if :

$$t_n \geq a_n \qquad ...(67)$$

Q.121 In applying SPR test procedure, the sequence { T_n } has one to one function of statistic - t defined as :

$$t = \frac{\overline{X}_n}{\sqrt{\sum_i (X_i - \overline{X}_n)^2}}$$

and it has "non-central t" distribution.

Q.122 A sequential procedure may also be used for the estimation of a function of the "Parameters".

Q.123 In general, it is not easy to determine the "most appropriate estimator" or to formulate a rule to terminate the "Sampling work."

Q.124 A sequential test procedure is defined by a sequence $\delta = \{\psi_n, T_n\}$, where ψ_n represents the stopping rule and T_n as the estimator to be used in case if the experiment stops at the nth stage.

Q.125 Let

$$\psi_n (X_1, X_2, \ldots\ldots, X_n) = \begin{cases} 1 \text{ if } N = n \\ 0 \text{ if } N \neq n \end{cases}$$

and in case N = n, the sequential procedure requires that the parametric function $\lambda(\theta)$ will be estimated by :

$$t_n = T_n(x_1, x_2, x_3, \ldots\ldots, x_n) \qquad \ldots(69)$$

Q.126 The sequential procedure will be terminated with probability 1.0 $\forall\ \theta \in \Omega\theta$ if :

$$\sum_{1 \le n < \infty} \psi_n = 1 \qquad \ldots(70)$$

Q.127 Let T_N be an unbiased estimator of parametric function $\lambda(\theta)$ then under certain regularity conditions to be held, Wolfowitz has given the following result :

For all $\theta \in \Omega\ \theta$, we have :

$$\text{Var}_\theta (T_N) \ge \frac{(\lambda^1(\theta))^2}{E_\theta\left[\frac{\partial}{\partial\theta} \log f_{\theta, N} (X_1, X_2, \ldots\ldots, X_n\right]^2}$$

Q.128 In sequential test procedure we have the following result :

$P_\theta\ [N(S) > n\ ] \to 0$ as $n \to \infty\ \forall\ \sigma$.

6.2 Fill in the Blanks

Q.1 A procedure to make an inference about the unknown values of parameters of one or more than one distribution in which the sample size n is random variable, is known as (..........).

Q.2 The principal feature of sequential procedure is a ".........." which provides a rule under which one decides at each stage of sampling either to stop or continue the sampling work.

Q.3 The decision about stopping or continuing the (..........) work, is taken in the light of observations already obtained.

Q.4 A rule which tells us about the type of action (decision) to make after the sampling work is stopped is known as (..........).

Q.5 A rule which tells us when to stop the (..........) in a sequential procedure is known as stopping rule.

Q.6 Let there be a sequence of r.v.'s say $X_1, X_2, \ldots\ldots X_m$. Then m^{th} value x_m is an observation of X*m*. Let $X_m = (x_1, x_2, \ldots\ldots, x_m)$ and let χ_m be the set of all values of X_m. Then stopping rule is equivalent to a set $A_m \in \chi_m$ for (m = 1, 2, 3, ...) such that the (..........) is to be stopped as soon as $X_m \in A_m$.

Q.7 Action rule specifies that for each m and for every possible vaule x_m of X_m, which decision has to be taken if (..........) is stopped after m observative taken.

Q.8 Suppose we have to test a hypothesis H_0 about the parameter θ then action rule will dictate which of two actions a_0 (acceptance of H_0) and a_1 (rejection of H_0) is to be taken after the (..........) is stopped.

Q.9 Let us consider two sets A_{0m} and A_{1m} $(m = 1, 2, \ldots\ldots)$ where A_{0m} & A_{1m} are disjoint sets of χ_m. Then take action 'a_0' if (..........)

Q.10 In Q.(9) take action 'a_1' if (..........).

Q.11 To judge the relative merits of two or more sequential tests, the two criterion say (..........) and A.S.N. function are considered in the study.

Q.12 The O.C. function is defined as the probability of (..........) when θ_0 is the true value of the parameter θ and is the function of θ.

Q.13 The O.C. function is closely related to the notion of the (.........) in the traditional theory of test of hypothesis.

Q.14 For any θ not consistent with H_0, the power is defined as :

Power of the test = (..........).

where $L(\theta)$ denotes the O.C. function.

Q.15 If θ is consistent with H_0 then $L(\theta)$ denotes the probability of taking (..........).

Q.16 If θ is consistent with H_0 then $(1-L(\theta))$ denotes the probability of type (.........).

Q.17 If θ is (..........) with H_0, then O.C. function is considered more favourable for higher value of $L(\theta)$.

Q.18 If θ is (..........) with H_0, then O.C. function is considered less favourable for higher value of $L(\theta)$.

Q.19 Let N be the number of observations actually taken according to sequential procedure. Then we call the (..........) procedure favourable if it requires a small value of N on the average.

Q.20 The average value of N, (the number of observations taken in a sequential procedure) is known as the (..........) of the sequential procedure.

Q.21 The A.S.N. say 'N' depend on (..........) θ and is denoted by $E_\theta(N)$.

Q.22 Every sequential procedure is associated with an .... (O.C.) function.

Q.23 Every sequential procedure is associated with an ".........." (AS.N.) function.

Q.24 The O.C. function in a sequential procedure describes the extent of taking (..........) about the objective of the test.

Q.25 The ASN function represents the price to be paid in reaching a decision in terms of number of observations required by a (.........) test procedure.

Q.26 The sequential probability ratio test (S.P.R.T.) developed by (..........) is the "Best sequential test procedure."

Q.27 If the null hypothesis $H_0 : \theta = \theta_0$ is true then for a positive integer m, the "Probability or Probability density" that the observations $(x_1, x_2, \ldots.., x_m)$ are required to take a decision in (..........) is given as :

$$f_{0m} = \prod_{i=1}^{m} f_{\theta_0}(x_i)$$

Q.28 In a situation described in Q.(27), if H_0 is false and H_1 is true then the probability for the sample $(x_1, x_2, \ldots.., x_m)$ required to take a decision in SPST is given as :

$f_{1,m} =$ ....

Q.29 The SPRT testing H_0 against H_1 is defined in term of the ratio λ written as :

$\lambda =$ ....

Q.30 Let there be two constents A and B such that $0 < B < 1$ and $A > 1$. Then in applying SPRT to continue by taking next observation as long as, we have the inequality written as ....

Q.31 In a problem described in Q.(30), stop to take further observations as soon and take the decision as accept H_0 if ....

Q.32 In Q.(31) take the decision to reject H_0 if ....

Q.33 Let us define $\log \lambda = \sum_{i=1}^{m} Z_i$ where $Z_i = \log \frac{(f_1(x_i)}{f_0(x_i)}$. Then we continue the (..........) test by taking more observations as long as :

$$\log B < \sum^{m} Z_i < \log A.$$

Q.34 In Q.(33), stop to take further observations and accept H_0 if $\sum_{i=1}^{m} Z_i$ ....

Q.35 In Q.(33), stop to take further observations and accept H_1 if $\sum_{i}^{m} Z_i$ is ....

Q.36 Let α' be the probability of rejecting H_0 when it is true, then we can write :

P_{θ_0} (A_1) = ...

Q.37 Let β' be the probability of accepting of H_0 when H_0 is false then we can write :

P_{θ_1} (θ) = ...

Q.38 In sequential test procedure, we have the result that :

B ≥ ....

Q.39 In sequential test procedure, we have the result that :

A ≤ ....

Q.40 The results in Q.(38, 39) jointly determines the constant (A & B) in terms of error (– –, –).

Q.41 Let the probability of error I and error Ii be denoted in 'usual' and 'sequential probability ratio test' as (α, β) and (α', β') respectively then we have the following relation :

(1 – α) β′ ≤ (..........) .....

Q.42 As considered the notations in Q.(41), we can write the relation as :

(1 – β) α' ≤ (..........) .....

Q.43 In SPR test, we can write the relation among (α', α, β) as :

$\alpha' \leq$ ....

Q.44 In SPRT, we can write the relation among (β', α, β) as :

$\beta' \leq$ ....

Q.45 When in usual test of hypothesis, the probability of wrong decisions are denoted by (α, β) then for practical purpose we may write the relation as :

A = ...

Q.46 As in Q.(45), for practical purpose, can write the relation as :

B = ...

Q.47 In SPRT, the statistician generally make study about (..........) function L(θ) of the test over the entire parametric space.

Q.48 Let us consider the following hypothesis

H : f_θ is the true pd.f. of x against.

H* : g_θ is the true p.d.f. of X.

Then under the SPRT, for testing H against H*, we continue to take observations as long as :

$$B^{h(\theta)} < \prod_{i=1}^{m} \frac{g_{\theta.}(x_i)}{f_\theta(x_i)} < \text{..........}$$

where h(θ) is the function of θ.

Q.49 For the problem discussed in Q.(48) we accept H if :

$$\prod_{i=1}^{m} \frac{g_\theta(x_i)}{f_\theta(x_i)} \leq \text{..........}$$

Q.50 For the problem discussed in Q.(48), we reject H and accept H* if :

$$\prod_{i=1}^{m} \frac{g_\theta(x_i)}{f_\theta(x_i)} \geq \text{.........}$$

Q.51 Consider ($A^{h(\theta)} > 0$, α' & β') as defined earlier, we can write :

$A^{h(\theta)} \leq$ ....

Q.52 Consider ($A^{h(\theta)} > 0$, α' & β') as defined earlier, we can write :

$B^{h(\theta)} \geq$ ....

Q.53 When the excess of $\prod_{i=1}^{\infty} \frac{g_\theta(x_i)}{f_\theta(x_i)}$ over the boundary A and B at the termination of the experiment is negligible, the equality sign holds in an approximate sense and we can write :

$A^{h(\theta)} \cong$ ....

Q.54 In situation discussed in Q.(53), we can write :

$B^{h(\theta)} \cong$ ....

Q.55 In applying SPRT procedure, since $\alpha' = 1 - L(\theta)$,

$$L(\theta) \cong \frac{A^{h(\theta)} - 1}{(\ldots\ldots\ldots\ldots\ldots)}$$

Q.56 Considering $A = (1 - \beta) / \alpha$ and $B = \beta / (1 - \alpha)$, we can write O.C. function $L(\theta)$ in applying SPRT procedure as :

$$L(\theta) = \frac{(\ldots\ldots\ldots\ldots)^h - 1}{\left(\frac{1-\beta}{\alpha}\right)^h - \left(\frac{\beta}{1-\alpha}\right)^h}$$

where $h(\theta) = h$ is considered as parameter :

Q.57 In SPRT procedure, one can draw the O.C. function easily by plotting a sufficiently large number of co-ordinates (....., ...) on graph paper.

Q.58 Let $Z_1, Z_2, \ldots..$ be identically distributed r v's. Then an approximate formula for the ASN function is given as :

$$E_\theta\,(N) = \frac{L(\theta) \log B + [\,1 - L(\theta)\,] \log A}{(\ldots\ldots\ldots\ldots)}$$

Q.59 Consider a class of sequential tests of a given strength (α, β). Then a test is said to be (..........) than the other test of the class if the former needs a smaller number of observations on the average than that of latter.

Q.60 If there exists a test in SPRT procedure for which $E_{\theta_0}(N)$ & $E_{\theta_1}(N)$ are smaller than that of any other test in the class, then the former test may be called as an (..........).

Q.61 Let us denote the minimum value of $E_{\theta_0}(N)$ by $N_0(\alpha, \beta)$ and minimum value of $E_{\theta_1}(N)$ by $N_1(\alpha, \beta)$ in the class of tests. Then the efficiency of a given test in the class under H_0 is defined by the ratio λ_0 written as :

$$\lambda_0 = \frac{N_0\,(\alpha, \beta)}{(\ldots\ldots\ldots\ldots)}$$

Q.62 In Q.(61), the efficiency of any test in the class under H_1 is defined by the ratio λ_1 written as :

$$\lambda_1 = \frac{..........}{E_{\theta_0}(N)}$$

Q.63 The efficiencies (λ_0 & λ_1) defined in Q.(61, 62) under H_0 and H_1 respectively lie in the limit as :

.... $< (\lambda_0, \lambda_1) <$ ....

Q.64 An optimum test in SPRT procedure has the efficiency say 'e'. Under H_0 and H_1, then e is equal to ....

Q.65 The statistician Wald has proved that under SPRT procedure the efficiency e under H_0 & H_1 is very nearly equal to ....

Q.66 If the value θ lie as ($\theta_0 < \theta < \theta_1$), the ASN value $E_\theta(N)$ of SPRT may exceed the sample size n of an equivalent (..........) test with the same errors (I & II) probabilities i.e. $E_\theta(N) \geq n$.

Q.67 The theoretical and empirical studies indicate that the distribution of 'N' in SPRT procedure is (..........).

Q.68 Suppose $H_0 : \theta = \theta_0$ is tested against $H_1 : \theta > \theta_0$. Then zone for (..........) of H_0 may be said to constitute the single value θ_0.

Q.69 The degree of preference to reject H_0 in SPRT procedure generally (..........) with the (..........) of θ in the domain $\theta > \theta_0$.

Q.70 We can find a value $\theta_1 > \theta_0$ such that the (.........) of H_0 is an error of practical importance whenever $\theta > \theta_1$.

Q.71 When θ lies as ($\theta_0 < \theta < \theta_1$) then the (..........) of H_0 is an error of no importance.

Q.72 The $\theta > \theta_1$ may be said to constitute the zone of preference of (..........) of θ_0.

Q.73 The limit ($\theta_0 < \theta < \theta_1$) constitute the zone of (..........).

Q.74 If X~ (Binomial or Poisson or Normal) distribution, the SPRT of strength (α, β) to test $H_0 : \theta = \theta_0$ v.s. $H_1 : \theta = \theta_1$. satisfies the required (..........).

Q.75 The L(θ) monotonically (..........) in θ for $\theta > \theta_1$.

Q.76 Suppose we have to test $H_0 : \theta = \theta_0$ v.s. $H_1 : \theta < \theta_0$, then the zone for the (..........) of H_0 may be said to constitute the single value θ_0.

Q.77 The degree of preference to reject H_0 generally (.........) with the (.........) of θ in the domain $\theta < \theta_0$.

Q.78 The $\theta < \theta_0$ may be said to constitute the zone of preference of (.........) of H_0.

Q.79 For $\theta < \theta_0$, the L(θ) monotonically (.........) as θ decreases.

Q.80 Suppose we have to test $H_0 : \theta \leq \theta_0$ v.s. $H_1 : \theta \geq \theta_0$, then we impose the following restriction :

$1 - L(\theta)$ .... if $\theta < \theta_0$.

Q.81 In Q.(80), the following restriction is imposed :
$L(\theta) <$ .... if $\theta > \theta_1$.

Q.82 For the problem of testing composite H_0 against composite H_1, we consider an approach as suggested by ....

Q.83 Let $X \sim N(\mu, \sigma^2)$ where both (μ & σ^2) are unknown. Suppose we have to test $H_0 : \mu = \underset{0}{\mu}$ v.s. $H_1 : \mu \neq \underset{0}{\mu}$ and $|\mu - \underset{0}{\mu}|/\sigma$ is small then (........) of H_0 will not be a serious matter.

Q.84 The importance of error committed in accepting H_0 will (.........) with the (..........) of $|\mu - \underset{0}{\mu}|/\sigma$.

Q.85 In Q.(83). we can find a value $\delta > 0$ such that the acceptance of H0 is considered as "am error of practical importance" only when

$$\frac{|\mu - \underset{0}{\mu}|}{\sigma} \text{ .... } \delta$$

Q.86 For the problem discussed in Q.(83), Wald has chosen the weight function 6_0 under H_0 of σ such that :

$$6_0(\sigma) = \begin{cases} \dfrac{1}{C} \text{ if } \sigma \text{ .... } C \\ 0, \quad \text{otherwise where } C \to \infty \end{cases}$$

Q.87 As in Q.(86), Wald has chosen the weight function 6_1 under H_1 of σ such that :

$$6_1(\sigma) = \begin{cases} \text{..........} & \text{for } \sigma \leq C \\ 0, & \text{otherwise} \end{cases}$$

Q.88 For fixed 'C' let (g_{0m}, g_{1m}) be the probability under (H_0 & H_1) respectively Wald has shown that the probability ratio $\lambda =$ .... for $C \to \infty$, is strictly increasing function of $\dfrac{|\bar{x} - \underset{0}{\mu}|\sqrt{n}}{s}$.

Q.89 The test procedure stated in Q.(88), is called as the (..........) and is illustrated as follow :

Continue to take observation as long as :

$$B < \frac{g_{1n}}{g_{0n}} < A.$$

Q.90 As illustrated in Q.(89), stop to take observation as soon as the condition mentioned in Q.(89) fails and (........) if :

$$\frac{g_{1n}}{g_{0n}} < B.$$

Q.91 As illustrated in Q.(89), we reject H_0 if $\frac{g_{1n}}{g_{0n}}$ ....

Q.92 Let us consider the sequence of statistics T_1, T_2, ... such that $T_n = t_n$ $(x_1, x_2, ..., x_n)$. For each n, T_n has a distribution that depends on θ through (.........) corresponding to the sequence $\{ T_n \}$.

Q.93 Let X~ N(μ, σ²) where both (μ, σ²) are unknown. Then we can find a (..........) t test the :

$$H_0 : \frac{\mu}{\sigma} = \lambda_0 \text{ vs } H_1 : \frac{\mu}{\sigma} = \lambda_1.$$

Q.94 Let X~ N(μ, σ²). Consider a new r.v. Y_i defined as $Y_i = X_i / | X_i |$, $i = 1, 2,$ ....

Then the distribution of each Y_i depends on (.........) alone.

Q.95 The test advised to apply on the sequence $\{ Y_i \}$ is the (........)

Q.96 In Q.(94) the SPRT may be expressed in terms of the (..........).

Q.97 As illustrated in Q.(93), the SPRT procedure will be based on the probability ratio η defined as :

$$\eta = \frac{f_{\lambda 1m}(x_1, x_2, ..., x_n)}{.....(x_1, x_2, ..., x_n)}$$

Q.98 In case if $\lambda_0 < \lambda_1$, the ratio η defined in Q.(97) is monotonically (........) function of t_n.

Q.99 If $\lambda_0 < \lambda_1$, the SPRT procedure may be stated as :

Continue (..........) as long as :

$b_n < t_n < a_n$

where (a_n, b_n) are two constants such that $\theta < b_n < 1$, and $a_n > 1$.

Q.100 In applying SPRT procedure, the sampling work is stopped as soon as either inequality fails and accept H_0 if t_n ....

Q.101 In Q.(99), the SPRT procedure will be stopped and we reject H0 : if t_n........

Q.102 In applying SPRT procedure, the sequence $\{T_n\}$ has one to one function of "student's-t" defined as :

$$t = \frac{\overline{x}_n}{\sqrt{\sum^{n} (x_i - \overline{x}_n)^2}}$$

and this t has (........) distribution.

Q.103 A sequential procedure may also be applied to estimate a function of (..........).

Q.104 In general, it is not easy to determine the "most appropriate estimator" or to formulate a rule to terminate the (..........).

Q.105 A sequential test procedure is defined by a sequence $\delta = \{ \psi_n, T_n \}$, where ψ_n represents the stopping rule and T_n represents as an estimator to be used in case if the experiment is stopped at the .... stage.

Q.106 Let ψ_n denotes the stopping rule in SPRT procedure. Then sequential procedure will be terminated with probability 1.0 $\forall\ \theta \in \Omega_\theta$ if :

$$\sum_{1 \le n < \infty} \psi_n = \text{.........}$$

Q.107 In sequential test procedure, we have the result that :

$P\theta\ [\ N\ (S) > n\ ] \rightarrow 0$ as $n \rightarrow$ ....

6.3 Multiple Choice Questions and Answers

Q.1 A procedure to make an inference about the unknown values of parameters of one or more than one distribution in which the sample size n is random variable is known as :

(a) Randomized procedure (b) Sequential procedure

(c) Experimental procedure (d) Sampling procedure.

Q.2 The principal feature of sequential procedure is a (..........) which provides a rule under which one decides at each stage of sampling either to stop or continue the sampling work.

(a) Sampling scheme (b) Randomized scheme

(c) Experimental scheme (d) Estimation theory.

Q.3 The decision about the stopping or continuing the (..........) is taken in the light of observations already obtained.

(a) Randomized work (b) Experimental work

(c) Sampling work (d) Estimation work

Q.4 A rule which tells us about the type of action (decision) to make after the sampling work is stopped is known as :

(a) Estimation theory (b) Sampling rule

(c) Decision theory (d) Stopping rule

Q.5 A rule which tells us when to stop the (..........) in a sequential procedure is known as stopping rule.

(a) Sampling work (b) Experimental work

(c) Decision making (d) Probability theory

Q.6 Let there be a sequence of r.v.'s say $X_1, X_2, \ldots.., X_m$. Then m^{th} value x_m is an observation on Xm. Let Xm = $(x_1, x_2, \ldots.., x_m)$ and let χ_m be the set of all values of X_m. Then stopping rule is equivalent to a set $A_m \in \chi_m$ for (m = 1, 2, 3, ...) such that the (..........) is to be stopped as soon as $X_m \in A_m$.

(a) Experimental work (b) Decision making

(c) Sampling work (d) Estimation work

Q.7 An action rule specifies that for each m and for every possible value x_m of X_m, which decision has to be taken if (..........) is stopped after m observations taken.

(a) Experimental work (b) Randomized work

(c) Decision making (d) Sampling work

Q.8 Suppose we have to test a hypothesis H_0 : about the parameter θ then action rule dictate which of two actions a_0 (acceptance of H_0) and a_1 (rejection of H_0) is to be taken after the (..........) is stopped.

(a) Randomization, (b) Sampling work

(c) Experimental work (d) Decision making

Q.9 Let us consider two sets A_{0m} and A_{1m} (m = 1, 2, ...) where A_{0m} & A_{1m} are disjoint sets of X_m. Then take action a_0 if (..........) :

(a) $x_m \in A_{0m}$ (b) $x_m \notin A_{0m}$

(c) $x_m \notin A_{1m}$ (d) Both (a) & (C)

Q.10 In Q.(9), we take action a_1 if (.........).

(a) $x_m \in A_{1m}$ (b) $x_m \notin A_{0m}$

(c) $x_m \notin A_{1m}$ (d) Both (a) & (b)

Q.11 To judge the relative merits of two or more sequential test, the two criterion say (.........) and (..........) functions are considered in the study.

(a) Operating character (OC) (b) Average sample number (ASN)

(c) Power function (d) Both (a) & (b)

Q.12 The O.C. function is defined as the probability of (........) when θ_0 is the true value of the parameter θ and is the function of θ.

(a) Accepting H_0 (b) Rejecting H_0

(c) Rejecting H_1 (d) Both (a) & (c)

Q.13 The O.C. function is closely related to the notion of the (..........) in the traditional theory of test of hypothesis.

(a) Power function (b) Probability of correct decision

(c) Type two error (d) Both (a) & (b)

Q.14 For any θ not consistent with H_0, the power is defined as :

Power of the test = (..........)

(a) $L(\theta)$ (b) $1-L(\theta)$

(c) $1-L^2(\theta)$ (d) $L^2(\theta)$

where $L(\theta)$ denotes the O.C. function.

Q.15 If θ is consistent with H_0 then $L(\theta)$ denotes the probability of taking (..........).

(a) False decision (b) Right decision

(c) Correct decision (d) Both (b) & (c)

Q.16 If θ is consistent with H_0 then $(1-L(\theta))$ denotes the probability of type (.........).

(a) One error (b) Two error

(c) Error I (d) Both (a) & (c)

Q.17 If θ is (........) with H_0, then O.C. function is considered more favourable for higher value of $L(\theta)$.

(a) Consistent (b) False

(c) In consistent (d) Both (b) & (c)

Q.18 If θ is (..........) with H_0, then O.C. function is considered less favourable for higher value of $L(\theta)$.

(a) Consistent (b) Inconsistent

(c) Insufficient (d) Unbiased

Q.19 Let N be the number of observations actually taken according the sequential procedue. Then we call the (..........) procedure favourable if it requires a small value of N on the average.

(a) Randomized (b) Estimation

(c) Sequential (d) Non-sequential

Q.20 The average value of N is known as the (..........) of sequential procedure.

(a) Sample size (b) Power

(c) Efficiency (d) A.S.N.

Q.21 The A.S.N. say 'N' depends on (..........) θ and is denoted by $E_\theta.(N)$.

(a) Population parameter (b) Population constant

(c) Both (a) & (b) (d) Population coefficient $\rho\ yx$

Q.22 Every sequential procedure is associated with an (.........), O.C. function.

(a) Operating character (b) Power

(c) Estimating (d) Efficiency

Q.23 Every sequential procedure is associated with an (.........) ASN function.

(a) Power (b) Average sample number

(c) Estimating (d) Likelihood

Q.24 The O.C. function in a sequential procedure describes the extent of taking (..........) about the objective of the test.

(a) Correct decision (b) Wrong decision

(c) Power (d) Estimate

Q.25 The A.S.N. function represents the price to be paid in reaching a decision in terms of number of observations required by a (..........) test procedure:

(a) Non-sequential (b) Likelihood ratio

(c) Sequential (d) Randomized

Q.26 The sequential probability ratio test (S.P.R.T.) developed by (........) is the "Best sequential test procedure."

(a) Wald (b) J. Neyman

(c) C.R. Rao (d) Prof. R.A. Fisher

Q.27 If null hypothesis $H_0 : \theta = \theta_0$ if true then for a positive integer m, the "Probability or Probability density" that the observations $(x_1, x_2, ..., x_m)$ are required to take a decision in (.........) is given as :

$$f_{0m} = \prod_{i}^{m} f_{\theta_0}(x_i)$$

(a) S.P.R.T. (b) Randomized test.

(c) Non-sequential test (d) Traditional test

Q.28 In a situation described in Q.(27), if H_0 is false and H_1 is true then the probability for the sample $(x_1, x_2, ..., x_m)$ required to take decision in S.P.R.T. is given by :

$f_{1m} =$ ....

(a) $\prod_{i}^{m} f_{\theta_1}(x_i)$ (b) $\prod_{i}^{m} f_{\theta_0}(x_i)$

(c) $f_{\theta_0}(\bar{x})$ (d) $f_{\theta_1}(\bar{x})$

Q.29 The S.P.R.T. testing H_0 against H_1 is defined in term of ratio λ written as :

λ = ...

(a) f_{0m} / f_{1m} (b) f_{1m} / f_{0m}

(c) α/β (d) β/α

Q.30 Let there be two constants A & B such that $0 < B < 1$, and $A > 1.0$. Then in applying SPRT to continue it by taking next observation as long as we have the inequality written as :

(a) $B < \frac{f_{1m}}{f_{0m}} < A$, (b) $A < \frac{f_{1m}}{f_{0m}} < B$

(c) $B < \frac{f_{0m}}{f_{1m}} < A$, (d) $A < \frac{f_{0m}}{f_{1m}} < B$

Q.31 In a problem discussed in Q.(30), stop to take further observations as soon and take the decision as accept H_0 if :

(a) $\frac{f_{1m}}{f_{0m}} \leq A$, (b) $\frac{f_{1m}}{f_{0m}} \leq B$

(c) $\frac{f_{0m}}{f_{1m}} \leq B$, (d) $\frac{f_{0m}}{f_{1m}} \leq A$

Q.32 In Q.(31), take the decision to reject H_0 if :

(a) $\frac{f_{1m}}{f_{0m}} \geq A$, (b) $\frac{f_{1m}}{f_{0m}} \geq B$

(c) $\frac{f_{0m}}{f_{1m}} \geq A$. (d) $\frac{f_{0m}}{f_{1m}} \geq B$.

Q.33 Let us define $\log \lambda = \sum_{i}^{m} Z_i$ where $Z_i = \log (f_1(x_i) / f_0(x_i))$. Then we continue the (..........) test by taking more observations as long as :

$$\log B < \sum^{m} Z_i < \log A$$

(a) Randomized (b) S.P.R.

(c) Traditional (d) Likelihood ratio

Q.34 In Q.(33), stop to take further observations and accept H_0 if $\sum_{i}^{m} Z_i$ is :

(a) $\le \log B$, (b) $\ge \log B$,

(c) $\le \log A$, (d) $\ge \log A$.

Q.35 In Q.(33) stop to take further observations and accept H_1 if $\sum_{i}^{m} Z_i$ is :

(a) $\ge \log B$ (b) $\ge \log A$

(c) $\le \log A$ (d) $\le \log B$

Q.36 Letl α' be the probability of rejecting H_0 when it is true, then we can write : $P_{\theta_0}(A)$ equal to :

(a) α' (b) $(1-\alpha')$

(c) $(\alpha'/2)$ (d) $(1-\frac{\alpha'}{2})$

Q.37 Let β' be the probability of accepting H_0 when H_0 is false, then we can write that : $P_{\theta_1}(\theta_0) =$ ....

(a) β' (b) α'

(c) $(1-\beta')$ (d) $(1-\alpha')$

Q.38 In sequential test procedure, we have the result that :

$B \ge$ ....

(a) $\frac{\beta'}{(1-\alpha')}$ (b) $\frac{\alpha'}{(1-\beta')}$

(c) $\frac{1-\alpha'}{\beta'}$ (d) $\frac{1-\beta'}{\alpha'}$

Q.39 In sequential test procedure we have the result that $A <$ ....

(a) $\frac{1-\beta'}{\alpha'}$ (b) $\frac{\alpha'}{1-\beta'}$

(c) $\frac{1-\alpha'}{\beta'}$ (d) $\frac{\beta'}{1-\alpha'}$

Q.40 Let the probability of error I and error II be denoted in traditional and sequential probability ratio test as (α, β) and (α', β') respectively then we have the following relation , $(1-\alpha)\,\beta' \le$.....

(a) $\frac{1-\alpha'}{\beta}$ (b) $\frac{1-\beta}{\alpha'}$

(c) $\frac{1-\alpha'}{\beta'}$ (d) $\frac{1-\beta'}{\alpha'}$

Q.41 As considered the notation in Q.(40), we can write relation as : $(1-\beta)\,\alpha' \leq$ ....

(a) $\frac{1-\beta'}{\alpha}$ (b) $\frac{1-\alpha}{\beta'}$

(c) $(1-\alpha')\,\beta$ (d) $(1-\beta)\,\alpha$

Q.42 In SPRT, we can write the relation among (α', α, β) as (α' ....).

(a) $\leq \frac{\alpha}{1-\beta}$ (b) $\leq \frac{1-\beta}{\alpha}$

(c) $\geq \frac{\alpha}{1-\beta}$ (d) $\geq \frac{1-\beta}{\alpha}$

Q.43 In SPRT, we can write the relation among (β', α, β) as : ($\beta' \leq$ .....).

(a) $\frac{\beta}{1-\alpha}$ (b) $\frac{1-\alpha}{\beta}$

(c) $\frac{1-\beta}{\alpha}$ (d) $\frac{\alpha}{1-\beta}$

Q.44 When in traditional test of hypothesis, the probability of wrong decisions are denoted by (α, β), then for practical purpose, we may write the relation as :
A = ...

(a) $\frac{\alpha}{1-\beta}$, (b) $\frac{1-\beta}{\alpha}$

(c) $\frac{\beta}{1-\alpha}$ (d) $\frac{1-\alpha}{\beta}$

Q.45 In Q.(45), we can write the relation that : B = ...

(a) $\frac{1-\alpha}{\beta}$ (b) $\frac{\beta}{1-\alpha}$

(c) $\frac{\alpha}{1-\beta}$ (d) $\frac{1-\beta}{\alpha}$

Q.46 In S.P.R.T. the statistician generally makes study about the (.........) function L(θ) of the test over the entire parametric space.

(a) A.S.N. (b) Power

(c) Estimating (d) O.C.

Q.47 Let us consider the following hypothesis to test : $H_0 : f_\theta$ is the true p.d.f. of X against $H_1 : g(\theta)$ is the true p.d.f. of X.

Then under the S.P.R.T. for testing H_0 vs H_1, we continue to take observations as long as :

$$B^{h(\theta)} < \prod_{i}^{m} \frac{g_\theta(x_i)}{f_\theta(x_i)} < \text{..........}$$

where h(θ) > 0 is the function of θ.

(a) $B^{h(\theta)}$ (b) $1-A^{h(\theta)}$

(c) $A^{h(\theta)}$ (d) $1-B^{h(\theta)}$

Q.48 For the problem discussed in Q.(47), we accept H_0 if :

$$\prod_{i=1}^{m} \frac{g_\theta(x_i)}{f_\theta(x_i)} \leq \text{..........}$$

(a) $B^{h(\theta)}$ (b) $A^{h(\theta)}$

(c) $1-B^{h(\theta)}$ (d) $1-A^{h(\theta)}$

Q.49 For the problem discussed in Q.(47), we reject H_0 and accept H_1 if :

$$\prod_{i}^{m} \frac{g_\theta(x_i)}{f_\theta(x_i)} \geq \text{..........}$$

(a) $B^{h(\theta)}$ (b) $1-A^{h(\theta)}$

(c) $1-B^{h(\theta)}$ (d) $A^{h(\theta)}$

Q.50 Consider ($A^{h(\theta)} > 0$, α' and β') as defined earlier, one can write :

$A^{h(\theta)} \leq$ ....

(a) $\frac{1-\beta'}{\alpha'}$ (b) $\frac{1-\alpha'}{\beta'}$

(c) $\frac{\alpha'}{1-\beta'}$ (d) $\frac{\beta'}{1-\alpha'}$

Q.51 Considering ($B^{h(\theta)}$, α' and β') as defined earlier, one can write :

$B^{h(\theta)} \geq$ ....

(a) $\frac{1-\alpha'}{\beta'}$ (b) $\frac{\beta'}{1-\alpha'}$

(c) $\frac{\alpha'}{1-\beta'}$ (d) $\frac{1-\beta'}{\alpha'}$

Q.52 When the excess of $\prod_{i}^{m} \frac{g_\theta(x_i)}{f_\theta(x_i)}$ over the boundary A and B at the termination of the experiment is negligible, the equality sign holds in an approximate sense and we can write :

$A^{h(\theta)} \cong$ ....

(a) $\frac{1-\alpha'}{\beta'}$ (b) $\frac{\alpha'}{1-\beta'}$

(c) $\frac{\beta'}{1-\alpha'}$ (d) $\frac{1-\beta'}{\alpha'}$

Q.53 In situation discussed in Q.(52), we can write :

$B^{h(\theta)} \cong$ ....

(a) $\frac{\beta'}{1-\alpha'}$ (b) $\frac{1-\alpha'}{\beta'}$

(c) $\frac{\alpha'}{1-\beta'}$ (d) $\frac{1-\beta'}{\alpha'}$

Q.54 In applying S.P.R.T. procedure, since $\alpha' = 1 - L(\theta)$, we can write :

$$L(\theta) \cong \frac{A^{h(\theta)} - 1}{(\ldots\ldots\ldots\ldots\ldots\ldots)}$$

(a) $B^{h(\theta)} - A^{h(\theta)}$ (b) $A^{h(\theta)} - B^{h(\theta)}$

(c) $A^{h(\theta)} / B^{h(\theta)}$ (d) $B^{h(\theta)} / B^{h(\theta)}$

Q.55 Considering $A = (1-\beta) / \alpha$ and $B = \beta\,(1-\alpha)$, we can write O.C. function $L(\theta)$ in applying S.P.R.T. procedure as :

$$L(\theta) = \frac{(\ldots\ldots\ldots)^h - 1}{\left(\frac{1-\beta}{\alpha}\right)^h - \left(\frac{\beta}{1-\alpha}\right)^h}$$

(a) $\frac{1-\beta}{\alpha}$ (b) $\frac{\alpha}{1-\beta}$

(c) $\frac{\beta}{1-\alpha}$ (d) $\frac{1-\alpha}{\beta}$

Q.56 In SPRT procedure, one can draw the O.C. function easily by plotting the sufficiently large number of co-ordinates (....., ...) on graph paper.

(a) $L(\theta), \theta,$ (b) $\theta, L(\theta)$

(c) $\theta, 1-L(\theta)$ (d) $\theta, 1+L(\theta)$

Q.57 Consider a class of sequential test of a given strength (α, β). Then a test is said to be (..........) than the other tests of the class if the former needs a smaller number of observation on the average then that of latter.

(a) More efficient (b) Unbiased

(c) M.P. test (d) Consistent

Q.58 If there exists a test in S P R T procedure for which $E_{\theta_0}(N)$ and $E_{\theta_1}(N)$ are smaller than that of any other test in the class, then the former test may be called as :

(a) Consistent test (b) Optimum test

(c) Unbiased test (d) Sufficient test

Q.59 Let us denote the minimum value of $E_{\theta_0}(N)$ by $N_0(\alpha, \beta)$ and minimum value of $E_{\theta_1}(N)$ by $N_1(\alpha, \beta)$ in the class of test. Then the efficiency of a given test in the class. Under H_0 is given by the ratio λ_0 written as :

$\lambda_0 =$ ....

(a) $\frac{N_0(\alpha, \beta)}{E_{\theta_0}(N)}$ (b) $\frac{E_{\theta_0}(N)}{No(\alpha, \beta)}$

(c) $\frac{N_1(\alpha, \beta)}{E_{\theta_1}(N)}$ (d) $\frac{E_{\theta_1}(N)}{N_1(\alpha, \beta)}$

Q.60 In Q.(59), the efficiency of any test in the class under H_1 is defined by the ratio λ_1 written as :

$\lambda_1 =$ ....

(a) $\frac{E_{\theta_1}(N)}{N_1(\alpha, \beta)}$ (b) $\frac{N_1(\alpha, \beta)}{E_{\theta_1}(N)}$

(c) $\dfrac{N_0(\alpha, \beta)}{E_{\theta_0}(N)}$ (d) $\dfrac{E_{\theta_0}(N)}{N_0(\alpha, \beta)}$

Q.61 The ratios (λ_0 & λ_1) defined in Q.(59, 60) under H_0 & H_1 lie in the limit written as :

.... < (λ_0 & λ_1) < ....

(a) (0, 1) (b) (1, 0)

(c) $(\frac{1}{2}, 1)$ (d) $(0, \frac{1}{2})$

Q.62 An optimum test in SPRT procedure has the efficiency say '*e*' under H_0 & H_1, then *e* is equal to :

(a) $\frac{1}{2}$, (b) 1.0

(c) 100% (d) Both (b) & (c)

Q.63 The statistician 'Wald' has proved that under SPRT procedure, the efficiency '*e*' under H_0 & H_1 is very nearly equal to :

(a) 1.0 (b) $\frac{1}{2}$

(c) 100% (d) Both (a) & (c)

Q.64 If the value θ lies as ($\theta_0 < \theta < \theta_1$), the ASN value $E_\theta(N)$ of SPRT may exceed the sample size *n* of an equvalent (..........) test with the same erroros (I & II) probabilities.

(a) Non-sequential (b) Traditional

(c) Randomized (d) Both (a) & (b)

Q.65 The theoretical and empirical studies made indicate that the distribution of 'N' is SPRT procedur is :

(a) Negatively skewed (b) Positively skewed

(c) Uniformally distributed (d) Binomially distributed

Q.66 Suppose $H_0 : \theta = \theta_0$ is tested against $H_1 : \theta > \theta_0$. Then the zone for (..........) of H_0 may be said to constitute the single value θ_0.

(a) Acceptance (b) Rejection

(c) Non-significance (d) Both (a) & (c).

Q.67 The degree of preference to reject H_0 in SPRT procedure generally (..........) with the (..........) of θ in domain $\theta > \theta_0$.

(a) Increases, Increase (b) Decreases, Increase

(c) Increases, Decrease (d) Decreases, Decrease

Q.68 We can find a value $\theta_1 > \theta_0$ such that the (.........) of H_0 is an error of practical importance whenever $\theta > \theta_1$.

(a) Acceptance (b) Non-significance

(c) Rejection (d) Both (a) & (b)

Q.69 When θ lies as $(\theta_0 < \theta < \theta_1)$, then the (.........) of H_0 is an error of no importance.

(a) Acceptance (b) Rejection

(c) Non-significance (d) Both (a) & (c)

Q.70 The $\theta \geq \theta_1$ may be said to constitute the zone of preference of (..........) of θ_0.

(a) Rejection (b) Acceptance

(c) Non-acceptance (d) Both (a) & (c)

Q.71 The $L(\theta)$ monotonically (..........) in θ for $\theta \geq \theta_1$.

(a) Decreases (b) Increases

(c) Remains constant (d) None

Q.72 Suppose we have to test $H_0 : \theta = \theta_0$ v.s. $H_1 : \theta < \theta_0$ then the zone for the (.........) of H_0 may be said to constitute the single value θ_0.

(a) Acceptance (b) Non-significance

(c) Both (a) & (b) (d) None

Q.73 The degree of preference to reject H_0 generally (..........) with the (..........) of θ in the domain $\theta < \theta_0$.

(a) Increases, Decrease (b) Increases, Increase

(c) Decreases, Increase (d) Decreases, Decrease

Q.74 The $\theta < \theta_0$ may be said to constitute the zone of preference of (..........) of H_0 :

(a) Rejection (b) Significant

(c) Both (a) & (b) (d) Acceptance

Q.75 For $\theta < \theta_0$, the $L(\theta)$ monotonically (..........) as θ decreases.

(a) Decreases (b) Increases

(c) Remains constant (d) None

Q.76 Suppose we have to test $H_0 : \theta = \theta_0$ v.s. $H_1 : \theta \geq \theta_0$, then we impose the following restriction :

$1 - L(\theta)$ .... if $\theta > \theta_1$.

(a) $\geq \alpha$, (b) $\leq \alpha$,

(c) $\leq \beta$, (d) $\geq \beta$.

Q.77 In Q.(76), the following restriction is imposed.

$L(\theta) \leq$ .... if $\theta \geq \theta_1$.

(a) β (b) α,

(c) $(1-\beta)$, (d) $(1-\alpha)$

Q.78 For the problem of testing composite H_0 v.s. composite H_1, we can consider an approach in SPRT suggested by :

(a) Wald (b) Prof. R. A. Fisher

(c) Crame Rao (d) J. Neyman

Q.79 Let X~ N(μ, σ^2) where both (μ & σ^2) are unknown. Suppose we have to test $H_0 : \mu = \mu_0$ v.s. $H_1 : \mu \neq \mu_0$ and $|\mu - \mu_0| / \sigma$ is small then (..........) of H_0 will not be a serious matter.

(a) Acceptance (b) Rejection

(c) Non-significance (d) Both (a) & (c)

Q.80 The importance of error committed in accepting H_0 will (..........) with the (..........) of $|\mu - \mu_0| / \sigma$.

(a) Increase, Increase (b) Increase, Decrease

(c) Decrease, Increase (d) Decrease, Decrease.

Q.81 In Q.(79), we can find a value $\delta > 0$, such that the acceptance of H_0 is considered as "An error of practical importance" only when :

$$\frac{|\mu - \mu_0|}{\sigma} \text{ .... } \delta.$$

(a) $\geq$ (b) $\leq$

(c) $=$ (d) None

Q.82 For the problem discussed in Q.(79), Wald has chosen the weight function 6_0 under H_0 of σ such that :

$$6_0(\sigma) = \begin{cases} \frac{1}{c} \; if \; \sigma \text{ .... } C \\ 0, \; otherwise \; where \; c \to \infty \end{cases}$$

(a) $\leq$ (b) $\geq$

(c) $=$ (d) None.

Q.83 As in Q.(82), Wald has chosen the weight function 6_1 under H_1 of σ such that : $6_1(\sigma) = \begin{cases} \text{..........} \; for \; \sigma \leq c \\ 0 \;, \; otherwise \end{cases}$

(a) $\frac{1}{c}$ (b) $\frac{1}{2c}$

(c) c, (d) 2c.

Q.84 For fixed 'c', let (g_{0m}, g_{1m}) be the probability under $(H_0 \& H_1)$ respectively. Wald shown that the probability ratio λ = ... for c → ∞ , is strictly increasing function of $\frac{|\bar{x} - \mu_0| \sqrt{n}}{s}$:

(a) $\frac{g_{1m}}{g_{0m}}$ (b) $\frac{1-g_{1m}}{1-g_{0m}}$

(c) $\frac{g_{0m}}{g_{1m}}$ (d) $\frac{1-g_{0m}}{1-g_{1m}}$

Q.85 The test procedure stated in Q.(84) is known as the (..........) and is illustrated as follows. Continue to take observation as long as :

$$B < \frac{g_{1m}}{g_{0m}} < A.$$

(a) Sequential Z test (b) Sequential t test

(c) Sequential randomized test (d) Student's test

Q.86 As illustrated in Q.(85), stop to take observation as soon as the condition mentioned in Q.(85) fails and (..........) if : $\frac{g_{1m}}{g_{0m}} \le B$.

(a) Accept H_0 (b) Reject H_0

(c) Declare non-significant. (d) Both (a) & (c)

Q.87 As illustrated in Q.(85), we reject H_0 if $\frac{g_{1m}}{g_{0m}}$ ....

(a) $\ge A$ (b) $\le A$

(c) $\ge B$ (d) $\le B$.

Q.88 Let us consider the sequence of statistics T_1, T_2, T_3, ... such that $T_n = t_n(x_1, x_2, ..., x_n)$. For each n, T_n has a distribution that depends on θ through λ(θ) alone. Then the hypothesis $H_0 : \lambda(\theta) = \lambda_0$ V.S. H1 : $\lambda(\theta) = \lambda_1$ is tested through (..........) corresponding to the sequence $\{T_n\}$.

(a) Student't test (b) S.P.R. test

(c) Z=test (d) Randomized test

Q.89 Let X~ N(μ, σ²) where both (μ, σ²) are unknown. Then we can find a (..........) to test the :

$$H_0 : \frac{\mu}{\sigma} = \lambda_0 \text{ vs. } H_1 : \frac{\mu}{\sigma} = \lambda_1.$$

(a) Sequential test (b) Student's t test
(c) Z-test (d) Randomised test.

Q.90 Let X~ N(μ, σ²). Consider a new r.v. Y_i defined as $y_i = \frac{X_i}{|X_i|}$, i = 1, 2, ... n then the distribution of each Y_i, depends on (..........) alone :

(a) $\frac{\mu}{\sigma}$ (b) $\frac{\sigma}{\mu}$

(c) $\frac{\overline{x}}{\sigma}$ (d) $\frac{\sigma}{\sqrt{n}}$

Q.91 The test advised to apply on the sequence $\{Y_1\}$ is the (..........).
(a) Student's t test (b) S.P.R. test
(c) Normal test (d) F ratio test

Q.92 In Q.(90), the S.P.R.T. may be expressed in terms of the (..........).
(a) Student's t statistic (b) Z statistic
(c) F statistic (d) χ^2 statistic

Q.93 In case if $\lambda_0 < \lambda_1$, the ratio η defined as $\eta = f_{\lambda 1}(x_1, x_2, \ldots\ldots\ldots. x_n)/\lambda_{0m}(x_1, x_2, \ldots.. x_n)$ is monotonically (..........) function of t_n.
(a) Increasing (b) Decreasing
(c) Non-decreasing (d) Both (a) & (c)

Q.94 In applying SPRT procedure, the sampling work is stopped as soon as either inequality fails and accept H_0 if t_n is
(a) $\leq b_n$ (b) $> b_n$,
(c) $> a_n$, (d) $< a_n$
where (a_n, b_n) are two constants such that $0 < b_n < 1$, and $a_n > 1$.

Q.95 In Q.(93), the SPRT procedure will be stopped and we reject H_0 if t_n.
(a) $\leq a_n$ (b) $\geq a_n$
(c) $\geq b_n$ (d) $\leq b_n$

Q.96 In applying S.P.R.T. procedure, the sequence $\{T_n\}$ has one to one function of "student's t" defined as $t_n = \frac{\bar{x}}{\sqrt{\sum^{n}(x_i - \bar{x}_n)^2}}$

and this 't_n' has (..........) distribution.

(a) central t (b) non-central t

(c) central χ^2 (d) Randomised t.

Q.97 A sequential procedure may also be applied to estimate a function of (.........).

(a) Probability (b) Population mean.

(c) Parameters (d) Population variance.

Q.98 In general, it is not easy to terminate the "Most appropriate estimator" or to formulate a rule to terminate the (..........).

(a) Sampling work (b) Experimental work

(c) Randomization (d) Decision making

Q.99 A sequential test procedure is defined by a sequence $\{\Psi_n, T_n\}$ where Ψ_n represents the stopping rule and T_n represents as (..........) to be used in case if the experiment is stopped at the nth stage.

(a) An estimator (b) An estimate

(c) Test statistic (d) Parameter

Q.100 Let Ψ_n denotes the stopping rule in S.P.R.T. procedure. Then sequential procedure will be terminated with probability 1.0 $\forall$ $\theta \in \Omega$ θ if

$$\sum_{1 < n < \infty} \Psi_n = \text{..........}$$

(a) > 1.0 (b) < 1.0

(c) 1.0, (d) None

6.4 Key Answers

Write True/False

Q	Ans.	Q	Ans.	Q	Ans.	Q	Ans.
1.	True	33.	True	65.	True	97.	True
2.	True	34.	True	66.	True	98.	True
3.	True	35.	True	67.	True	99.	True
4.	True	36.	True	68.	True	100.	True
5.	True	37.	True	69.	True	101.	True
6.	True	38.	True	70.	True	102.	True
7.	True	39.	True	71.	True	103.	True
8.	True	40.	True	72.	True	104.	True
9.	True	41.	True	73.	True	105.	True
10.	True	42.	True	74.	True	106.	True
11.	True	43.	True	75.	True	107.	True
12.	True	44.	True	76.	True	108.	True
13.	True	45.	True	77.	True	109.	True
14.	True	46.	True	78.	True	110.	True
15.	True	47.	True	79.	True	111.	True
16.	True	48.	True	80.	True	112.	True
17.	True	49.	True	81.	True	113.	True
18.	True	50.	True	82.	True	114.	True
19.	True	51.	True	83.	True	115.	True
20.	True	52.	True	84.	True	116.	True
21.	True	53.	True	85.	True	117.	True
22.	True	54.	True	86.	True	118.	True
23.	True	55.	True	87.	True	119.	True
24.	True	56.	True	88.	True	120.	True
25.	True	57.	True	89.	True	121.	True
26.	True	58.	True	90.	True	122.	True
27.	True	59.	True	91.	True	123.	True
28.	True	60.	True	92.	True	124.	True
29.	True	61.	True	93.	True	125.	True
30.	True	62.	True	94.	True	126.	True
31.	True	63.	True	95.	True	127.	True
32.	True	64.	True	96.	True	128.	True

Fill in the Blanks.

Q.	Ans.
1.	Sequential procedure
2.	Sampling scheme
3.	Sampling
4.	Sampling rule
5.	Sampling work
6.	Sampling work
7.	Sampling work
8.	Sampling work
9.	$x_m \in A_{0m}$
10.	$x_m \in A_{1m}$
11.	O.C. function
12.	Accepting H_0
13.	Power function
14.	$1 - L(\theta)$
15.	Correct decision
16.	One error
17.	Consistent
18.	Inconsistent
19.	Sequential
20.	Average sample number
21.	Parameter
22.	Operating character
23.	Average sample number
24.	Correct decision
25.	Sequential
26.	Wald
27.	SPRT
28.	$\prod^{m} f_{\theta_1}(x_i)$
29.	f_{1m}/f_{0m}

Q.	Ans.
30.	$B < \dfrac{f_{1m}}{f_{0m}} < A$
31.	$\dfrac{f_{1m}}{f_{0m}} \leq B$
32.	$f_{1m}/f_{0m} \geq A$
33.	Sequential probability ratio
34.	$\leq \log B$
35.	$\geq \log A$
36.	α'
37.	β'
38.	$\alpha'/(1-\alpha')$
39.	$(1-\beta')/\alpha'$
40.	(α', β')
41.	$\leq (1-\alpha')\,\beta$
42.	$\leq (1-\beta')\,\alpha$
43.	$\leq \alpha/(1-\beta)$
44.	$\leq \beta/(1-\alpha)$
45.	$= \alpha/(1-\beta)$
46.	$= \beta/(1-\alpha)$
47.	Operating characteristic
48.	$A^{h(\theta)}$
49.	$B^{h(\theta)}$
50.	$A^{h(\theta)}$
51.	$(1-\beta')/\alpha'$
52.	$\beta'(1-\alpha')$
53.	$(1-\beta')/\alpha'$
54.	$\beta'/(1-\alpha')$
55.	$A^{h(\theta)} - B^{h(\theta)}$
56.	$(1-\beta)/\alpha$
57.	$(\theta, L(\theta))$

Q.	Ans.	Q	Ans.
58.	$E_\theta(Z_1)$	83.	Acceptance
59.	More efficient	84.	Increase, Increase
60.	Optimum test	85.	$\geq$
61.	$E_{\theta_0}(N)$	86.	$\leq$
62.	$N_1(\alpha, \beta)$	87.	$\frac{1}{2C}$
63.	(0, 1)	88.	g_{1m}/g_{0m}
64.	1.0	89.	Sequential *t* test
65.	1.0	90.	Accept H0
66.	Non-sequential	91.	$\geq A$
67.	Negatively skewed	92.	SPRT
68.	Acceptance	93.	Sequential test
69.	Increases, Increase	94.	μ/σ
70.	Acceptance	95.	SPRT
71.	Acceptance	96.	Student's statistic
72.	Rejection	97.	$f_{\lambda 0m}$
73.	Indifference	98.	Increasing
74.	Assumptions	99.	Sampling work
75.	Decreases	100.	$\leq b_n$
76.	Acceptance	101.	$\geq a_n$
77.	Increases, Increase	102.	Non-central *t*
78.	Rejection	103.	Parameters
79.	Decreases	104.	Sampling work
80.	$\leq \alpha$	105.	n^{th}
81.	β	106.	1.0
82.	Wald	107.	∞

Multiple Choice Questions and Answers

Q.	Ans.	Q.	Ans.	Q.	Ans.	Q.	Ans.
1.	(b)	6.	(c)	11.	(d)	16.	(d)
2.	(a)	7.	(d)	12.	(d)	17.	(a)
3.	(c)	8.	(b)	13.	(d)	18.	(b)
4.	(d)	9.	(d)	14.	(c)	19.	(b)
5.	(a)	10.	(d)	15.	(d)	20.	(d)

Q.	Ans.	Q.	Ans.	Q.	Ans.	Q.	Ans.
21.	(c)	41.	(a)	61.	(a)	81.	(a)
22.	(a)	42.	(a)	62.	(d)	82.	(a)
23.	(b)	43.	(a)	63.	(d)	83.	(b)
24.	(a)	44.	(a)	64.	(d)	84.	(a)
25.	(c)	45.	(b)	65.	(a)	85.	(b)
26.	(a)	46.	(d)	66.	(d)	86.	(d)
27.	(a)	47.	(c)	67.	(a)	87.	(a)
28.	(a)	48.	(a)	68.	(d)	88.	(b)
29.	(b)	49.	(d)	69.	(d)	89.	(a)
30.	(a)	50.	(a)	70.	(d)	90.	(a)
31.	(b)	51.	(b)	71.	(a)	91.	(b)
32.	(a)	52.	(c)	72.	(c)	92.	(a)
33.	(b)	53.	(a)	73.	(b)	93.	(d)
34.	(a)	54.	(b)	74.		94.	(a)
35.	(b)	55.	(a)	75.	(a)	95.	(b)
36.	(a)	56.	(b)	76.	(b)	96.	(b)
37.	(a)	57.	(a)	77.	(a)	97.	(c)
38.	(a)	58.	(b)	78.	(a)	98.	(a)
39.	(a)	59.	(a)	79.	(d)	99.	(a)
40.	(a)	60.	(b)	80.	(a)	100.	(c)

❑❑❑

CHAPTER - 7

Decision Theory

7.1 Write True/False

Q.1 All the problem of statistical inference may be considered as the problem of "Decision making in the face of uncertainty."

Q.2 The problem of point estimation is the problem if deciding which value in the "parameter space" Ω_θ is the true value of the parameter.

Q.3 The problem of statistical hypothesis testing is to decidie which one of H_0 and H_1 is true.

Q.4 The problem of interval estimation is to decide which of the possible class of intervals includes the true value of θ.

Q.5 In each problem of statistical inference, the decision is taken in the light of sample values $x_1, x_2, ..., x_n$ drawn from the population under study.

Q.6 Suppose a manufacturer of steel has to choose one out of seven methods of producing steel. He will like to choose the method that would lead to the "maximum net profit."

Q.7 The basic theory of "Elements of decision theory" has been pioneered by the statisticison "A. Wald."

Q.8 The first element of the decision making problem is the "Set of alternative values or forms" which may be assumed by the particular characteristic of real world.

Q.9 The whole set of the possible values of the parameter θ is denoted by Ω_θ.

Q.10 In the theory of decision making, the Ω_θ is known as the "Space of the possible states of nature."

Q.11 In the theory of "Decision making" a particular decision is denoted by 'd' and the whole set of decisions is denoted by 'D'.

Q.12 In the theory of decision making, the 'D' is known as "Decision space" relevent to the problem under study.

Q.13 When θ is the true state of nature then in taking decision 'd', we may or may not commit an error.

Q.14 While making a decision about unknown parameter θ, a loss may incurred by as if we take the decision 'd' when true state of nature is θ.

Q.15 In most of the decision making problems, it is possible to represent the loss incurred by us "Numerial."

Q.16 The loss incurred in decision making is denoted by $L(\theta, d)$.

Q.17 The function L which is dependent on two variables θ & d, is said to be the "Loss function" related to our decision problem.

Q.18 Any decision procedure dictates that which decision out of many should be taken about the problem considered under study.

Q.19 A decision procedure sets up a correspondance between each element of sample space, "R_n" and some element d of decision space "D".

Q.20 A decision procedure may be considered as a mapping (function) whose domain is 'R_n' and range is 'D'.

Q.21 A term "Decision function" or "Decision rule" is used to mean a particular procedure and is denoted by δ.

Q.22 If $x = \{ x_1, x_2, ..., x_n \}$ is the observed set of n values (sample values) then $\delta(x)$ denotes the decision dictated by δ.

Q.23 A specified decision rule may in some sense be regarded as the "Most desirable" in a given set of rules say 'Δ'.

Q.24 A criteria for a good decision rule should be based on the loss function $L(\theta, \delta(x))$.

Q.25 We can not ensure that a given decision rule based on a observed sample values $(x_1, x_2, ..., x_n)$ would always lead to the smallest loss even if the true value of θ were known.

Q.26 The expected value of $L(\theta, \delta(x))$ for given (θ, δ) but looked upon as a measurable function of the r.v.'s $X_1, X_2, ..., X_n$, is called the "risk" of using $\delta(x)$ for θ being the true value and is denoted by $r(\theta, \delta)$.

Q.27 The function $r(\theta, \delta)$ defined as :

$$r(\theta, \delta) = E_\theta [L(\theta, \delta(x)] \qquad ...(1)$$

for varying $\theta \in \Omega\ \theta$, is said to be the risk function of the decision rule.

Q.28 The performance of any decision rule δ is judged in the light of behaviour of its risk function $r(\theta, \delta)$ for varying θ.

Q.29. If there exists a decision rule δ in Δ such that :

$r(\theta, \delta) \leq r(\theta, \delta') \ \forall \ \theta \in \Omega \ \theta.$ ...(2)

for any rival decision $\delta' \in \Delta$, then δ is said to be the "Uniformaly best decision rule."

Q.30 A decision rule $\delta \varepsilon \Delta$ is called admissible if $\exists$ no other rule $\delta' \in \Delta$ with uniformally smaller risk.

Q.31 A decision rule d is admissible if $\exists$ no other rule $\delta' \in \Delta$ such that :

$r(\theta, \delta') < r(\theta, \delta) \ \forall \ \theta \in \Omega_\theta.$ ...(3)

and $r(q, d') < r(q, d)$ for some $\theta \in \Omega_\theta$.

Q.32 A decision rule δ is called a "Minimax rule" if it provides the maximum protection against the highest possible risk.

Q.33 A decision rule $\delta_0 \in \Delta$ is called a "Minimax rule" in the class Δ among all rules in D, if it minimises the "Maximum risk."

Q.34 The decision rule $\delta 0 \in \Delta$ is minimax if :

$$\underset{\theta}{Sup}\ r(\theta, \delta_0) \leq \underset{\theta}{Sup}\ r(\theta, \delta') \qquad ...(4)$$

for whatever is the other rule $\delta' \in \Delta$.

Q.35 The "Minimax apprach" in decision theory is also known as "Pessimist approach."

Q.36 A "minimax apprach" is called a "Pessimist apprach" because this approach always thinks of the worst but tries to avert it if possible.

Q.37 It is not true that we will necessarily get a decision rule $\delta \in \Delta$ which has the smallest maximum risk.

Q.38 For selecting a best decision rule, θ itself is considered as a r.v. with a specified distribution say 6 (continuous or discrete). This p.d.f. 6 of θ is known as "Prior distribution of θ."

Q.39 A decision rule $\delta_6 \in \Delta$ is called a "Bayes rule" in the class D if among all the rules $\in \Delta$, it has the "smallest average risk."

Q.40 The rule $\delta 6$ is a Bayes rule if :

$r(6, \delta_6) \leq r(6, \delta)$ ...(5)

for whatever the other rule $\delta \in \Delta$ may be.

Q.41 The risk function for a decision rule $\delta \in \Delta$ is defined as :

$$r(6, \delta) = \int L(\theta, \delta(x))\ f_{\theta\,(x)}\, dx. \qquad ...(6)$$

Q.42 The average risk function is defined as :

$$\text{Average } r(6, \delta) = \int_{\Omega_\theta} \int L(\theta, \delta(x))\ f_\theta(x)\ 6(\theta)\ dx\ d_\theta. \qquad ...(7)$$

Q.43 In Q.(42), we assume that r(6, δ) admits a change in the order of integration and we can write :

$$r(6, \delta) = \int_{\Omega_\theta} \int L(\theta, \delta(x))\, f_\theta(x)\, 6(\theta)\, d_\theta . dx. \qquad \ldots(8)$$

Q.44 Let $6_X(\theta)$ be the function of θ and is interpreted as the conditional p.d.f. of θ for given x as the observed value of r.v.x. Then $6_X(\theta)$ is called the probability density function of "Posterior distribution of θ corresponding to "Prior distribution" 6.

Q.45 The average risk of δ(*x*) may be expressed as :

$$r(6, \delta) = \int_X [\int_\theta L(\theta, \delta(x))\, 6_X(\theta)\, d_\theta]\, 6(x)\, dx. \qquad \ldots(9)$$

Q.46 In Q.(45) the integral within the square brackets may be looked upon as the conditional expected loss given that *x* is the observed value of X.

Q.47 A Bayes rule may not exist for a simple reason that the minimum of the conditional expected loss may not exist for each X.

Q.48 In case of Bayes rule being not exits, we may use another decision rule which is called an "∈ - Bayes decision rule."

Q.49 A class C of decision rules is said to be complete if for any rule δ not in C, there is a rule δ′ in C which is uniformally better than δ.

Q.50 A class C of decision rules δ(*x*), said to be complete if :

$$r(\theta, \delta') < r(\theta, \delta) \ \forall\ \theta \in \Omega\ \theta. \qquad \ldots(10)$$

and r(θ, δ') < r (θ, δ) for some θ

for a δ′ in C but not δ in C.

Q.51 A complete class C of decision rules δ(*x*) is "Minimal" if it does not contain a complete subclass of rules.

Q.52 A Bayes decision rule δ(*x*) is admissible if it is unique except for risk equivalence.

Q.53 A decision function δ_0 will be said to be "Bayes-solution" in the wide sense ∃ a sequence 6_k (k = 1, 2, ...) of prior distribution such that δ_0 is a Bayes solution related to the sequence i.e. if :

$$\lim_{k\to\infty} \left[\inf_{\delta} r(6_k, \delta_0) - r(6_k, d_\theta) \right] = 0. \qquad \ldots(11)$$

Q.54 Wald has shown that under very general conditions, the totality of "Bayes solution" in wide sense constitute a complete class.

Q.55 Wald has given a greater importance to "Minimax rules."

Q.56 Bayes rules are the only means through which the "Minimax rules" can be deduced.

Q.57 Let (δ6 & δ6') be the Bayes rules for a decision problem w.r. to prior distributions (6, 6′). Then 6 is called the "Least favourable" prior distribution if:

$$r(6, \delta 6) > r(6', \delta_{6'}). \quad \ldots(12)$$

Q.58 Let there (6, δ6) be the prior distribution and Bayes rule corresponding to (θ & 6) then δ_6 will be a "minimax rule" if

$$r(6, \delta_6) > \underset{\theta}{Sup}\ r(\theta, \delta 6). \quad \ldots(13)$$

Q.59 In Q.(58), the distribution 6 will be said to be Least favourable" prior distribution if:

$$r(6, \delta_6) > \underset{\theta}{Sup}\ r(\theta, \delta 6). \quad \ldots(14)$$

Q.60 Let δ be a Bayes decision rule whose risk r (θ, δ6) is constant over parameter space Ωθ. then δ6 is minimax.

Q.61 Let { 6_k, k = 1, 2, 3, ... } be a sequence of prior distributions, {δ_k, k = 1, 2, ... } be the sequence of Bayes rules w.r. to 6_k and there exists a decision rule δ_0 such that :

$$\lim_{k \to \infty} r(6_k, \delta_k) = \underset{\theta}{Sup}\ r(\theta, \delta_0) \quad \ldots(15)$$

then $\delta_0(x)$ is minimax decision rule.

Q.62 Let λ(θ) be a real valued function of parameter θ and is to be estimated from the observed sample (xi, i = 1, 2, 3, ..., n). Then a decision rule or decision function δ(x) is the estimator of λ(θ).

Q.63 Suppose L(θ, d) denotes the loss occured in decision d ∈ Δ. in estimating the parameter function λ(θ). Let us define loss function L(θ, d) as :

$$L(\theta, d) = c(\theta).w\ [\ |\ d - \lambda(\theta)\ |\ ] \quad \ldots(16)$$

where c(θ) is the positive and w is strictly increasing function of absolute error | d – λ(θ) |. Then a quadratic loss function is defined as :

$$L^*(\theta, d) = c(\theta)\ [\ d - \lambda(\theta)\ ]^2. \quad \ldots(17)$$

Q.64 Let D = { d_0, d_1 } be the decision space corresponding to hypothesis testing $H_0 : \theta \in \Omega\ \theta_0$ v.s. $H_1 : \theta \in \Omega\ \theta_1$. Then hypothesis testing is a two decisions (d_0, d_1) problem.

Q.65 Let us suppose that D = { d_0, d_1 } denotes the correct decisions for ($\theta \in \Omega\ \theta_0$ & $\theta \in \Omega\ \theta_1$) respectively. Then we define loss function L(θ, d_0) given as :

$$L(\theta, d_0) = \begin{cases} 0 \; if \; \theta \in \Omega \; \theta_0 \\ a(\theta) \; if \; \theta \in \Omega \; \theta_1 \end{cases} \quad \ldots(18)$$

for $a(\theta) > 0$.

Q.66 In relation to the Q.(65), the loss function $L(\theta, d_1)$ is defined as given below:

$$L(\theta, d_1) = \begin{cases} b(\theta) \; if \; \theta \in \Omega \; \theta_0 \\ 0 \; if \; \theta \in \Omega \; \theta_1 \end{cases} \quad \ldots(19)$$

for $b(\theta) > 0$.

Q.67 Corresponding to loss functions defined in equations. (18, 19), the risk function $r(\theta, \delta)$ with decision rule δ and critical region w, is defined as :

$$r(\theta, \delta) = b(\theta) \int_w f_\theta(x)\, dx, \text{ if } \theta \in \Omega \; \theta_0. \quad \ldots(20)$$

Q.68 As in Q.(67), the risk function $r'(\theta, \delta)$ can also be defined as :

$$r'(\theta, \delta) = a(\theta) \int_w f_\theta(x)\, dx \quad \text{if } \theta \in \Omega \; \theta_1 \quad \ldots21)$$

Q.69 The nature of a Bayes solution (test) may be considered as similar to that of a M.P. test for a simple null hypothesis H_0 against a simple alternative H_1.

Q.70 In case of applying Bayes test, we consider a sort of weighted average densities under (H_0 & H_1) corresponding to critical regions say (W_0 & W_1). respectively.

Q.71 Suppose we have to estimate a real valued function $\lambda(\theta)$ by means of an interval say :

$d \cong (d_1, d_2) \in I$.

where I is a finite or infinite inverval containing all possible values of $\lambda(\theta)$. Then the general form of a decision function is given as :

$$\delta(x) \cong [\, d_1(x), \delta_2(x) \,]. \quad \ldots(22)$$

Q.72 We define a simple loss function $L(\theta, d)$ given as :

$$L(\theta, d) = \begin{cases} 0, \; if \; d_1 \le r(\theta) \le d_2 \\ 1, \; otherwise \end{cases} \quad \ldots(23)$$

Q.73 The risk function $r(\theta, \delta)$ of a decision function $\delta(\underset{\sim}{x})$ is defined as :

$$\begin{aligned} r(\theta, \delta(\underset{\sim}{x})) &= E\theta.\, L[\theta, \delta(x)] \\ &= 1 - P_\theta [\, \delta_1(x) < r(\theta) < \delta_2(\underset{\sim}{x}) \,]. \quad \ldots(24) \end{aligned}$$

Q.74 A game in the mathematical sense is just as a triplet (Ω_θ, D, L) in context of elements of decision theory.

Q.75 The loss function $L(\theta, D)$ may be negative and this negative value may be interpreted as a gain.

Q.76 In a two persons game, the two persons try to simultaneously to maximise their winning i.e. to minimise their loss.

Q.77 Any function $d(\underset{\sim}{x})$ that maps the sample sapce S into a class D of all non-randomised decision rules or function provided the risk function $r(\theta, d)$ exists and is finite $\forall\ \theta \in \Omega\ \theta$.

Q.78 If θ is a random variable then conditional p.d.f. of θ given $Y = y$ is :

$$y(\theta/y) = \frac{g(y, \theta)}{g_1(y)} = \frac{h(\theta).\,l(\theta/y)}{g_1(y)}, g_1(y) > 0.$$

Q.79 In context of Bayesian statistic, the p.d.f. $h(\theta)$ is called "Prior p.d.f. of θ."

Q.80 In context of Bayesian statistic, the conditional p.d.f. $l(y/\theta)$ is called the "Posterior p.d.f. of θ."

Q.81 In Q.(78), $h(\theta)$is the p.d.f. of θ prior to the observation of y has been made where as $l(\theta/y)$ is the p.d.f. of θ after the observation y has been made.

7.2 Fill in the Blanks

1. All the problems of statistical inference may be considered as the problem of (..........) in the face of uncertainity.
2. The problem of point estimation is the problem of deciding which value in the (..........) say Ω_θ, is the true value of parameter.
3. The problem of statistical (..........) is to decide which one of H_0 and H_1 is true.
4. The problem of (..........) is to decide which of the possible class of intervals includes the true value of θ.
5. In each problem of (..........), the decision is taken in the light of sample values $(x_1, x_2, ... x_n)$ drawn from the population under study.
6. Suppose a manufacturer of steel has to choose one out of seven methods of producing steel. He will like to choose the method that would lead to the (..........) net profit.
7. The basic theory of elements of (..........) has been pioneered by the statistician 'A Wald.'
8. The whole set of the possible values of the (..........) say θ is denoted by Ω_θ.

9. In the theory of (..........), the Ω_θ is known as the "Space of the possible states of nature."
10. In the theory of decision making, a particular decision is denoted by 'd' and the whole set of (..........) is denoted by 'D'.
11. In the theory of decision making, the 'D' is known as (..........) relevant to the problem under study.
12. When θ is the true state of nature then in taking the (..........) 'θ', we may or may not commit an error.
13. While making a decision about an unknown (..........) sat 'θ', a loss may incurred by us if we take the decision 'd' when true state of native is θ.
14. The loss incurred in a decision making is denoted by..........
15. The function L which is dependent on two variables θ & d, is said to be the (..........) related to our decision.
16. Any (..........) dictates that which decision out of many decisions should be taken about the problem under study.
17. A decision procedure sets up a correspondance between each element of (..........) 'R^n' and some element 'd' of decision space "D".
18. A decision procedure setsup a correspondance between each element of sample space 'R^n' and some element 'd' of (..........) say 'D'.
19. A decision procedure may be considered as a mapping (function) whose (..........) is 'Rn' and range is 'D'.
20. A decision procedure may be considered as a mapping (function) whose domain is 'R^n' and (..........) is 'D'.
21. The term (..........) or "Decision rule"" is used to mean a particular procedure and is denoted by 'δ'.
22. If (x) is the observed set of n values (sample values) then (..........) denotes decision dictated by δ.
23. A specified decision rule may in some sense be regarded as the (..........) in a given set of rules say 'D'.
24. A criteria for a good decision rule should be based on the (..........) say $L(\theta, \delta(x))$.
25. We can not ensure that a given decision rule based on a observed sample values (x) would always lead to (..........) even if true value of θ were known.
26. The expected value of $L(\theta, \delta(x))$ for give (θ, δ) but considered as a (..........) of r.v.s (x) is called 'risk' of using $\delta(x)$, for θ being true value and the risk is denoted by r (.........).

27. The function r(θ, δ) defined as :

    $r(\theta, \delta) = E_\theta$ [L (θ. δ(*x*)] for varying ∀ θ ∈ Ω θ, is known as (..........) of the decision rule.

28. The performance of any decision rule δ is judged in the light of behaviour of its (.........). r (θ, δ) for varying θ.

29. If there exists a decision rule δ in Δ such that r(θ, δ) < r (θ, δ') ∀ θ ∈ Ω θ.

    for an arbitrary rule δ' ∈ Δ, then 'δ' is known an (.........).

30. A decision rule δ ∈ Δ is known as (..........) if ∃ no other rule δ' ∈ Δ with uniformally smaller risk.

31. A decision rule δ is called a (...........) if it provides the maximum protection against the highest possible risk.

32. A decision rule $\delta_0 \in \Delta$ is called 'Minimax rule' in the class D if among all rules in D, it minimises the "..........".

33. A decision rule $\delta_0 \in \Delta$ is minimax if

    $$\underset{\theta}{Sup}\ r(\theta, \delta_0) < \underset{\theta}{Sup}\ (..........)$$

    for whatever be the other rule δ' ∈ Δ.

34. The (..........) in decision theory is also known as "Pessimist approach."

35. A "Minimax approach" in decision theory is also known as (..........).

36. A "Minimax approach" is also called "Pessimist approach" because this apparoach always thinks of the worst but tries to avert it if possible.

37. It is not true that we will necessarily get a decision rule δ ∈ Δ which has the (..........) maximum risk.

38. For selecting a best decision rule, θ itself is considered as a r.v. with a specified distribution say 6 (continuous or discrete). This p.d.f. 6 of θ is known as (..........).

39. A decision rule δ6 ∈ Δ is called a (.........) in the class Δ if among all the rules ∈ Δ, it has the smallest average risk.

40. A rule δ6 is a Bayes rule if :

    r(6, δ6) ....

    for whatever the other rule δ ∈ Δ may be.

41. The risk function for a decision rule δ ∈ Δ. is defined as :

    r (6, δ) = ...

42. Let 6_x (θ) be the function of θ and is interpreted as the conditional distribution of θ for given x as observed value of r.v. X. Then $6_x(\theta)$ is

called the p.d.f. of (..........) corresponding to prior distribution 6.

43. Let $6_x(\theta)$ be the function of θ and is interpreted as the conditional p.d.f. of θ for given x as observed value of r v X. Then $6_x(\theta)$ is known as the p.d.f. of "Posterior distribution of θ corresponding to (.........) 6.

44. A (..........) may not exist for a simple reason that the minimum of the conditional expected loss may not exist for each x.

45. In case if Bayes rule does not exist, we may use an other rule which is called an (.........).

46. A class C of decision rules is said to be complete if for any rule δ not in C, there exists a rule δ' in C which is (..........) than δ.

47. A class C of decision rules $\delta(x)$ is said to be (..........) if :

$r(\theta, \delta') \leq r(\theta, \delta) \ \forall \ \theta \in \Omega_\theta$.

and $r(\theta, \delta') < r(\theta, \delta)$ for some θ.

for a δ' being in C but δ not being in C.

48. A complete class C of decision rules $\delta(x)$ is (..........) if it does not contain a complete subclass of rules.

49. A Bayes decision rule $\delta(x)$ is (..........) if it is unique except for risk equivalenc.

50. A decision function δ_0 will be said to be (..........) solution in the wide sense $\exists$ a sequence 6_k ($k = 1, 2, ...$) of prior distributions such that δ_0 is a Bayes solution related to the sequence i.e. if :

$$\lim_{k\to\infty} \left[\inf_{\delta} r(6_k, \delta_0) - r(6_k, \delta_0) \right] = 0.$$

51. Wald has shown that under very general conditions, the totality of (..........) in a wide sence constitutes a complete class.

52. Wald has shown a great importance to (..........) rules.

53. Bayes rules are the only means through which the (..........) rules can be deduced.

54. Let ($\delta 6$ & $\delta 6'$) be the Bayes rules for a decision problem w.r. to prior distributions (6, 6'). Then 6 is called the (.........) prior distribution if :

$r(6, \delta_6) \geq r(6', \delta_6')$.

55. Let (6, & $\delta 6$) be the prior distribution and Bayes rule corresponding to (θ, 6), then $\delta 6$ will be a (..........) if :

$$r(6, \delta 6) > \sup_{\theta} r(\theta, \delta 6).$$

56. In Q.(55), the distribution 6 will be said to be (..........) prior distribution if :

$r(6, \delta 6) > \underset{\theta}{Sup}\ r(\theta, \delta 6)$.

57. Let δ be the Bayes decision rule whose risk r (θ, δ6) is constant over parameter space Ωθ. then δ6 is (..........).

58. Let { 6_k, $k = 1, 2,$ .... } be a sequence of prior distributions, { δk, $k =$ 1, 2, ... } be the sequence of Bayes rules w.r. to 6_k. and there exists a decision rule δ_0 such that :

$$\lim_{k \to \infty} r(6_k, \delta_k) = \underset{\theta}{Sup}\ r(\theta, \delta_0)$$

then $\delta_0(x)$ is known as (.........) rule.

59. Let λ(θ) be a real valued function of parameter θ and is to be estimated from the observed sample (x). Then a decision rule or decision function $\delta(x)$ is the (..........) of λ(θ).

60. Let L(θ, d) be the loss occured in decision d ∈ Δ in estimating the parametric function λ(θ). Let us define loss function as :

$L(\theta, d) = c(\theta)\ w(\,|\,d - \lambda(\theta)\,|\,)$

where c(θ) is positive and w is strictly increasing function of absolute error | d – λ(θ) |. Then a (..........) is defined as :

$L^*(\theta, d) = C(\theta)\,(\,d - \lambda(\theta)\,)^2$.

61. Let D = { d_0, d_1 } be the (..........) corresponding hypothesis testing

$H_0 : \theta \in \Omega_{\theta_0}$ v.s. $H_1 : \theta \in \Omega_{\theta_1}$.

Then hypothesis is a two decision (d_0, d_1) problem.

62. Let D = { d_0, d_1 } be the correct decisions for ($\theta \in \Omega\,\theta_0$, & $\forall\ \theta \in \Omega\theta_1$) respectively. Then we define loss function L (θ, d0) as given :

$$L(\theta, d_0) = \begin{cases} 0 \text{ } if \text{ } \theta \in .... \\ a(\theta) \text{ } if \text{ } \theta \in .... \end{cases}$$

for a(θ) > 0.

63. In relation to the Q.(62), the loss function L (θ, d_1) is defined as given below :

$$L(\theta, d_1) = \begin{cases} b(\theta) \text{ } if \text{ } \theta \in .... \\ 0 \quad if \text{ } \theta \in .... \end{cases}$$

for b(θ) > 0.

64. Corresponding to the loss functions defined in Q.(62, 63) the risk function r (θ, δ) with decision rule δ and critical region w is defined as :

$r(\theta, \delta) = b(\theta)$ .... if $\theta \in \Omega_{\theta_0}$.

65. In Q.(64), the risk function $r'(\theta, \delta)$ can also be defined as :

$$r'(\theta, \delta) = a(\theta) \int_{w^e} f_\theta(x)\, dx \quad \text{if ....}$$

66. The nature of a Bayes test (solution) may be considered as similar to that of a (..........) for a simple H_0 vs simple H_1.

67. In ase of applying Bayes test, we consider a sort of weighted average densities under (H_0 & H_1) corresponding to (..........) say (W_0 & W_1) respectively.

68. Suppose we have to estimate a real valued function $\lambda(\theta)$ of parameter θ by means of an interval say :

$d = (d_1, d_2) \in I$.

where I contains all possoble values of (..........). Then the general form of a decision function is given as :

$\delta(\underset{\sim}{x}) = (\delta_1(\underset{\sim}{x}), \delta_2(\underset{\sim}{x}))$.

69. We define a simple loss function $L(\theta, d)$ given as $L(\theta, d) =$

$$\begin{cases} 0 \text{ if } \ldots\ldots < r(\theta) < \ldots\ldots \\ 1 \quad , \quad \text{otherwise} \end{cases}$$

where $d = (d_1, d_2) \in I$.

70. The risk function $r(\theta, \delta)$ of a decision function $\delta(x)$ is defined as :

$r(\theta, \delta(x)) =$ (....................).

71. A game in the mathematical sense is just as a triplet (Ω_θ, D, L) in the context of elements of (..........).

72. Let the loss function $L(\theta, D)$ may be negative, then this negative value may be interpreted as a (..........).

73. In a two persons game, they may try to (..........) simultaneously their loss.

74. In context of Bayesian statistic, the p.d.f. of (θ) say $h(\theta)$ is called (..........) of θ.

75. In context of Bayesian statistic, the conditional p.d.f. say $f(y/\theta)$ is called (..........) of θ.

76. In Q.(74 & 75), the p.d.f.'s say $h(\theta)$ and $f(y/\theta)$ are the probability density functions (.........) and after to the observations y have been taken respectively.

7.3 Multiple Choice Questions and Answers

1. All the problems of statistical inference may be considered as the problem of (..........) in the face of uncertainity :

(a) Decision making (b) Decision theory
(c) Decision procedure (d) Decision function

Q.2 The problem of point estimation is the problem of deciding which value in the (.........) say $\Omega\theta$ is the true value of parameter.

(a) Sample space (b) Parameter space
(c) Decision space (d) Decision function

Q.3 The problem of statistical (.........) is to decide which one of (H_0, H_1) is true.

(a) Decision making (b) Hypothesis testing
(c) Decision procedure (d) Null hypothesis

Q.4 The problem to decide which one of the class of intervals includes the true value of parameter θ, is known as :

(a) M.L. estimation (b) Point estimation
(c) Interval estimation (d) Minimum value estimation

Q.5 The problem of decision taken about the population parameter θ on the basis of sample values (x) drawn from the population under study is known as :

(a) Statistical inference (b) Statistical decision
(c) Both (a) & (b) (d) None

Q.6 Suppose a manufacturer of steel has to chose one out of seven methods of producing steel. He will like to choose the method that would lead to (..........) net profit.

(a) Maximum (b) Minimum
(c) Minimax (d) None

Q.7 The basic theory of elements of (..........) has been pioneered by the statistion "A. Wald".

(a) Decision making (b) Decision theory
(c) Decision function (d) Both (a) & (b)

Q.8 The whole set of all possible values of (..........) say $\Omega\theta$ is known as parameter space.

(a) Parameter (b) Statistic
(c) Population constant (d) Both (a) & (c)

Q.9 In the theory of (.........), the $\Omega\theta$ is known as the "Space of possible states of nature."

(a) Decision making (b) Decision theory

(c) Decision procedure (d) Decision space

Q.10 In the theory of decision making, a particular decision is denoted by 'd' and the whole set of (.........) is denoted by "D".

(a) Decision (b) Inference

(c) Both (a) & (b) (d) Decision constant.

Q.11 In the theory of decision making, the 'D' is known as (..........) relevent to the problem under-study.

(a) Decision space (b) Decision function

(c) Decision procedure (d) Decision theory

Q.12 When θ is the true state of nature, then in taking (.........) 'd', we may or may not commit an error .

(a) Decision space (b) Decision

(c) An estimate (d) An estimator

Q.13 When making a decision about an unknown (..........) say 'θ', a loss may incurred by us, if we take decision 'd' when true state of nature is θ.

(a) Parameter (b) Population constant

(c) Both (a) & (b) (d) Estimate

Q.14 The loss incurred in decision making is denoted by (..........).

(a) $L(\theta, d)$ (b) $L(\theta, \delta(x))$

(c) $L(\theta, \Delta)$ (d) $L(d_1, d_2)$

Q.15 The function L which is dependent on two variables θ and d, is known as the (..........) related to our decision.

(a) Loss function (b) An estimate

(c) Amount of risk (d) Risk function

Q.16 Any (..........) dictates that which decision out of many decisions should be taken about the problem under study.

(a) Decision theory (b) Decision procedure

(c) Decision making (d) Loss function

Q.17 A decision procedure sets up a correspondance between each element of (..........) say 'R^n' and some elements 'd' of decision space "D".

(a) Sample space (b) Parameter space

(c) Population (d) Decision making

Q.18 A decision procedure sets up a correspondance between each elements of sample space 'R^n' and some elements 'd' of (.........) say 'D'.

(a) Sample space (b) Decision space

(c) Decision function (d) Decision procedure

Q.19 A decision procedure may be considered as a mapping (function) whose (..........) is 'R^n' and range is 'D'.

(a) Domain (b) Range

(c) Parameter space (d) None

Q.20 A decision proceudre may be considered as a mapping (function) whose domain is 'R^n' and (..........) is 'D'.

(a) Domain (b) Range

(c) Parameter space (d) Estimate

Q.21 The term (..........) or "decision rule" is used to mean a particular procedure and is denoted by 'δ'.

(a) Decision function (b) Decision space

(c) Both (a) & (b) (d) Decision rule

Q.22 If x is the observed set of n values (sample values) then the decision dictated by δ is denoted by :

(a) $D(x)$ (b) $\delta(\underset{\sim}{x})$

(c) $L(\theta, \delta)$ (d) None

Q.23 A specified decision rule may in some sense be regarded as the (..........) in a given sets of rules say 'D' :

(a) Most desirable (b) Most appropriate

(c) Both (a) & (b) (d) Most efficient

Q.24 A criteria for a good decision rule should be based on the (.........) say $L(\theta, \delta(x))$.

(a) Loss function (b) Risk function

(c) Decision space (d) Decision procedure

Q.25 We can not ensure that a decision rule based on a observed sample values (x) would always lead to (.........) even if true value of θ were known.

(a) Smallest loss (b) Smallest risk

(c) Both (a) & (b) (d) Maximum loss

Q.26 The expected value of $L(\theta, \delta(x))$ for given (θ, δ) but considered as a (.........) of r.v.'s (x) is called 'risk' of using $\delta(x)$ for true value θ.

(a) Measurable function (b) Parametric function
(c) Sample estimator (d) Sample Statistic

Q.27 The function r(θ, δ) defined as :
$r(\theta, \delta) = E_\theta [L(\theta, \delta(x)]$ for varying $\theta \in \Omega\theta$ is known as ....
(a) Loss function (b) Risk function
(c) Parametric function (d) Decision function

Q.28 The performance of any decision rule δ is judged in the light of behaviour of its (..........). r(θ, δ) fo varying θ.
(a) Loss function (b) Risk function
(c) Parametric function (d) Decision function

Q.29 If there exists a decision rule δ in Δ such that $r(\theta, \delta) \leq r(\theta, \delta') \ \forall \ \theta \in \Omega\theta$ for an arbitrary $\delta' \varepsilon \theta$, then 'δ' is known as :
(a) Unbiased rule (b) Unbiased estimate
(c) Unbiased procedure (d) Uniformally best decision rule

Q.30 A decision rule $\delta \in \Delta$ is known as (..........) if ∃ no other rule $\delta' \in \Delta$ with uniformally smaller risk.
(a) Admissible (b) Not admissible
(c) Adjusted (d) Appropriate

Q.31 A decision rule δ is called (..........) if it provides maximum protection against the highest possible risk.
(a) Decision function (b) Minimax rule
(c) Minimax decision (d) Decision making

Q.32 A decision rule $\delta_0 \in \Delta$ is called "Minimax rule" in the class Δ if among all rules in Δ, it minimises the (.........).
(a) Maximum risk (b) Maximum gain
(c) Minimum loss (d) Minimum risk

Q.33 A decision rule $\delta_0 \in \Delta$ is minimax if :

$$\sup_\theta r(\theta, \delta_0) \ \ldots\ldots\ldots \ \sup_\theta r(\theta, \delta)$$

(a) $\leq$ (b) $\geq$
(c) $=$ (d) $>$

Q.34 The (..........) in decision theory is also known as "Pessimist approach."
(a) Traditional approach (b) Procedural approach
(c) Minimax approach (d) Theoretical approach

Q.35 A "Minimax approach" in decision theory is also known as (..........).

(a) Pessimist approach (b) Procedural approach
(c) Theoretical approach (d) Traditional approach

Q.36 A (minimax approach) is also called (Pessimist approach) because this approach always thinks of the (.........) but tries to avert it.

(a) Best (b) Worst
(c) Better (d) Unimportent

Q.37 It is not true that we will necessarily get a decision rule $\delta \in \Delta$ which has the (..........). maximum risk.

(a) Smallest (b) Least
(c) Both (a) & (b) (d) Adopted

Q.38 For selecting a best decision rule, θ itself is considered as a r.v. with a specified distribution say 6. (continuous or discrete). This p.d.f. 6 of θ is known as :

(a) Prior distribution (b) Posterior distribution
(c) Truncated distribution (d) Sampling distribution

Q.39 A decision rule $\delta_6 \in \Delta$, is called a (..........) in the class Δ if among all the rules $\in \Delta$, it has the smallest average risk.

(a) Admissible rule (b) Bayes rule
(c) Minimax rule (d) Decision space

Q.40 A rule δ_6 is a Bayes rule if :

$r(6, \delta_6)$ ....

(a) $\leq r(6, \delta)$ (b) $\geq r(6, \delta)$
(c) $= r(6, \delta)$ (d) $> r(6, \delta)$

Q.41 The risk function for a decision rule $\delta \in \Delta$, is defined as :

$\leq r(6, \delta) =$ ....

(a) $\int L(\theta, \delta(x))\, f_{\theta_0}(x)\, dx$ (b) $\int L(\theta, \delta(x))\, f_{\theta}(x)\, dx$

(c) $\int L(\theta, \delta(x))\, f_{\theta_0}(x)\, dx$ (d) None.

Q.42 Let $6_X(\theta)$ be the function of θ and is interpreted as the conditional distribution of θ for given x as observed value of X. Then $6_x(\theta)$ is known as the p.d.f. of (..........) corresponding to prior distribution 6.

(a) Posterior distribution of θ. (b) Prior distribution of θ.
(c) Truncated distribution of θ. (d) Sampling distribution of θ.

Q.43 In Q.(42), the $6_x(\theta)$ is known as p.d.f. of "Posterior distribution" of θ

corresponding to (..........) 6.

(a) Truncated distribution (b) Prior distribution

(c) Sampling distribution (d) None

Q.44 A (..........) may not exist for a simple reason that the minimum of conditional expected loss may not exist for each x.

(a) Bayes rule (b) Minimax rule

(c) Procedural rule (d) Traditional rule

Q.45 In case if Bayes rule does not exist, we may use an other rule which is known as :

(a) $\in$-Bayes decision rule (b) Minimax rule

(c) Procedural rule (d) Admissible rule

Q.46 A class C of decision rules is said to be complete if for any rule δ not in C, there exists a rule δ' in C which is (..........) than δ.

(a) More efficient (b) More powerful

(c) Uniformally better (d) Uniformally unbiased

Q.47 A class 'C' of decision rules $\delta(x)$ is said to be (..........) if :

$r(\theta, \delta') < r(\theta, \delta) \ \forall \ \theta \in \Omega \ \theta$

and $r(\theta, \delta') < r(\theta, \delta)$ for some θ

for δ' being in C but δ not being in C.

(a) Incomplete (b) Sufficient

(c) Insufficient (d) Complete

Q.48 A complete class 'C' of decision rules $\delta(x)$ if it does not contain a complete subclass.

(a) Minimal (b) Sufficient

(c) Complete (d) Admissible

Q.49 A Bayes decision rule $\delta(x)$ is (..........) if it is unique except for risk equivalence.

(a) Sufficient (b) Minimal

(c) Complete (d) Admissible

Q.50 A decision function δ_0 will be said to be (..........) solution in the wide sense $\exists$ a sequence 6_k (k = 1, 2, ...) of prior distributions such hat δ_0 is a Bayes solution related to the, ie. if :

$$\lim_{k \to \infty} [\inf r(6_k, \delta_0) - r(6_k, \delta_\theta] = 0$$

(a) Bayes solution (b) Bayes procedure

(c) Minimax solution (d) Best solution

Q.51 Wald has shown that under a very general condition the totality of (.........) in a wide sence constitutes a complete class.

(a) Bayes procedure (b) Bayes solution

(c) Best solution (d) Minimax solution

Q.52 Wald has shown a great importance to (..........) rules.

(a) Bayes rule (b) Admissible rule

(c) Traditional rule (d) Minimax rule

Q.53 Bayes rules are the only means through which the (..........) rules can be deduced.

(a) Minimax (b) Procedural

(c) Admissible (d) Traditional

Q.54 Let (δ6 & δ6') be the Bayes rules for a decision problem w.r. to prior distributions (6, 6'). Then 6 is called as (..........) prior distribution if :

r (6, δ6) > r (6', δ6').

(a) Most favourable (b) Least favourable

(c) Admissible (d) Acceptable

Q.55 Let r (6, & δ6) be the prior distribution and Bayes rule corresponding to r (θ, 6), then δ_6 will be a (..........) if :

$$r(6, \delta 6) \geq \sup_{\theta} r(\theta, \delta 6).$$

(a) Admissible rule (b) Less favourable rule

(c) Minimax rule (d) More favourable rule

Q.56 In Q.(55), the distribution 6 will be said to be (..........) prior distribution if :

$$r(6, \delta 6) \geq \sup_{\theta} r(\theta, \delta 6).$$

(a) Least favourable (b) Most favourable

(c) Minimax rule (d) Acceptable rule

Q.57 Let δ be the Bayes decision rule whose risk r (θ, δ_6) is constant over parameter space Ωθ, then δ6 is (..........).

(a) Admissible (b) Minimax

(c) Less admissible (d) More admissible

Q.58 Let { 6_k, k = 1, 2, ... } be a sequence of prior distributions, {δ_k, k = 1, 2, ...} be the sequence of Bayes rules w.r. to 6_k. and there exists a decision rule δ_0 such that :

$$\lim_{k\to\infty} r(6_k, \delta_k) = \underset{\theta}{Sup}\ r(\theta, \delta_0)$$

then $\delta_0(x)$ is known as :

(a) Minimax decision rule (b) More desirable rule

(c) Least desirable rule (d) Acceptable rule

Q.59 Let $\lambda(\theta)$ be the real valued function of parameter θ and it is to be estimated from the observed sample (x). Then a decision rule or decision function $\delta(x)$ is the (..........) of $\lambda(\theta)$.

(a) Parameter (b) Estimator

(c) Estimate (d) Statistic

Q.60 Let $L(\theta, d)$ denotes the loss incurred in deicsion $d \in \Delta$ in estimating the parametric function $\lambda(\theta)$. Let us defined loss function as :

$L(\theta, \delta) = C(\theta)\ W(|d - \lambda(\theta)|)$

where $C(\theta)$ is positive and W is strictly increasing function of absolute error $|d - \lambda(\theta)|$. Then a (..........) function is defined as :

$L^*(\theta, d) = C(\theta) . (d - \lambda(\theta))^2$.

(a) Quadratic loss (b) Linear loss

(c) Admissible loss (d) Considerable loss

Q.61 Let $D = (d_0, d_1)$ be the (.........) corresponding to hypothesis testing $H_0 : \theta \in \Omega\ \theta_0$ V.S. $H_1 : \theta \in \Omega\ \theta_1$. Then hypothesis is a two decision (d_0, d_1) problem.

(a) Sample space (b) Decision space

(c) Parameter space (d) Confidence interval

Q.62 Let $\Delta = \{ d_0, d_1 \}$ be the correct decisions for $(\theta \in \Omega_{\theta_0}\ \&\ \theta \in \Omega_{\theta_1})$ respectively. Then we define loss function $L(\theta, d_0)$ as given below :

$$L(\theta, d_0) = \begin{cases} 0 \text{ if } \theta \in .... \\ a(\theta) \text{ if } \theta \in .... \end{cases}$$

for $a(\theta) > 0$

(a) $\Omega\theta_0, \Omega\theta_1$ (b) $\Omega\theta_1, \Omega\theta_0$

(c) $\Omega\theta, \Omega\theta_0$ (d) $\Omega\theta_1, \Omega\theta$

Q.63 In relation to Q.(62), the loss function $L(\theta, d_1)$ is defined as :

$$L(\theta, d_1) = \begin{cases} b(\theta) \text{ if } \theta \in .... \\ 0 \text{ if } \theta \in .... \end{cases}$$

for $b(\theta) > 0$

(a) $\Omega\theta_1, \Omega\theta_0$ (b) $\Omega\theta_0, \Omega\theta_1$

(c) $\Omega\theta_0, \Omega\theta,$ (d) $\Omega\theta_1, \Omega\theta.$

Q.64 Corresponding to loss functions defined in Q.(62, 63), the risk function $r(\theta, \delta)$ with decision rule δ and critical region 'w' is defined as :

$r(\theta, \delta) = b(\theta)$ .... if $\theta \in \Omega_{\theta_0}$.

(a) $\int_w f_\theta(x)\, dx$ (b) $\int_{w1} f_\theta(x)\, dx$

(c) $\int_w f_{\theta_0}(x)\, dx$ (d) $\int_{w1} f_{\theta_1}(x)\, dx$

Q.65 In Q.(64), the risk function $r(\theta, \delta)$ can also be defined as :

$r(\theta, \delta) = a(\theta) \int_{w^c} f_\theta(x)\, dx$ if :

(a) $\theta \in \Omega\, \theta_1$ (b) $\theta \in \Omega\, \theta_0$

(c) $\theta \notin \Omega\, \theta_0$ (d) Both (a) & (c)

Q.66 The nature of a Bayes test (solution) may be considered as similar to that of a (..........) for a simple H_0 vs. simple H_1.

(a) M.P. test (b) U.M.P. test

(c) M.P. unbiased test (d) None

Q.67 In case of applying Bayes test, we consider a sort of weighted average densities under (H_0 & H_1) w.r. to (..........) say (W_0 & W_1) respectively.

(a) Acceptance regions (b) Critical regions

(c) Central regions (d) Confidence interval

Q.68 Suppose we have to estimate a real valued function $\lambda(\theta)$ of parameter θ by means of an interval say $d = (d_1, d_2) \in I$. where I contains all possible values of (..........). Then a general form of a decision function is given as :

$\delta(x) = (\ \delta_1(x),\ \ \delta_2(x)\ )$

(a) $\lambda(\theta_0)$ (b) $\lambda(\theta_1)$

(c) $\lambda(\theta)$ (d) None

Q.69 We define a simple loss function $L(\theta, d)$ given as :

$$L(\theta, d) = \begin{cases} 0 \; if \; (.......... < r\,(\theta, d) < ....) \\ 1, \qquad\qquad otherwise \end{cases}$$

(a) (d_2, d_1) (b) $\left(\frac{d_1}{2}, \frac{d_2}{2}\right)$

(c) (d_1, d_2) (d) $\left(\frac{d_2}{2}, \frac{d_1}{2}\right)$

Q.70 The risk function r (θ, δ) of a decision function δ(x) is defined as :
r (θ, δ(x)) = (.........).

(a) E_{θ}. L [θ, δ(x)] (b) E_{θ} L [θ_0, δ(x)]

(c) E_{θ} L [θ_1, δ(x)] (d) $E\theta_1$ L [θ_1, δ(x)]

Q.71 A game in the mathematical sense is just as a triplet (Ωθ, D, L) in the context of elements of (.........).

(a) Decision theory (b) Decision space

(c) Decision procedure (d) Both (a) & (c)

Q.72 Let the loss function L(θ, D) may be negative, then this negative valule may be interpreted as a (..........).

(a) Gain (b) Profit

(c) Both (a) & (b) (d) Loss

Q.73 In a two persons game, they may try to (..........) simultaneously their loss.

(a) Minimise (b) Maximise

(c) Average (d) Eliminate

Q.74 In context of Bayesian statistic the conditional p.d.f. say f(y/θ) is known as (.........) of θ.

(a) Posterior p.d.f. (b) Prior p.d.f.

(c) Truncated p.d.f. (d) Discreate p.d.f.

Q.75 In context of Bayesian statistic, the p.d.f. of θ say h (θ) is known as :

(a) Posterior p.d.f. (b) Prior p.d.f.

(c) Truncated p.d.f. (d) Continuous p.d.f.

7.4 Key Answers

Write True/False

Q.	Ans.	Q.	Ans.	Q.	Ans.	Q.	Ans.
1.	True	21.	True	41.	True	61.	True
2.	True	22.	True	42.	True	62.	True
3.	True	23.	True	43.	True	63.	True
4.	True	24.	True	44.	True	64.	True
5.	True	25.	True	45.	True	65.	True
6.	True	26.	True	46.	True	66.	True
7.	True	27.	True	47.	True	67.	True
8.	True	28.	True	48.	True	68.	True
9.	True	29.	True	49.	True	69.	True
10.	True	30.	True	50.	True	70.	True
11.	True	31.	True	51.	True	71.	True
12.	True	32.	True	52.	True	72.	True
13.	True	33.	True	53.	True	73.	True
14.	True	34.	True	54.	True	74.	True
15.	True	35.	True	55.	True	75.	True
16.	True	36.	True	56.	True	76.	True
17.	True	37.	True	57.	True	77.	True
18.	True	38.	True	58.	True	78.	True
19.	True	39.	True	59.	True	79.	True
20.	True	40.	True	60.	True	80.	True
						81.	True

Fill in the Blanks

Q.	Ans.	Q.	Ans.
1.	Decision making	7.	Decision theory
2.	Parameter space	8.	Parameter
3.	Hypothesis testing	9.	Decision making
4.	Interval estimation	10.	Decisions
5.	Sequential inference	11.	Decision space
6.	Maximum	12.	Decision

Q.	Ans.
13.	Parameter
14.	L(θ, d)
15.	Loss function
16.	Decision procedure
17.	Sample space
18.	Decision space
19.	Domain
20.	Range
21.	Decision function
22.	$\delta(x)$
23.	Most desirable
24.	Loss function
25.	Smallest loss
26.	Measurable function, (θ, δ)
27.	Risk function
28.	Risk function
29.	Uniformally best decision rule
30.	Admissible
31.	Minimax rule
32.	Maximum risk
33.	$\leq$ r (θ, δ')
34.	Minimax approach
35.	Pessimist approach
36.	Pessimist approach
37.	Smallest
38.	Prior distribution
39.	Bayes rule
40.	$\leq$ r (6, δ)
41.	$\int L(\theta, \delta(x))\, f_\theta(x)\, dx$
42.	Posterior distribution of θ
43.	Prior distribution
44.	Bayes rule
45.	∈-Bayes decision rule
46.	Uniformally better
47.	Complete
48.	Minimal
49.	Admissible
50.	Bayes solution
51.	Bayes solution
52.	Minimax rule
53.	Minimax
54.	Least favourable
55.	Minimax rule
56.	Least favourable
57.	Minimax
58.	Minimax decision
59.	Estimator
60.	Quadratic loss function
61.	Decision space
62.	$\Omega\theta_0$, $\Omega\theta_1$
63.	$\Omega\theta_0$, $\Omega\theta_1$
64.	$\int_w f_\theta(x)\, dx$
65.	$\theta \in \Omega\theta$
66.	M P test
67.	Critical region
68.	$\lambda(\theta)$
69.	$(d1, d_2)$
70.	Eθ L [θ, $\delta(x)$]
71.	Decision theory
72.	Grain
73.	Minimise
74.	Prior p.d.f.
75.	Posterior p.d.f. A θ
76.	Prior to.

Multiple Choice questions and Answers

Q.	Ans.	Q.	Ans.	Q.	Ans.
1.	(a)	26.	(a)	51.	(b)
2.	(b)	27.	(b)	52.	(d)
3.	(b)	28.	(b)	53.	(a)
4.	(c)	29.	(d)	54.	(b)
5.	(c)	30.	(a)	55.	(c)
6.	(a)	31.	(b)	56.	(a)
7.	(d)	32.	(a)	57.	(b)
8.	(d)	33.	(a)	58.	(a)
9.	(a)	34.	(c)	59.	(b)
10.	(c)	35.	(a)	60.	(a)
11.	(a)	36.	(b)	61.	(b)
12.	(b)	37.	(c)	62.	(a)
13.	(c)	38.	(a)	63.	(b)
14.	(a)	39.	(b)	64.	(a)
15.	(a)	40.	(a)	65.	(b)
16.	(b)	41.	(b)	66.	(a)
17.	(a)	42.	(a)	67.	(b)
18.	(b)	43.	(b)	68.	(c)
19.	(a)	44.	(a)	69.	(c)
20.	(b)	45.	(a)	70.	(a)
21.	(a)	46.	(c)	71.	(d)
22.	(a)	47.	(d)	72.	(c)
23.	(c)	48.	(a)	73.	(a)
24.	(a)	49.	(d)	74.	(a)
25.	(c)	50.	(a)	75.	(b)

❑❑❑

Chapter - 8

Linear Estimation

8.1 Write True/False

Q.1 Any linear function of sample observations which is used to estimate the population parameter or parametric function,is called the "Linear estimator."

Q.2 The sample mean $\overline{Y}_n = \frac{1}{n}\sum_{i=1}^{n} y_i$ is a linear estimator of the population mean $\overline{Y}_N = \frac{1}{N}\sum_{i}^{N} y_i$.

Q.3 The sample mean square error (sample variance)

$s^2 = \frac{1}{(n-1)}\sum_{i=1}^{n}(x_i - \overline{x}_n)^2$ is quadratic estimator of population variance

$\sigma_x^2 = \frac{1}{N}\sum_{i=1}^{N}(x_i - \overline{X}_N)^2$.

Q.4 A linear estimator t_n which has the maximum concentration about the unknown true parametric value, is known as "Minimum Variance Unbiased Linear Estimator."

Q.5 A linear estimator having minimum sampling variance in the class of linear unbiased estimators is known as "Best Linear Unbiased Estimator", (B.L.U.E.).

Q.6 Let $Y_1, Y_2, ..., Y_n$ be n independent r.v.'s with common variance σ^2. Let :

$E(Y_i) = a_{i_1} B_1 + a_{i_2} B2 + ... + a_{i_p} Bp$ ...(1)

$V(Y_i) = \sigma^2 \forall i = 1, 2, ..., n$

and $\text{cov}(Y_i, Y_j) = 0$ for $i \neq j$

where B_i (i = 1, 2, ... p < *n* are p unknown parameters and (a_i)'s) are known constant. Then model (i) is known as Gauss Markov" linear model.

Q.7 Let (Y, β, A) be expressed as column vectors and matrix written as :

$$Y = \begin{pmatrix} Y_1 \\ Y_2 \\ \cdot \\ \cdot \\ \cdot \\ Y_n \end{pmatrix}, \quad B = \begin{pmatrix} B_1 \\ B_2 \\ \cdot \\ \cdot \\ \cdot \\ B_p \end{pmatrix}, \quad A = \begin{pmatrix} a_{11} & a_{12} & \cdots\cdots & a_{1p} \\ a_{21} & a_{22} & \cdots\cdots & a_{2p} \\ \cdot & & & \\ \cdot & & & \\ \cdot & & & \\ a_{n1} & a_{n2} & \cdots\cdots & a_{np} \end{pmatrix}$$

and Y = AB ...(2)

Then A is known as "Design matrix."

Q.8 The model (2) may be written as :

$E(Y) = A\beta$

$D(Y) = \sigma^2 I.$ ...(3)

where D(Y) is the dispersion matrix and I is an identity matrix of order nxn.

Q.9 The model (3) may be written in an alternative way as :

$Y = AB + e$

$E(e) = \underset{\sim}{0}$ and $D(e) = \sigma_I^2.$ ...(4)

Q.10 The unknown parameters β's in the model are called "Effect."

Q.11 The model in which all the effects (unknown parameters) are fixed, are known as "Fixed effect model" or model I.

Q.12 If in the model 1,one of the $\frac{B}{j}$'s is constant with $a_{ij} = 1$, for that J and all i = 1, 2, ... *n*. then such effect is called a "General effect or an Additive constant."

Q.13 If in the model 1, $a_{ij's}$ are indicator variables then such model is known as "ANOVA" model.

Q.14 If in the model 1, the $a_{ij's}$ are independent variables (also called as regressor variables) then the model is called as "Regression model".

Q.15 If in the model 1, the $a_{ij's}$ are of types "Indicator variables" and "Independent variables" then such model is known as "ANCOVA" model.

Q.16 A linear function of the parameters $\beta_1, \beta_2, \ldots\ldots \beta_p$, is called a "Parametric function."

Q.17 A parameteric function $l'\beta = \sum_{i}^{p} li\ \beta_i$ for which an unbiaed linear estimator exists, is known as "Estimable function."

Q.18 If there exists a vector C = (C1, C2, ..., Cn)' of constant such that :

$$E(C'Y) = l'\beta \qquad \ldots(5)$$

which is independent of parameters then the parametric function $l'\beta$ i said to be "Linearly estimable" or an "Estimable function."

Q.19 A necessary and sufficient condition for the parametric function $l'\beta$ to be linearly estimable is :

$$\text{rank (A)} = \text{rank}\begin{pmatrix} A \\ I' \end{pmatrix} \qquad \ldots(6)$$

where $\begin{pmatrix} A \\ I' \end{pmatrix}$ is the matrix obtained from A by adjoining the row vector l'.

Q.20 If the r(A) = p, then every linear function of the parameters is "Linearly estimable."

Q.21 Every linearly estimable parametric function is of the form b'Aβ i.e.

$$b'A\beta = b_1\ E\ (Y_1) + b_2\ E\ (Y_2) + \ldots\ldots\ldots + b_n\ E\ (Y_n) \qquad \ldots.(7)$$

where $b' = (b_1, b_2, \ldots\ldots, b_n)$.

Q.22 A linear function of (Yi, i = 1, 2, ..., n) is said to belong to error if its expected value Vanishes independently of parameters.

Q.23 If $d' = (d_1, d_2, \ldots\ldots, d_n)$ is a row vector then d'y belongs to error if :

$$E(d'Y) = d'\,A\beta = 0\ \forall\ \beta \qquad \ldots(8)$$

i.e. if $d'A = \underset{\sim}{O}'$ or $A'd = \underset{\sim}{O}$.

Q.24 A necessary and sufficient condition for the linear function d'Y to belong to error is that d′ is orthogonal to the vector space V(A') generated by the row vector of A'_{pxn}.

Q.25 Let the rank of V(Á) be no < p. The vector space V(β) which is orthogonal to the space V(Á) is called the "Error space."

Q.26 The error space V(β) has the rank equal to

$n_e = (n - n_0)$. ...(9)

Q.27 The space V(A′) which contains the coefficient vectors of best estimators of the estimable function, is known as "estimation space".

Q.28 The estimation space V(A′) has the rank equal to no.

Q.29 Let C′Y be the linear function of the random variables $Y_1, Y_2, \ldots\ldots, Y_n$. Such that C ∈ V(A′) and E(C'Y) = l'β.

Then C'Y has the minimum variance among all linear unbiased estimators of l'β.

Q.30 The best estimator of any estimable function l'β must be of the form q′A′Y. where q' = $(q_1, q_2, \ldots\ldots, q_p)$ is a row vector and satisfies the equation

q' A ' A = 1′ ...(10)

Q.31 Let li'b(i = 1, 2, ..., k) be k estimable parametric functions and let

Ti = Ci Y for i = 1, 2, ..., k ...(11)

be their estimators. Then the best estimator of :

$\sum_{i}^{k} b_i \, l_i' \beta$ is T written as :

$$T = \sum_{i=1}^{k} b_i \, T_i \, . \qquad \ldots(12)$$

Q.32 Let $\beta_1 = \beta_1, (Y),\ \beta_2 = \beta_2 (Y), \ldots\ldots\ldots\ldots, \beta_P = \beta_P(Y)$ be a set of measurable function of Y such that the values $\beta_1, \beta_2, \ldots\ldots, \beta_P$ minimise the SS. of deviations of $Y_1, Y_2, \ldots\ldots, Y_n$ from their expectations i.e.

S = (Y – Aβ)′ (Y – Aβ) ...(13)

and is called a set of "Least square estimator" of the unknown parameters β_i (i = 1, 2, ..., p).

Q.33 The best estimator of the estimable linear function l'β is T given as :

T = l'β ...(14)

where (β_i, i = 1, 2, ..., p) are the L.S. estimators of (Bi, i = 1, 2, ..., p).

Q.34 The least square estimator of $l'\beta$ is identical with the B.L.U.E. of $l'\beta$.

Q.35 Let rank (A) = p, the number of unknown parameters. Usually p < n, this is known as the situation of "Full rank").

Q.36 Let C be such that rank (C) = rank (A). Then C is "Non-singular."

Q.37 In Q.(36), the inverse of the matrix C say $C^{-1} = (c^{w})$ exists and the unique solution of β is given by :

$$\beta = C^{-1} A'Y \qquad ...(15)$$

Q.38 The estimators β given in (15) is unbiased for the parameter β.

Q.39 The dispersion matrix D(β) is given by :

$$D(\beta) = \sigma^2 C^{-1} \qquad ...(16)$$

Q.40 In Q.(37), the estimator β_j is "Unique linear unbiased estimator" β_j for j = 1, 2, ..., p.

Q.41 In Q.(39), the sampling variance of the estimator β_j is given by :

$$V(\beta_j) = \sigma^2 C^{jj'} \qquad ...(17)$$

Q.42 In Q.(39) the covariance of the estimator (β_j, β_j') will be given by :

$$COV.(\beta_j, \beta_{j'}) = \sigma^2 C^{jj'} \qquad ...(18)$$

Q.43 Let A be any mxn matrix. Then the generalised inverse of the matrix A denoted by A^- is an *n x m* matrix such that for any arbitrary vector y for which the equations :

$$AX = Y \qquad ...(19)$$

are consistent.

Q.44 The generalised inverse A^- is not unique but A^- exists.

Q.45 For the generalised inverse A^-, the identity given as :

$$AA^-A = A \qquad ...(20)$$

holds.

Q.46 Let $(l_1'\beta \;\&\; l_2'\beta)$ be two least square (L.S.) estimators of two estimable linear functions $(l_1'\beta \;\&\; l_2'\beta)$ respectively. Then

$V(l_1' \beta) = \sigma^2 \, l_1' \, C^- \, l_1$

where C^- is the g inverse of $C = A^*A$ ...(21)

Q.47 In Q.(46), we have the result that :

$COV\,(l_1' \beta, l_1' \beta) = \sigma^2 l_1' \, \overline{C} \, l_2$ ...(22)

Q.48 Let us suppose that the parameters are subject to a set of consistent linear restrictions :

$P'\beta = L$, with rank $(P) = (p - s)$ ...(23)

Then appropriate linear model is :

$E(Y) = A\beta$

Q.49 The dispersion matrix of Y is given as :

$D(Y) = \sigma^2 I.$ ...(24)

Q.50 The minimum variance unbiased estimator (M.V.U.E.) of $l'\beta$ is $l'\beta$. where β is a solution of normal equations obtained by minimising $(Y - A\beta)'(Y - A\beta)$ subject to restriction $P'\beta = L$.

Q.51 Using λ as a vector of Lagrangian multipliers, the normal equations are written as :

$$\left.\begin{aligned} A'A\,\beta + P\lambda &= A'\,Y \\ \text{and } P'\beta &= L \end{aligned}\right\} \qquad \text{...(25)}$$

Q.52 Let us consider a more general form of the model as :

$E(Y) = A\beta$ ...(26)

$D(Y) = \sigma^2\beta$

where σ^2 is an unknown positive quantity and B is a known matrix. Then B is symmetrical and the variables are correlated.

Q.53 In Q.(52), we assume that the matrix B is non-singular i.e. $|B| \neq 0$.

Q.54 In Q.(52), the L.S. estimator may be found by minimising the S.S. written as :

$S.S. = (Y - A\beta)' \, B^{-1} \, (Y - A\beta)$ ...(27)

Q.55 The S.S. written in equation (27) is called the "Weighted S.S.).

Q.56 In case of correlated variables as stated in Q.(52), the L.S. estimators of the β's are then found by minimising the "Weighted S.S.."

Q.57 In Q.(52), if variables are uncorrelated but do not have constant σ^2 then B will be diogonal matrix.

Q.58 Let us consider the problem of linear estimation with ordered observation in a distribution depending on location and scale parameters only. Then the location and scale parameters (μ & σ) need not to be the "Mean and standard deviation" of the variable Y.

Q.59 The set of linear functions $C_1^{'}$ Y, $C_2^{'}$ Y, ...,,, $C_k^{'}$ Y ($k \geq 2$) is said to form an "Independent set" if no linear function of this set depends upon the other.

Q.60 A single linear function C'Y is said to be dependent or independent according. as the coefficient vector C' is "Null or Non-null."

Q.61 A necessary and sufficient (n & s) condition that the set of linear functions $C_1^{'}$Y, $C_2^{'}$ Y, ...$C_k^{'}$ is an independent set, is the non-existance of numbers $b_1, b_2, \ldots, b_k$ not all zero such that :

$$\sum_{i}^{k} b_i \, c_i^{'} = \underset{\sim}{O} \qquad \ldots(28)$$

where '$\underset{\sim}{O}$' is the null (1 × n) row vector.

Q.62 A set of all linear functions dependent on the linear functions $C_1^{'}$ Y, $C_2^{'}$Y, ..., $C_k^{'}$ Y is called the "Linear set or Linear system" generated by $C_1^{'}$ Y, $C_2^{'}$Y, ..., $C_k^{'}$Y.

Q.63 The rank of the linear set is the number of independent linear functions in the generating set.

Q.64 The rank of the linear set is also known as the "Number of degree of freedom" of the linear set.

Q.65 If a linear function b'Y is orthogonal to each of the linear functions $C_1^{'}$ Y, $C_2^{'}$Y, ..., $C_k^{'}$ Y generating a linear set L. then we say that the set $C_1^{'}$ Y, $C_2^{'}$Y, ..., $C_m^{'}$Y is orthogonal to the linear set L.

Q.66 The d.f. carried out by all (B.L.U.Es) of estimable parametric function, is equal to the rank of "Estimation space."

Q.67 The d.f. carried out by all possible linear functions belonging to the error space is equal to the rank of the "Error space" i.e. $n_e = (n - n_0)$. ...(29)

Q.68 If a set of linear functions $C_1^{'}$ Y, $C_2^{'}$Y, ..., $C_k^{'}$ Y carries r.d.f. then the set of linear functions orthogonal to the given set carries (n – r) d.f.

Q.69 Consider a linear function $C'Y = \sum_{i=1}^{n} C_i Y_i$. Then the quantity s^2 defined as :

$$s^2 = \left(\sum^{n} C_i Y_i\right)^2 / \left(\sum^{n} C_i^2\right) = \frac{(C'Y)^2}{C'C} \qquad \ldots(30)$$

is known as "S.S. due to linear function C'Y."

Q.70 The S.S. due to the linear function C'Y defined in Q.69 possess single d.f.

Q.71 The S.S. (C'Y) is equal to the square of the length of projection of Y' on the coefficient vector C'.

Q.72 The expectation of the S.S. due to linear function C'Y is given as :

$$E\,(\text{S S due to } C'Y) = \sigma^2 + \frac{(C'E(Y))^2}{C'C} \qquad \ldots(31)$$

Q.73 From linear functions $C_1'Y, C_2'Y, \ldots\ldots\ldots, C_k'Y$ with r.d.f. choose any r mutually orthogonal linear functions $b_1'Y, b_2'Y, \ldots\ldots, b_r'Y$. Then the total S.S. corresponding to these r chosen mutually orthogonal linear function carries r.d.f.

Q.74 The total SS due to r chosen mutually orthogonal linear functions is "Unique" and is quite independent of the choice of r mutually orthogonal linear function.

Q.75 If all the linear functions of the set belongs to error then :

$$E(S^2) = r\,\sigma^2 \qquad \ldots(32)$$

where S^2 is the S.S. due to set of linear functions carrying r.d.f.

Q.76 Let there be two linear sets with (r_1, r_2) d.f. such that every linear function from the first set is orthogonal to every linear function of the next set. If (S_1^2, S_2^2) are the S.S. corresponding to two sets then S.S due to joint set having $(r_1 + r_2)$ d.f. will be given as :

$$S^2 = S_1^2 + S_2^2 \qquad \ldots(33)$$

Q.77 The generalisation of the problem discussed in Q.(76) holds good for K such sets.

Q.78 Let the number of estimable linear parametric function be n_0 and hence corresponding d.f. is equal to n_0. Then the rank of estimation space is equal to n_0.

Q.78 Let us suppose that the coefficient vectors of the best estimators lie in the estimation space V(A'). with rank (V(A')) = n_0. Hence the exact number of independent best estimators in the estimation space is equal to n_0.

Q.79 The S.S. due to the totality of all best estimators is equal to the "Square of the projection of Y" on V(A').

Q.80 Let Z' be the projection of Y'. Then we have

$$Y' = \beta' A' + e' = \beta' A' + e'$$

and $$Z' = \beta' A'$$

where e′ is the projection of Y′ on the error space.

Q.81 The S.S. due to the set of all B.L.U.E.s carrying n_0 d.f. is S_0^2 given as :

$$S_0^2 = Z' Z = Z' Y$$

$$= \beta' A'Y. \qquad ...(35)$$

$$= Y' A.\beta$$

Q.82 The S.S. due to all the linear functions belonging to the error is called the "S.S. due to error."

Q.83 All the linear functions belonging to error generates the "Error space" of rank $n_e = (n - n_0)$.

Q.84 The conditional variance of Y given a fixed value of X = x is denoted by $\sigma^2_{Y.x}$.

Q.85 In the general regression model, the conditional mean of the regressed variable Y for the given ($x_1, x_2, \ldots\ldots, x_p$) is given as :

$$E(Y/x_1, x_2, \ldots\ldots, x_p) = \beta_0 + \beta_1 x_1 + \beta_2 x_2 + \ldots\ldots\ldots\ldots + \beta_p x_p \qquad ...(36)$$

Q.86 For the two regressor variables (X_1, X_2), the non linear regression model is written as :

$$E(Y/x_1, x_2) = \beta_0 + \beta_1 x_1 + \beta_1 x_1^2 + \beta_2 x_2 \qquad ...(37)$$

Q.87 The non-linear regression model written in equation (37) covers both the "Multiple linear regression" and "Curvilinear or Polynomial regression."

Q.88 Gauss Mavkov linear regression model with regression coefficients β's as the parameters and the matrix "A" of the known values of the regressor variable X_{px1} is expressed as :

$$\left.\begin{array}{l} E(Y) = Xb \\ D(Y) = \sigma^2 I \end{array}\right\} \quad \text{...(38)}$$

Q.89 In the Gauss Markov linear model written in equation 38, the matrix 'X' X' is "Non-singular."

Q.90 The rank of vector X of regressor variable is equal to the number of parameters say $(\beta_0, \beta_1, \beta_2, \ldots\ldots, \beta_p)$ i.e. rank (X) = (p + 1) ...(39)

Q.91 In the regression model written in Q.(88), the Unique solution of the normal equations :

$$X'X\beta = X'Y$$

is given as :

$$\beta = (X'X)^{-1} X'Y. \quad \text{...(40)}$$

Q.92 In Q.(91), the dispersion matrix of β is given as :

$$D(\beta) = \sigma^2 (X'X)^{-1}. \quad \text{...(41)}$$

Q.93 An unbiased estimator of σ^2 is given as :

$$\sigma^2 = \frac{SSE}{(n-p-1)} \quad \text{...(42)}$$

with $SSE = Y'Y - \beta' X' Y$.

Q.94 Y_{ij} denotes the life in the hours of Jth bulb prepared by i[th] method (i = 1, 2, ..., $p, j = 1, 2, \ldots\ldots, n_i$). Let us consider that :

$$\left.\begin{array}{l} E(Y_{ij}) = \mu_i \\ V(V_{ij}) = \sigma^2 \end{array}\right\} \quad \text{...(43)}$$

where μ_i is the population mean life in hours for bulb prepared by i[th] method. Then estimate of μ_i is given as :

$$\mu_i = \frac{1}{n_i} \sum_{j=1}^{n_i} Y_{ij} = \overline{Y_i}.$$

Q.95 For the problem discussed in Q.(94), the B.L.U.E. of any parametric function $L = \sum_{i}^{p} l_i \mu_i$ is given as

$$L = \text{est}\left(\sum_{i}^{p} l_i \mu_i\right) = \sum_{i}^{p} l_i \overline{Y}_{i.}$$

Q.96 In the problem discussed in Q.(94), the rank of the estimation space is given as :

Rank n_e (Estimation Space) = $(n - p)$.

Q.97 For the regression model discussed in Q.(94), the S.S. due to error given as :

$$\text{S.S.E.} = \sum_{i}^{p} \sum_{j=1}^{n_i} (Y_{ij} - \overline{Y}_{i.})^2 \qquad \ldots(45)$$

with d.f. = $(n-p)$.

Q.98 An unbiased estimate of σ^2 in the model (43) is given as :

$$\sigma^2 = \frac{1}{(n-p)} \sum_{i=1}^{p} \sum_{j}^{n_i} (Y_{ij} - \overline{Y}_{i.})^2 . \qquad \ldots(46)$$

8.2 Fill in the Blanks

1. Any linear function of sample observations which is used to estimate the 'population parameter' or 'parametric function' is called the (..........).

2. The sample mean $\overline{Y}_n = \frac{1}{n} \sum_{i}^{n} Y_i$ is a linear estimator of (.........) $\overline{Y}_N$.

3. The sample mean square (sample variance) $s^2 = \frac{1}{n-1} \sum^{n} (Y_i - \overline{Y}_n)^2$ is a (..........) of population variance $\sigma_y^2 = \frac{1}{N} \sigma_y^2 \frac{1}{N} \sum_{i}^{N} (y_i - \overline{Y}_N)^2$.

4. A linear estimator t_n which has the maximum concentration about the

unknown true parameteric value, is known as (..........) linear estimator.

5. A linear estimator having minimum sampling variance in the class of linear unbiased estimators, is known as (..........).

6. Let $Y_1, Y_2, \ldots\ldots, Y_n$ be independent r.v.'s with common variance σ_y^2. Let

$$E(Y_i) = \sum_j^p a_{ij}\ \beta_j \text{ and } V(Y_i) = \sigma_y^2\ \forall\ i = 1, 2, \ldots\ldots n. \text{ and } \operatorname{cov}(Y_i, Y_j) = 0$$

for $i \neq j$. where β_j (i = 1, 2, ..., p) are p unknown parameters and (a_{ij}'s) are known constant. Then the model written as :

$$E(Y_i) = \sum_j^p a_{ij}\ \beta_j \qquad \ldots(1)$$

is known as (..........) linear model.

7. Let (Y, β, A) be expressed as column vectors and matrix written as :

$$Y = \begin{bmatrix} Y_1 \\ Y_2 \\ \cdot \\ \cdot \\ \cdot \\ Y_n \end{bmatrix},\ \beta = \begin{bmatrix} \beta_1 \\ \beta_2 \\ \cdot \\ \cdot \\ \cdot \\ \beta_p \end{bmatrix},\ A = \begin{bmatrix} a_{11} & a_{12}, & \cdots & a_{1p} \\ a_{21} & a_{22} & \cdots & a_{2p} \\ \cdot & & & \\ \cdot & & & \\ \cdot & & & \\ a_{n1} & a_{n2} & \cdots & a_{np} \end{bmatrix}$$

and $Y = A\beta$ ...(2)

Then 'A' is known as (..........).

8. The model (2) in Q.(7) may be written as :

$$E(Y) = A\beta \text{ and } D(Y) = \sigma^2_{I} \qquad \ldots(3)$$

Then D(Y) is known (..........) matrix and 'I' is an identity matrix of order $n \times n$.

9. The model (3) is Q.(8) may be written in an alternative way as :

$Y = Ab + e$ ...(4)

E(e) = ..., and D(e) = (..........).

10. The unknown parameters β's in the model are called (..........).
11. The model in which all the effects (Unknown parameters) are fixed, are known as "..........'' or model I.
12. In the model 1, one of the β_j's is constant with aij = 1, for that j and all i = 1, 2, ... n. Then such effect is known as (..........) or an additive constant.
13. If in the model 1, a_{ij}'s are (..........) variables then such model is known as (..........) model.
14. If in the model 1, the a_{ij}'s are independent variables (also called regressor variables) then the model is called (..........).
15. If in model 1, the a_{ij}'s are of types "Indicator variables" and "Independent variables" then such model is known as (.........) model.
16. A linear function of the parameters $\beta_1, \beta_2, \ldots, \beta_p$ is known as (..........).
17. A parametric function $l'\beta = \sum_{i=1}^{p} l_i \beta_i$ for which an unbiased linear estimator exists, is known as an (..........).
18. Let $C = (C_1, C_2, \ldots, C_n)'$ be a vector of constants such that $E(C'Y) = l'\beta$.

    which is independent of parameters. Then the parametric fuction $l'\beta$ is said to be (..........) or estimable function.
19. A n & s condition for the parameteric function $l'\beta$ to be (.........) is :

    $$\text{rank }(A) = \text{rank}\begin{pmatrix} A' \\ l \end{pmatrix}$$

    where $\begin{pmatrix} A' \\ l \end{pmatrix}$ is the matrix obtained from A. by adjoining the row vector l'.
20. If the rank (A) = (..........) then every linear function of the parameters is linearly estimable.
21. A linear function of Y_i (i = 1, 2, ..., n) is said to belong to 'error' if its (..........) vanishes independently of parameters.
22. Let d' = $(d_1, d_2, \ldots, d_n)$ be a row vector then d'Y belongs to (..........) if :

    $$E(d'Y) = d'A\beta = 0 \ \forall \ \beta.$$

    ie if $d'A = \underset{\sim}{O}'$ or $A'd = \underset{\sim}{O}$.

23. A n & s condition for the linear function d'Y to belong to error is that d' is (..........) to the vector space V(A′) generated by the row vector of A'_{pxn}.

24. A n & s condition for the linear function d'Y to belong to error is that d' is orthogonal to the (..........) V(A′) generated by row vector of A'_{pxn}.

25. Let the rank V(A′) be $n_0 < p$. The vector space V(β) which is orthogonal to the space V(A′) is called the (..........).

26. The error space V(β) has the rank n_0 equal to (..........).

27. The space V(A′) which contains the coefficient vector of best estimators of the estimable function, is known as (..........).

28. Let C'Y be the linear function of the random variables $Y_1, Y_2, ... Y_n$ such that C ∈ V(A') and E(C'Y) = $l'\beta$. Then C'Y has the (..........) variance among all linear unbiased estimators of $l'\beta$.

29. The best estimator ofany estimable function l'β must be of the form q' A'Y where $q' = (q_1, q_2, ..., q_p)$ is a (..........) vector and satisfies the equation :

q' A' A = l'

30. Let $l_i'\beta$ (i = 1, 2, ..., k) be k estimable parametric functions and let :

$T_i = C_i' Y$ (i = 1, 2, ..., k)

be their best estimators. Then the best estimator of

$\sum_{i}^{k} b_i l_i' \beta$ is T written as :

T = ...

31. Let { $\beta_i = \beta_i'$ (Y), i = 1, 2, ..., p} be a set of measurable function of Y.

such that the values (β_i, i = 1, 2, ..., p) minimise the S.S. of deviations of (Y_i, i = 1, 2, ..., n) from their expectations. Then the set {β_i Y, i = 1, 2, ... p) is called a set of (..........) of the unknown parameters (β_i, i = 1, 2, ..., p).

32. Let T be the test estimator of estimable linear function $l'\beta$ which is written as :

T = ...

where ($\hat{\beta}_i$, $i = 1, 2, \ldots\ldots, p$) are the L S estimators of (β_i, $i = 1, 2, \ldots\ldots, p$).

33. The least square estimator of $l'\beta$ is identical with the (..........) of $l'\beta$.

34. Let rank (A) = p, the number of unknown parameters. This is known as the situation of (.........).

35. Let C be such that rank (C) = rank (A). Then C is (.........) matrix.

36. In Q.(35), the inverse of the matrix C say $C^{-1} = (C^{W})$ exists and the unique solution of β is given as :

    β = (.........).

37. The estimators β given in Q.(36) are (..........) for the parameters β.

38. The dispersion matrix D(β) is given as :

    D(β) = (.........).

39. In Q.(38), the (.........) of the estimator β_j, is given as :

    $V(\beta_j) = \sigma^2 C^{jj}$.

40. In Q.(38), the (..........) of the estimators (β_j, β_j) will be given as :

    $COV(\beta_j, \hat{\beta}_{j'}) = \sigma^2 C^{jj}$.

41. Let A be any *mxn* matrix. Then the (..........) of the matrix A denoted by A^- is an *nxm* matrix such that for any arbitrary vector Y, the equations Ax = Y are consistent.

42. For the generalised inverse A^-, the following identity holds :

    $A A^- A$ = ...

43. Let ($l_1' \beta$ & $l_2' \beta$) be two least square estimator of two estimable linear functions ($l_1' \beta$ & $l_2' \beta$) respectively. Then

    $V(l_1' \beta)$ = ...

44. In Q.(43), we have the result that

    $COV(l_1' \beta, l_2' \beta)$ = ...

45. The M.V.U.E. of $l'\beta$ is $l'\beta$ where β is the solution of normal equations obtained by minimising the function (..........)' (..........) subject to restriction $P'\beta = L$.

46. Using λ as a vector of (..........), the normal equations are written as :

$$A' A\beta + P\lambda = A'Y$$

and $\quad P'\beta = L$

47. Let us consider a more general form of the model as :

$E(Y) = A\beta$

$D(Y) = \sigma^2 B$

where $\sigma^2 > 0$ (Unknown) and B is known matrix. Then B is (.........) and variables are correlated.

48. In Q.(47), the L.S. estimator may be found by minimising the S.S. written as :

S.S. = (..........)′ (..........).

49. The S.S. given in Q.(48) is called as (..........).

50. In Q.(47), if variables are uncorrelated but do not have constant σ^2 then B will be a (.........) matrix.

51. Let us consider the problem of linear estimation with ordered observations in a distribution depending on location and scale parameters only. Then the location and scale parameters (μ, σ) need not to be the (......... and ....) of the variable Y respectively.

52. A single linear function $C'Y$ is said to be dependent or independent according as the coefficient vector C' is (......... or ....). respectively.

53. A *n* & *s* condition that the set of linear functions $C_1'Y, C_2'Y, \ldots, C_k'Y$ is an (.........) is the non-existant of number $b_1, b_2, \ldots, b_k$ not all zero such that :

$$\sum_{i=1}^{k} b_i C_i' = O'$$

where O' is the null $(1 \times n)$ row vector.

54. The rank of linear set of functions is equal to the number of (..........) in the generating set.

55. The rank of linear set of functions is also known as the (..........) of the linear set.

56. The d.f. carried out by all (B.L.U.E.'s) of a estimable parametric function, is equal to the rank of (..........).

57. The d.f. carried out by the all possible linear functions belonging to the error is equal to the rank of the (.........) i.e. ne = $(n - n_0)$.

58. If a set of linear functions $C_1' Y, C_2' Y, \ldots, C_k' Y$ carries r.d.f. then the set of linear functions orthogonal to a given set of n linear functions carries d.f. equal to (..........).

59. Consider a linear function $C'Y = \sum_i^n C_i Y_i$. Then the quantity s^2 defined as :

$$s^2 = \left(\sum_i^n C_i Y_i\right)^2 / \left(\sum^n C_i^2\right) = \frac{(C'Y)^2}{C'C}$$

is known as (.....................).

60. The S.S. due to linear function C'Y defined in Q.(59) has d.f. equal to ....

61. The expectation of the S.S. due to linear function C'Y is given as :
E (S.S. due to C'Y) = ...

62. From the linear functions $C_1' Y, C_2' Y, \ldots, C_k' Y$ with r.d.f., choose any r mutually orthogonal linear functions $b_1' Y, b_2', \ldots, b_k' Y$. Then total S.S. due to those r chosen mutually orthogonal linear functions have d.f. equal to ....

63. The total S.S. due to r chosen mutually arthogonal linear functions is 'Unique' and is quite independent of choice of r (..........) linear functions.

64. If all the linear functions of the set belong to the error then :
$E(S^2) = r\,\sigma^2$
where S^2 is the (..........) the set of linear functions carrying r.d.f.

65. Let there be two linear sets with (r_1, r_2) d.f. such that every linear function from the first set is orthogonal to every linear function of the next set. If (S_1^2, S_2^2) are the S.S. due to two sets then S.S. due to the joint set having $(r_1 + r_2)$ d.f. will be given as :
S^2 = (..........).

66. Let us suppose that coefficient vector of the best estimator lie in the (..........) V(A') with rank V(A') = n_0. Hence the exact number of independent best estimators in the estimation space is equal to n_0.

67. Let Z' be the projection of Y'. Then we have :

$$Y' = \beta' A' + e'$$

and $Z' = \beta' A'$

Then S.S. due to the set of all BLUE's carrying no. d.f. is S_0^2 and is given as :

S_0^2 = ...

68. The S.S. due to all the linear functions belonging to the error is called the S.S. (.........).

69. All the linear functions belonging to error generates the (.........). of rank $n_e = (n - n_0)$.

70. The conditional variance of Y given a fixed value X = x, is denoted by ....

71. In general regression model, the conditional mean of regressed variable Y for given $(x_1, x_2, ..., x_p)$ is given as :
$E(Y / x_1, x_2, ..., x_p)$ = ...

72. The non linear regression model written in (71), covers both the (..........) and (..........).

73. Gauss-Markov linear regression model with regression coefficients β's as parameters and matrix 'A' of known values of the (..........) X_{px1} is expressed as :
$E(Y) = X\beta$
$D(Y) = \sigma^2 I$

74. In Gauss-Markove linear model written in Q.(73), the matrix X'X is (..........).

75. The rank of vector X of regressor variable is equal to the number of parameters say $(\beta_0, \beta_1, \beta_2, ..., \beta_p)$ i.e. rank (X) = ...

76. In the regression model written in Q.(73), the unique solution of normal equations :
$X' X \beta = X' Y$ is given as

β = ...

77. In Q.(76), the dispersion matrix of β is given as :

$D(\beta)$ = ...

78. An unbiased estimator of σ^2 in the Q.(77) is given as :

σ^2 = ...

with S.S.E. = $Y' Y - \beta' X' Y$.

8.3 Multiple Choice Questions and Answers

Q.1 Any linear function of sample observations which is used to estimate the population parameter or parametric function, is known as :

(a) Linear estimator (b) Non-linear estimator

(c) Quadratic estimator (d) Point estimator

Q.2 The sample mean $\overline{Y}_n = \frac{1}{n}\sum_{i}^{n} y_i$ is the linear estimator of :

(a) Population mean (b) Population constant

(c) Population (d) None

Q.3 The sample mean square s^2 (sample variance) $s^2 = \frac{1}{n-1}\sum^{n}(y_i - \overline{y}_n)^2$

is a (..........) of population variance σ_y^2.

(a) Linear estimator (b) Non-linear estimator

(c) Quadratic estimator (d) Both (b) & (c)

Q.4 A linear estimator t_n which has the maximum concentration about the unknown true parametric value is known as :

(a) Minimum variance unbiased linear estimator

(b) Maximum variance unbiased linear estimator

(c) Best linear unbiased estimator

(d) Both (a) & (c)

Q.5 A linear estimator having minimum sampling variance in the class of linear estimators, is known as :

(a) Minimum variance unbiased linear estimator

(b) Best linear unbiased estimator (B.L.U.E.)

(c) Both (a) & (b)

(d) Consistent estimator

Q.6 Let $(Y_1, Y_2, ..., Y_n)$ be n independent r.v.'s with common variance σ^2.

Let $E(Y_i) = \sum_{j=1}^{p} a_{ij}\,\beta_j$ and $V(Y_i) = \sigma^2, \forall\ i = 1, 2, ..., n\}$

and COV $(Y_i, Y_i' =) = 0 \ \forall\ i \neq i'$

where B_j ($j = 1, 2, \ldots\ldots, p < n$) are p unknown parameters and (a_{ij}) are known constants. Then the model (1) is, known as :

(a) Consistent estimator (b) Gauss Markov linear model

(c) Sufficient estimator (d) Unbiased estimator.

Q.7 Let (Y, β, A) be expressed as colum vectors and matrix written as :

$$Y = \begin{bmatrix} Y_1 \\ Y_2 \\ . \\ . \\ . \\ Y_n \end{bmatrix}_{nx1}, \beta = \begin{bmatrix} \beta_1 \\ \beta_2 \\ . \\ . \\ . \\ \beta_p \end{bmatrix}_{px1}, A = \begin{bmatrix} a_{11} & a_{12} & \cdots & a_{1p} \\ a_{21} & a_{22} & \cdots & a_{2p} \\ . \\ . \\ . \\ a_{n1} & a_{n2} & \cdots & a_{np} \end{bmatrix}_{nxp}$$

and Y = A β. ...(2)

Then A is known as :

(a) Design matrix (b) Covariance matrix

(c) Variance matrix (d) Correlation matrix

Q.8 The model 2 in Q.(7) may be written as :

$E(Y) = A\,\beta$ and $D(Y) = \sigma^2\,I$ ...(3)

Then D(Y) is known as (..........) and I is an identity matrix of order $n\ x\ n$.

(a) Design matrix (b) Dispersion matrix

(c) Correlation matrix (d) Singular matrix

Q.9 Unknown parameters β's involved in the model are known as the :

(a) Effects (b) Cofactors

(c) Multipliers (d) Coefficients

Q.10 The model in which all the effects (Unknown parameters) are fixed, are known as (..........) or model I.

(a) Mixed effect model (b) Fixed effect model

(c) Random effect model (d) None

Q.11 In the model 1 given in Q.(6), one of the β's is constant with $a_{ij} = 1$, for that j and all i = 1, 2, ..., n. Then such effect is known as (..........) or an additive constant.

(a) General effect (b) Treatment effect

(c) Simple effect (d) Random effect

Q.12 If in model 1 written in Q.(6), the $a_{ij's}$ are the (.........), then such model is known as ANOVA model.

(a) Random variable (b) Indicator variable

(c) Continuous variable (d) Statastic variable

Q.13 In the model 1 of Q.(6), if a_{ij}'s are indicator variables, then such model is known as :

(a) ANOVA model (b) ANCOVA model

(c) Fixed effect model (d) Random model

Q.14 In the model 1, if a_{ij}'s are independent variables (also called regression variables) then the model is known as :

(a) ANOVA model (b) ANCOVA model

(c) Regression model (d) Mathematical model

Q.15 In the model 1, the a_{ij}'s of type "Indicator and independent" variables, then such model is known as :

(a) ANCOVA model (b) ANOVA model

(c) Regression model (d) Multiplicative model

Q.16 The linear function of the parameters $\beta_1, \beta_2, \ldots, \beta_p$ is known as :

(a) Statistical function (b) Parametric function

(c) Additive function (d) Regression function

Q.17 A parametric function $l' \beta = \sum_{i=1}^{p} l_i \beta_i$ for which an unbiased linear estimator exists, is known as :

(a) Estimable function (b) Parametric function

(c) Statistical function (d) Mathematical function

Q.18 Let $C = (C1, C_2, \ldots, C_n)'$ be a vector of constants such that :

$$E(C'Y) = l'\beta$$

which is independent of parameters. Then the parametric function is said to be (..).

(a) Linearly estimable function (b) Estimable function

(c) Both (a) & (b) (d) Statistical function

Q.19 A n & s condition for the parametric function $l'\beta$ to be (..........) is :

$$\text{rank}\,(A) = \text{rank}\begin{pmatrix} A \\ l' \end{pmatrix}$$

where $\begin{pmatrix} A \\ l' \end{pmatrix}$ is the matrix obtaind from 'A' by adjoining the row vector l'.

(a) Linearly estimable (b) Not linearly estimable
(c) Efficiently estimable (d) None

Q.20 If the rank(A) = (..........), then every linear function of the parameters is linearly estimable.

(a) No. of columns (b) No. of rows
(c) No. of parameters (d) No. of d.f.

Q.21 A linear function of Yi's (i = 1, 2, ..., n) is said to belong to 'error' if its (..........) vanishes independently of parameters.

(a) Expected value (b) Mean
(c) Both (a) & (b) (d) None

Q.22 Let $d' = (d_1, d_2, ..., d_n)$ be a row vector then d'Y belongs to (.........) if:
$E(d'Y) = d'A\beta = 0 \; \forall \; \beta$
i.e. iff $d'A = \underset{\sim}{O}'$ or $A'd = \underset{\sim}{O}$.

(a) Error space (b) Vector space
(c) Estimation space (d) None

Q.23 A. n & s condition for the linear function d' Y to belong to error is that d' is (..........) to the vector space V′(A) generated by the row vector of A_{nxp}.

(a) Independent (b) Dependent
(c) Orthogonal (d) None

Q.24 A n & s condition for the linear function d'Y to belong to 'Errir soace' is that d′ is orthogonal to the (..........) V(A') generated by row vector of A'_{pxn}.

(a) Vector space (b) Error space
(c) Estimation space (d) None

Q.25 Let the rank (A′) be $n_0 < p$. The vector space V(β) which is orthogonal to the space V(A′) is called the ....

(a) Vector space (b) Error space
(c) Estimation space (d) Parameter space

Q.26 The error space V(β) has the rank η_e equal to ....

(a) $(n-n_0)$ (b) n
(c) n_0 (d) p

Q.27 The space V(A′) which contains the coefficients vector of best estimators of the estimable functions is known as :

(a) Vector space (b) Error space

(c) Parameter space (d) Estimation space

Q.28 Let C′Y be the linear function of random variables $Y_1, Y_2, \ldots, Y_n$ such that C ∈ V(A′) and E(C'Y) = $l'\beta_j$. Then C′Y has the (..........) among all linear unbiased estimators of $l'\beta$.

(a) Minimum variance (b) Maximum variance

(c) Least variance (d) Both (a) & (c)

Q.29 The best estimator of any estimable function $l'\beta$ must be of the form q′ A′ Y where $q' = (q_1, q_2, \ldots, q_p)$ is a (..........) satisfies the equation : q′ A′ A = l'.

(a) Row vector (b) Column vector

(c) Unit vector (d) None

Q.30 Let $l_i'\,\beta$ $(i = 1, 2, \ldots, k)$ be k estimable parametric functions and let :

Ti = C_i' Y, (i = 1, 2, ..., k) be their best estimators. Then the best estimator of

$\sum_i^k b_i\, l_i'\, \beta$ is T written as :

T = ...

(a) $\sum_{i=1}^k b_i\, T_i$ (b) $\sum_i^k C_i'\, Y$

(c) $\sum_i^k b_i'\, C_i'\, Y$ (d) Both (a) & (c)

Q.31 Let $\{\beta_i - \beta_i(Y), i = 1, 2, \ldots, p\}$ be a set of measureable functions of Y such that the values (β_i, i = 1, 2, ..., p) minimise the S.S. of deviations of Y_j, $j = 1, 2, \ldots, p$) is known as a set of (..........) of the unknown parameters (βi, i = 1, 2, ..., p).

(a) Sufficient estimators (b) Consistent estimators

(c) Least square estimators (d) Biased estimators

Q.32 Let T be the best estimator of estimable linear function $l'\beta$, then T will be written as :

T = ...

(a) $l'\beta$ (b) $l'T$

(c) $l\beta$ (d) None

Q.33 The L.S. estimator of $l'\beta$ is identical with the (.........) of $l'\beta$.

(a) B.L.U.E. (b) Unbiased estimator

(c) Sufficient estimator (d) Consistent estimator

Q.34 Let rank (A) = p, the number of unknown parameters. This is known as the situation of (..........) rank.

(a) Half rank (b) Full rank

(c) Estimable rank (d) None

Q.35 Let C be a matrix such that rank (c) = rank (A). Then C is known as :

(a) Non-singular matrix (b) Symmetric matrix

(c) Singular matrix (d) Digonal matrix

Q.36 In Q.(35), the inverse of the matrix say $C^{-1} = (C^{ij})$ exists then unique solution of β is given as : $\hat{\beta}$ = ...

(a) $C^{-1}AY$, (b) CAY

(c) ACY (d) None

Q.37 The estimators β given in Q.(36) are (..........) of the parameters β.

(a) Biased estimators (b) Consistent estimators

(c) Sufficient estimators (d) Unbiased estimators

Q.38 The dispersion matrix D (β) is given as :

D (β) = ...

(a) σ^2C^{-1} (b) $C^{-1}\sigma^2$

(c) $C\sigma^2$ (d) Both (a) & (b)

Q.39 In Q.(38), the (..........) of the estimator β_j is given as :

$V(\beta_j) = \sigma^2C^{jj}$

(a) Sampling variance (b) Standard error

(c) Standard deviation (d) Variance

Q.40 In Q.(38), the (..........) of the estimators (β_j, β_j') will be written as :

$V^*(\beta_i, \beta_j) = \sigma^2 C^{ij}$

(a) Variance (b) Covariance
(c) Sampling variance (d) Standard deviation

Q.41 Let A be mxn matrix. Then the (..........) of the matrix a denoted by A^- is a *nxn* matrix such that for am arbitrary vector Y, the equations Ax = Y are consistent.

(a) Genarized inverse (b) Inverse
(c) Adjoint (d) None

Q.42 For the generalised inverse A^-, the folowing identity holds.

$A A^- A =$ ....

(a) A (b) (A^{-1})
(c) Both (a) & (b) (d) None

Q.43 Let ($l_1' \beta$ & $l_2' \beta$) be two L.S. estimators of two estimable linear functions ($l_1' \beta$ & $l_2' \beta$) respectively.

Then :

$V(l_1' \beta) =$ ....

(a) $\sigma^2 l_1' C^- l_1$ (b) $\sigma^2 l_1 C^- l_1'$
(c) $l_1' C^- l_1 \sigma^2$ (d) Both (a) & (c)

Q.44 In Q.(43), we have the result that :

$COV(l_1' \beta, l_2' \beta) =$ ....

(a) $\sigma^2 l_1' C^- l_2$ (b) $l_1' C^- l_2 \sigma^2$
(c) Both (a) & (b) (d) None

Q.45 The M.V.U.E. of $l'\beta$ is $1'\beta$ where β is the solution of normal equations obtained by minimising the function (.........)′ (..........). subject to restriction $P'\beta = L$

(a) $(Y - A\beta)' B^{-1} (Y - A\beta)$ (b) $(Y - A\beta)' B. (Y - A\beta)$
(c) $(Y - A\beta) B (Y - A\beta)'$ (d) None

Q.46 Using λ as a vector of (..........), the normal equations are written as :

$$A' A\beta + P\lambda = A'Y$$

and

$$P'\lambda = L$$

(a) Lagranjian multipliers (b) F-Yates multipliers
(c) Pearsonian multipliers (d) Fisher's multipliers

Q.47 Let us consider a more general form of the model written as :

$$E(Y) = A\beta$$

$$D(Y) = \sigma^2_\beta$$

where $\sigma^2 > 0$ (Unknown) and B is known matrix.

Then B is (..........) and variables are correlated.

(a) Symmetrical matrix (b) Asymmetrical matrix

(c) Digonal matrix (d) Identity matrix

Q.48 In Q.(47), the L.S. estimator may be found by minimising the S.S. written as :

S.S. = (..........)′ (..........).

(a) $(Y-A\beta)' B^{-1} (Y-A\beta)$ (b) $(Y-A\beta)' B. (Y-A\beta)$

(c) $(Y-A\beta) B^{-1} (Y-A\beta)'$ (d) None

Q.49 The S.S. written in Q.(48) is known as (..........).

(a) Weighted S S (b) Usual S S

(c) Minimum S.S. (d) Maximum S.S.

Q.50 In Q.(47), if variables are uncorrelated but do not have constant σ^2 then B will be a (.........) matrix.

(a) Digonal matrix (b) Symmetric matrix

(c) Singular matrix (d) Identity matrix

Q.51 Let us consider a problem of linear estimation with ordered observations in a distribution depending on location and scale parameters (μ, σ) only. Then the parameters (μ, σ) need not to be the (.......... and ....) of the variable Y. respectively.

(a) Mean and standard deviation

(b) Mean and variance

(c) Standard deviation and mean

(d) Variance and mean

Q.52 A single linear function C′Y is said to be dependent or independent according as the coefficient vector C′ is (.......... or ....) respectively.

(a) Null or non-null (b) Non-null or Null

(c) Scaler or Non-scaler (d) Vector or scaler.

Q.53 A n & s condition that the set of linear functions $C_1'Y, C_2'Y, \ldots, C_k'Y$ is an (.........) is the non-existence of number $b_1, b_2, \ldots, b_k$, not all zero such that :

$$\sum_{i}^{k} b_i' C_i' = O'$$

where O′ is the null $(1xn)$ row vector.

(a) Independent (b) Non-dependent

(c) Both (a) & (b) (d) Dependent

Q.54 The rank of linear set of functions is equal to the number of (..........) in the generating set.

(a) Independent linear function

(b) Non-dependent linear function

(c) Both (a) & (b)

(d) Dependent functions

Q.55 The rank of set of linear functions is also known as the (..........) of the linear set.

(a) Degree of freedom (b) No. of parameters

(c) The No of constants (d) None

Q.56 The d.f. carried out by all BLUE's of a estimable parametric function, is equal to the rank of (..........).

(a) Error space (b) Parametric function

(c) Non-parametric function (d) Estimation space

Q.57 The d.f. carried out by all possible linear functions belonging to error is equal to the rank of the (..........) i.e. $n_e = (n - n_0)$.

(a) Parametric space (b) Estimation space

(c) Error space (d) Vector space

Q.58 If a set of linear function C_1' Y, C_2' Y, ..., C_k' Y carries r. degree of freedom, then the set of linear functions orthogonal to a given set of n linear functions carries d.f. equal to (..........).

(a) $(n - r)$ (b) $(r - n)$

(c) n_e (d) $(n - n_0)$

Q.59 Consider a linear function $C'Y = \sum_{i}^{n} C_i' Y_i'$. Then the quantity s^2 defined as :

$$\hat{s}^2 = \left(\sum^{n} C_i Y_i\right)^2 / \left(\sum^{n} C_i^2\right) = (C'Y)^2 / C'C \text{ is known as } (..........).$$

(a) S.S. due to linear function C′Y
(b) S.S. due to linear function YC′
(c) S.S. due to non-linear function $C'Y^2$
(d) S.S. due to linear function Y′C′

Q.60 The S.S. due to linear function C'Y defined in Q.(59) has d.f. equal to :

(a) n (b) $n-1$
(c) 1 (d) none

Q.61 The expectation of S.S. due to linear function C′Y is given as :
E (S. S. due to C′Y) = ...

(a) $\sigma^2 + \dfrac{(C'E(Y))^2}{C'C}$ (b) $\sigma + \dfrac{(C'E(Y))^2}{C'C'}$

(c) $\sigma^2 + \dfrac{(C'E(Y))^2}{CC'}$ (d) $\sigma^2 + \dfrac{(C'E(Y))}{C'C}$

Q.62 From the linear functions C_1' Y, C_2' Y, ..., C_k' Y, with r.d.f., choose any r mutually orthogonal linear functions b_1'Y, b_2'Y, ..., b_r' Y. Then total S.S. due to these r chosen mutually orthogonal linear functions have d.f. equal to :

(a) $n-1$ (b) r
(c) $n-r$ (d) $r-1$

Q.63 The total S.S. due to r chosen mutually orthogonal linear functions is 'Unique' and is quite independent of choice of r (..........) linear functions.

(a) Dependent (b) Mutually dependent
(c) Mutually orthogonal (d) None

Q.64 If all the linear functions of the set belong to the error then
$E(S^2) = r\,\sigma^2$
where S^2 is the (..........) the set of linear functions carrying r.d.f.

(a) S.S. due to (b) M.S. due to
(c) Error S.S. of (d) Sampling variance

Q.65 Let there be two linear sets with (r_1, r_2) d.f. such that every linear function from the first set is orthogonal to every linear function of the next set. If (S_1^2, S_2^2) are the S.S. due to two sets then S.S. due to the joint set having $(r_1 + r_2)$ d.f. will be given as :
$S^2 = (\text{..........})$

(a) $S_1^2 + S_2^2$ (b) $S_2^2 + S_1^2$

(c) $S_1 + S_2$ (d) Both (a) & (b)

Q.66 Let us suppose that the coefficient vector of the best estimators lie in the (..........) V(A′) with rank (V (A′)) = n_0. Hence the exact number of the independent best estimators in the estimation space is equal to n_0.

(a) Estimation space (b) Error space

(c) Vector space (d) Parametric space

Q.67 Let Z' be the function of Y′. Then we have :

$Y' = \beta' A' + e'$ and $Z' = \beta' A'$

Then S.S. due to the set of all B.L.U.E.'s carrying n_0 d.f. is S_0^2 and is given as :

$S_0^2 = \text{..........}$

(a) $Y' A \beta$ (b) $Y' A \beta$

(c) $Y A \beta$ (d) $Y' A' \beta$

Q.68 The S.S. due to all linear functions belonging to the error is known as the S.S. (.........).

(a) Due to treatments

(b) Due to error (b)

(c) Due to estimators

(d) Due to interaction effects (d)

Q.69 All the linear functions belonging to the error generate the (.........) having rank $n_e = (n - n_0)$.

(a) Error space (b) Estimation space

(c) Vector space (d) Parameter space

Q.70 The conditional variance of Y for a given fixed value X = x, is denoted by ....

(a) $\sigma_{y.x}^2$ (b) σ_y^2

(c) $\sigma^2_{x.y}$ (d) σ^2_{yx}

Q.71 In general regression model, the conditional mean of regressed variable Y for given $(x_1, x_2, ..., x_p)$ is given as :
$E(Y / x_1, x_2, ..., x_p) =$ ....

(a) $\beta_0 + \sum_i^p \beta_i x_i^2$ (b) $\beta_0 + \sum_{i=1}^p \beta_i x_i$

(c) $\beta_0 + \sum^p \beta_i^2 x_i$ (d) $\beta_0 + \sum_i^p \beta_i^2 x_i^2$

Q.72 The non-linear regression model covers both the (.........) and (..........) parts.
(a) Multiple linear regression and curvilinear regression
(b) Multiple curvilinear regression and multiple linear regression
(c) Both (a) & (b)
(d) Only curvilinear regreesion.

Q.73 Gauss Markove linear regressionmodel with regression coefficients β's as parameters and matrix A of known values of the (........) X_{px1} is expressed as :

$$E(Y) = X\beta$$
$$D(Y) = \sigma^2 I$$

(a) Regression variables (b) Auxiliary variables
(c) Predictor variables (d) All the above

Q.74 InGauss Markove linear model written in Q.(73) the matrix X' X is (.........).
(a) Non-singular (b) Singular
(c) Digonal (d) Non-symmetrical

Q.75 The rank of vector X of regressor variable is equal to the number of parameters $(\beta_0, \beta_1, \beta_2, ..., \beta_p)$ i.e.
rank (X) = ...
(a) p (b) p + 1
(c) p – 1 (d) None

Q.76 In the regression written in Q.(73), the unique solution of normal equations :

X' X b = X' Y

is given as β = ...

(a) $(X'X)^{-1} X'Y$ (b) $(X'X) X'Y$

(c) $(X'X)^{-1} XY$ (d) $(X'X) Y X'$

Q.77 In Q.(76), the dispersion matrix of β is given as

$D(\beta)$ = ...

(a) $\sigma^2 (X'X)^{-1}$ (b) $(X'X)^{-1} \sigma^2$

(c) Both (a) & (b) (d) $(X'X)^{-1} \sigma$

Q.78 An unbiased estimator of σ^2 in the Q.(77) is given as :

σ^2 = ...

with S.S.E. = $Y'Y - \beta' X'Y$

(a) $SSE / (n-p-1)$ (b) $SSE / (n-p)$

(c) $SSE / (n-1)$ (d) $SSE / (p-1)$

8.4 Key Answers

Write True/False

Q.	Ans.	Q.	Ans.	Q.	Ans.	Q.	Ans.
1.	True	12.	True	23.	True	34.	True
2.	True	13.	True	24.	True	35.	True
3.	True	14.	True	25.	True	36.	True
4.	True	15.	True	26.	True	37.	True
5.	True	16.	True	27.	True	38.	True
6.	True	17.	True	28.	True	39.	True
7.	True	18.	True	29.	True	40.	True
8.	True	19.	True	30.	True	41.	True
9.	True	20.	True	31.	True	42.	True
10.	True	21.	True	32.	True	43.	True
11.	True	22.	True	33.	True	44.	True

Q.	Ans.	Q.	Ans.	Q.	Ans.	Q.	Ans.
45.	True	59.	True	73.	True	87.	True
46.	True	60.	True	74.	True	88.	True
47.	True	61.	True	75.	True	89.	True
48.	True	62.	True	76.	True	90.	True
49.	True	63.	True	77.	True	91.	True
50.	True	64.	True	78.	True	92.	True
51.	True	65.	True	79.	True	93.	True
52.	True	66.	True	80.	True	94.	True
53.	True	67.	True	81.	True	95.	True
54.	True	68.	True	82.	True	96.	True
55.	True	69.	True	83.	True	97.	True
56.	True	70.	True	84.	True	98.	True
57.	True	71.	True	85.	True		
58.	True	72.	True	86.	True		

Fill in the Blanks

Q.	Ans.	Q.	Ans.
1.	Linear estimator	15.	ANCOVA
2.	Population mean	16.	Parametric function
3.	Quadratic estimator	17.	Estimable function
4.	Minimum variance unbiased	18.	Linearly estimable
5.	Best linear unbiased estimator (B.L.U.E.)	19.	Linearly estimable
		20.	Number of parameters
6.	Gauss Markove	21.	Expected value
7.	Design	22.	Error
8.	Dispersion matrix	23.	Orthogonal
9.	O, σ^2 I	24.	Vector space
10.	Effect	25.	Error space
11.	Fixed effect model	26.	$(n - n_0)$
12.	General effect	27.	Estimation space
13.	Indicator, ANOVA	28.	Minimum
14.	Regression model	29.	ROW

Q.	Ans.
30.	$T = \sum_{i}^{k} b_i T_i$
31.	Least square estimators
32.	$l'\beta$
33.	BLUE
34.	Full rank
35.	Non-singular
36.	C^{-1} AY
37.	Unbiased
38.	$\sigma^2 C^{-1}$
39.	Sampling variance
40.	Co variance
41.	Generalized inverse
42.	= A
43.	$= \sigma^2 l_1' C^- l_1$
44.	$= \sigma^2 l_1' C^- l_2$
45.	$(Y - A\beta)'(Y - A\beta)$
46.	Lagrangian multipliers
47.	Symmetrical
48.	$(Y - A\beta)' B^{-1} (Y - A\beta)$
49.	Weighted S.S.
50.	Digonal
51.	Mean & Standard deviation
52.	Null or Non-Null
53.	Independent set
54.	Independent linear function
55.	Degree of freedom
56.	Estimation space

Q.	Ans.
57.	Error space
58.	$(n - r)$
59.	S.S. due to linear function C'Y
60.	1
61.	$\sigma^2 + \frac{(C'E(Y))^2}{C'C}$
62.	r
63.	Mutually orthogonal
64.	S.S. due to
65.	$S_1^2 + S_2^2$
66.	Estimation space
67.	$S_0^2 = Y'A\beta$
68.	Due to error
69.	Error space
70.	$\sigma^2_{y.x}$
71.	$\beta_0 + \sum_{i}^{p} \beta_i x_i$
72.	Multiple linear regression, curvilinear regression
73.	Regression variable
74.	Non-singular
75.	= (p + 1)
76.	$= (X'X)^{-1} X'Y$
77.	$\sigma^2 (X'X)^{-1}$
78.	$\frac{S.S.E.}{(n-p-1)}$

Multiple Choice Questions and Answers

Q.	Ans.	Q.	Ans.	Q.	Ans.	Q.	Ans.
1.	(a)	21.	(b)	41.	(a)	61.	(a)
2.	(a)	22.	(a)	42.	(a)	62.	(b)
3.	(d)	23.	(c)	43.	(d)	63.	(c)
4.	(d)	24.	(a)	44.	(c)	64.	(a)
5.	(b)	25.	(b)	45.	(a)	65.	(d)
6.	(b)	26.	(a)	46.	(a)	66.	(a)
7.	(a)	27.	(d)	47.	(a)	67.	(a)
8.	(b)	28.	(d)	48.	(a)	68.	(b)
9.	(a)	29.	(a)	49.	(a)	69.	(a)
10.	(b)	30.	(a)	50.	(a)	70.	(a)
11.	(a)	31.	(c)	51.	(a)	71.	(b)
12.	(b)	32.	(a)	52.	(b)	72.	(c)
13.	(a)	33.	(a)	53.	(c)	73.	(d)
14.	(c)	34.	(b)	54.	(c)	74.	(a)
15.	(a)	35.	(a)	55.	(a)	75.	(b)
16.	(b)	36.	(a)	56.	(d)	76.	(a)
17.	(a)	37.	(d)	57.	(c)	77.	(c)
18.	(c)	38.	(d)	58.	(a)	78.	(a)
19.	(a)	39.	(a)	59.	(a)		
20.	(c)	40.	(b)	60.	(c)		

❑❑❑

Chapter - 9

Distribution of Order Statistics

9.1 Write True/False

Q.1 In the subject matter of order statistic, the useful beta distribution is written as :

$$f(x) = \frac{1}{\beta(p, q)} x^{p-1} (1-x)^{q-1} \qquad ...(1)$$

for $0 < x < 1$, and $f(x) = 0$ else, where (p, q) are positive and real. Values.

Q.2 The multivariate analogue of $b(x, p, q)$ distribution is :

$$f(x_1, x_2, ..., x_k) = \frac{\Gamma(n_1 + n_2 + ... + n_{k+1})}{\Gamma(n_1)\ \Gamma(n_2)\ .....\ \Gamma(n_{k+1})} . \qquad ...(2)$$

$$x_1^{n_1}\ x_2^{n_2}\ .....\ x_k^{n_k}\ (1 - x_1 - .....- x_k)^{n_{k+1}}$$

for $x_i > 0$, $i = 1, 2, ..., k$, $\sum_{i=1}^{k} x_i \le 1$,

Q.3 The distribution defined in (2) is called as "Dirichlet distribution."

Q.4 In the distribution, we get the result that :

$$E(X_i) = \frac{n_i}{n_1 + n_2 + ... + n_{k+1}} \qquad ...(3)$$

Q.5 In Q.(2), the mathematical expectation of square of X_i is given as :

$$E\left(X_i^2\right) = \frac{n_i\ (n_i + 1)}{(n_1 + n_2 + ... + n_{k+1} + 1)\ (n_1 + n_2 + ... + n_{k+1})} . \qquad ...(4)$$

Q.6 In Q.(2), the variance of X_i is given as :

$$V(X_i) = \frac{n_i\ (n_1 + n_2 + ... + n_{k+1} - n_i)}{(n_1 + n_2 + ... + n_{k+1})\ (n_1 + ... + n_{k+1} + 1)} . \qquad ...(5)$$

Q.7 In Q.(2), the COV (X_i, X_j) are given as :

$$\text{COV}(X_i, X_j) = \frac{-n_i n_j}{(n_1 + n_2 + \ldots.. + n_{k+1})^2 (n_1 + n_2 + \ldots.. + n_{k+1} + 1)} \quad \ldots(6)$$

Q.8 For the sake of convenience, the distribution (2) is denoted by D(n_1, n_2, ..., n_{k+1}).

Q.9 Let X_1, X_2, ..., X_{k+1} be (k+1) be independent gamma variables with Xi having the distribution G (x_i, $\frac{1}{2}$, n_i). Then the variable Y_i defined as :

$$\text{Yi} = \frac{X_i}{X_1 + X_2 + \ldots.. + X_{k+1}}, \; i = 1, 2, \ldots., k.$$

has the distribution D (n_1, n_2, ..., n_k, n_{k+1}).

Q.10 Let X_1, X_2, ..., X_k follow Dirichlet distribution :
D (n_1, n_2, ..., n_k, n_{k+1}). Suppose that $Z_1 = X_1 + X_2 + \ldots.. + X_k$. $Z_2 = X_{k_{1+1}} + \ldots. + X_{k_1+k_2}$, ...,

$Z_k = X_{k_1} + \ldots. + K_{k-1+1} + \ldots.. + X_{k_1} + \ldots.. + k_k$.

where $k_1 + k_2 + \ldots.. + k_k < k$. Then ($Z_1$, ..., Z_k) has a s-variate Dirichlet distribution D(r_1, ..., r_k, r_{k+k}) where

$$r_1 = n_1 + \ldots.. + n_{k1}$$

$$r_2 = n_{k_{1+1}} + \ldots.. + n_{k1}+k_2$$

.

.

.

$$r_s = n_{k_1} + \ldots.. + k_{k-1} + 1 + \ldots.. n_{k1 + \ldots.. + ks}$$

$$r_{k+1} = n_{k1} + \ldots.. + k_{k+1} + \ldots.. + n_{k+1}$$

Q.11 Let us denote a random sample from continuous population by $x_{(1)}$, $x_{(2)}$....., $x_{(n)}$. Arrange these observations in ascending order of magnitude and denote this arranement by x_1, x_2, ..., x_n where $x_1 \leq x_2 \leq \ldots.. \leq x_n$. Then x_1, x_2, ..., x_n are called the order statistics of the sample.

Q.12 In order statistics $x_1 < x_2 \leq \ldots.. < x_i \leq \ldots.. \leq x_n$, the x_i is known as the ith order statistic.

Q.13 Let us denote by $F_{(x)}$ distribution function from which the random sample is drawn. As the population is assumed to be continous then we have :

$P_{(x_k = x_{k+1})} = 0$ for $k = 1, 2, \ldots, n-1$.

Q.14 The intervals $(-\infty, x_1), (x_1, x_2), \ldots (x_n, \infty)$ are called the "Sample blocks."

Q.15 The functions $F_{(x_1)}, F_{(x_2)} - F_{(x_1)}, \ldots, F_{(x_{n+1})} - F_{(x_n)}$ of these blocks are known as "Coverages."

Q.16 The coverage of the block denotes the probability of the population contained in that "Sample block."

Q.17 The study of sampling distribution of order statistics and coverages, is useful for a study of "Non parametric inference."

Q.18 If $x_1, x_2, \ldots, x_n$ are order statistics from a continuous distribution frunction $F_{(x)}$, then the distribution of random variables $F_{(x_1)}, F_{(x_2)}, \ldots, F_{(x_n)}$ is the Dirichlet distribution D(1, ..., 1 : 1).

Q.19 Let $x_1, x_2, \ldots, x_n$ be order statistics from a distribution function $F_{(x)}$ and propability density function $f_{(x)}$. Then the p.d.f. of order statistics is :

$$n!\ f_{(x_1)}\ f_{(x_2)} \ldots f_{(x_n)}\ \text{dx1 dx} d_{x_1}\ d_{x_2} \ldots d_{x_n}$$

Q.20 The distribution of x_k, $1 \le k \le n$ is given as :

$$\frac{\overline{|(n+1)}}{\overline{|(k)}\overline{|n-k+1)}} \left\{F_{(x_k)}\right\}^{k-1} \left\{1 - F_{(x_k)}\right\}^{n-k} f_{(x_k)}\ d_{x_k}$$

Q.21 The distribution of x_1, is denoted by $m(x_1)$ is given as :

$$m_{(x1)} \{ 1 - F_{(x_1)} \}^{n-1} f_{(x_1)}\ d_{x_1}$$

Q.22 The distribution of the largest sample values x_n is given as :

$$g(x_n) = n \{ \text{f}F_{(x_n)} \}^{n-1} f_{(x_n)}\ d_{x_n}$$

Q.23 If the parent distribution is rectangular in the range (0, β) then the distribution illustrated in Q.(21), reduces to the form :

$$m_{(x_1)} = \left(1 - \frac{x_i}{\beta}\right)^{n-1} \frac{d_{x_1}}{\beta}$$

Q.24 If the parent distribution is rectangular in the range (0, β), then the distribution illustrated in Q.(22), takes the form :

$$g_{(x_n)} = \frac{nx^{n-1}\, dx_n}{\beta^n}$$

Q.25 Let $x_1 \le x_2 \le \ldots \le x_n$ be an ordered sample of size n then the range R = $(x_n - x_1)$. Then generalising this concept, the difference between the mth

values from either side is called the mth range say R_m and mathematically, it is written as :

$$R_m = (x_{n-m+1} - x_m)$$

Q.26 The mth mid range is defined as :

$$M_m = \frac{1}{2}(x_{n-m+1} + x_m)$$

Q.27 Let R denotes the range of the ordered sample $x_1 \leq x_2 \leq x_3 < ..... < x_n$. and hence :

$$R = (x_n - x_1)$$

Then the expected value of R is given as :

$$E(R) = \int_{-\infty}^{\infty} [1 - \{F_{(x)}\}^n - \{1 - \{F_{(x)}\}^n] \, dx \;.$$

Q.28 Let us denoted the coverages as $U_1 = F_{(x_1)}$, $U_2 = F_{(x_2)} - F_{(x_1)}$ ..... Un $= F_{(x_n)} - F_{(x_{n-1})}$ where F(x) is continuous distribution. Then the joint distribution of U_1, U_2, ..., Un is D(1,..... 1 : 1) i.e. n-variate Dirchlet distribution.

Q.29 Any p ($\leq n$) coverages defined above has a Dirichlet distribution D(1, ..., 1 :, $n - p + 1$).

Q.30 The sum of any p of the coverage considered above, has a beta distribution $\beta_{(p,\, n-p+1)}$.

Q.31 If $w_1, w_2, ..., w_s$ are sums, without overlaping, of $K_1, K_2, ..., K_s$ coverages respectively, then the distribution of $w_1, w_2, ..., w_s$ is the s-variate Dirchlet distribution. D(K_1, ... K_s : $n - k_1 - ..... - K_s + 1$).

9.2 Fill in the Blanks

Q.1 In subject matter of order statistics, the useful beta distribution is written as :

$$f(x, p, q) = \frac{1}{\beta(p, q)} x^{p-1} \; (..........)$$

for 0 < x < 1, and $f(x) = 0$ else, where (p, q) are positive and real values.

Q.2 The multivariate analogue of $b(x, p, q)$ distribution is :

$$f(x_1, x_2, ..., x_k) = \frac{\overline{\overline{\quad -\,-\,-\,-\,- \quad}}}{\left| \left(\prod_{i}^{k+1} n_i \right) \right.} x_1^{n_1} \; x_2^{n_2} \; ..... \; x_k^{n_k} \; (1 - x_1 - ..... - x_k)^{n_{k+1}}$$

for $x_i > 0$, i = 1, 2, ..., k, $\sum_{i}^{k} x_i < 1$.

Q.3 The distribution defined in Q.(2) is known as (..........).

Q.4 In the distribution defined in Q.(2), we get

$E(X_i)$ = ...

Q.5 In Q.(2), the mathematical expectation of square of X_1 is given as :

$$E\,(X_i^2) = \frac{n_i\,(n_i + 1)}{\left(\sum_{i=1}^{k+1} n_i + 1 \right) (..........)}$$

Q.6 In Q.(2), the variance of X_i is given as :

$$V(Xi) = \frac{n_i\,(..........)}{\left(\sum_{i=1}^{k+1} n_i \right) \left(\sum_{i=1}^{k+1} n_i + 1 \right)}$$

Q.7 In Q.(2), the COV (X_i, X_j) are given as :

$$COV\,(X_i, X_j) = \frac{-n_i\; n_j}{(..........)^2\;(..........)}$$

Q.8 Consider a random sample $x_{(1)}, x_{(2)}, ..., x_{(n)}$ from a continuous population. Arrange these sample values in ascending order of magnitude as :

$x_1 \leq x_2 \leq ..... \; x_n$. Then $x_1, x_2, ..., x_n$ are known as :

(.....................) of the sample values.

Q.9 In ordered statistics $x_1 \leq x_2 \leq ..... \leq x_n$, the x_i is known as the (..........).

Q.10 Let $F_{(x)}$ denotes the distribution function from which a random sample is drawn. As the population is assumed to be continuous, then we have :

$P(x_k = x_{k+1})$ = ... for k = 1, 2, ..., n–1.

Q.11 The intervals $(-\infty, x_1), (x_1, x_2), ... (x_n, \infty)$ are known as the (.........).

Q.12 The study of sampling distribution of order statistics and coverages, is useful for a study of (..........) inference.

Q.13 Let $x_1, x_2, ..., x_n$ be the ordered statistics from a continuous distribution $F_{(x)}$. Then the distribution $(F_{(x_1)}, F_{(x_2)}, ..., F_{(x_n)}$ is known as (..........) and is noted as D(1, ... 1, : 1).

Q.14 The distribution of the largest sample values x_n is given as :

$g(x_n) = n \{ ....\}^{n-1} f(x_n)\, dx_n$

Q.15 Let $x_1 \leq x_2 \leq ..... \leq x_n$ be an ordered sample of size n then the range $R = (x_n - x_1)$. Then generalizing this concept, the difference between m^th^ value from either side is known as (..........) say R_m is written as :

$R_m = (x_{n-m+1} - x_m)$

Q.16 In Q.(15), the mth range R_m will be defined as :

R_m = (....................)

Q.17 In Q.(15), the mth mid range will be defined as :

$$M_m = \frac{1}{2} (.........).$$

9.3 Multiple Choice Questions and Answers

Q.1 In subject matter of order statistics, the useful beta distribution is written as :

$$f(x, p, q) = \frac{1}{\beta(p, q)} x^{p-1} (..........)$$

(a) $(1-x)^{q-1}$ (b) $(1-x)^{p-1}$

(c) x^{q-1} (d) $(1-x)^{p}$

Q.2 The multivariate analogue of b(x, p, q) distribution is :

$$f(x_1, x_2, ..., x_k) = \frac{..........}{\sqrt{\left(\prod_{i}^{k+1} n_i\right)}} \left(\prod_{i=1}^{k} x_i^{n_i}\right) (1 - \sum_{i}^{k} x_i)^{n_{k+1}}$$

for $x_i > 0$, $i = 1, 2, ..., k$, $\sum_{i}^{k} x_i < 1$.

(a) $\sqrt{\left(\sum_{i}^{k} n_i\right)}$ (b) $\sqrt{\left(\sum_{i=1}^{k+1} n_i\right)}$

(c) $\sqrt{\left(\sum_{i=1}^{k-1} n_i\right)}$ (d) None

Q.3 The distribution defined in Q.(2), is known as (..........).

(a) Multivariate distribution (b) Binomial distribution

(c) Dirichlet distribution (d) Truncated distribution

Q.4 In distribution defined in Q.(2), we get :

$E(X_i)$ = ...

(a) $n_i \Big/ \sum_{i=1}^{k+1} n_i$ (b) $n_i \Big/ \sum_{i=1}^{k} n_i$

(c) $(n_i + 1) \Big/ \sum_{i=1}^{k+1} n_i$ (d) $(n_i + 1) \Big/ \sum_{i=1}^{k} n_i$

Q.5 In Q.(2), the mathematical expectation of square of X_i is given as :

$$E\left(X_i^2\right) = \frac{n_i\ (n_i + 1)}{\left(\sum_{i=1}^{k+1} n_i + 1\right)(..........)}$$

(a) $\left(\sum_{i=1}^{k} n_i - n_i\right)$ (b) $\left(\sum_{i=1}^{k+1} n_i - n_i\right)$

(c) $\sum^{k} n_i + n_i$ (d) $\sum_{i=1}^{k+1} n_i + n$

Q.6 In Q.(2), the variance of X_i is given as :

$$V(X_i) = \frac{n_i\ (..........)}{\left(\sum_{i=1}^{k+1} n_i\right)\left(\sum_{i=1}^{k+1} n_i + 1\right)}$$

(a) $\left(\sum^{k+1} n_i - n_i\right)$ (b) $\left(\sum_{i=1}^{k} n_i - n_i\right)$

(c) $\left(\sum^{k-1} n_i - n_i\right)$ (d) $\left(\sum_{i}^{k+1} n_i + n_i\right)$

Q.7 In Q.(2), the COV (X_i, X_j) are given as :

$$COV(X_i, X_j) = \frac{-n_i\, n_j}{(..........)\,(..........)}$$

(a) $\left(\sum_{i=1}^{k+1} n_i\right)^2 \left(\sum_{i}^{k+1} n_i + 1\right)$ (b) $\left(\sum_{i=1}^{k+1} n_i\right)\left(\sum^{k+1} n_i + 1\right)$

(c) $\left(\sum_{i=1}^{k+1} n_i\right)\left(\sum^{k+1} n_i + 1\right)^2$ (d) $\left(\sum_{i=1}^{k} n_i\right)\left(\sum_{i=1}^{k} n_i + 1\right)$

Q.8 Consider a random sample $x_{(1)}, x_{(2)}, ..., x_{(n)}$ from a continuous population. Arrange these sample values in ascending order of magnitude as $x_1 \le x_2 \le ..... \le x_n$ Then $x_1, x_2, ..., x_n$ are known as (.........) of the sample values.

(a) Discrete statistic (b) Order statistic

(c) Continuous statistic (d) Alternative statistic

Q.9 In order statistics $x_1 \le x_2 \le ..... \le x_n$, the value x_i is known as the (..........).

(a) i^{th} order statistic (b) $(i + 1)^{th}$ order statistic

(c) $\left(\frac{i}{2}\right)^{th}$ order statistic (d) $\left(\frac{i+1}{2}\right)^{th}$ order statistic.

Q.10 Let $F_{(x)}$ denote the distribution function from which a random sample is drawn. As the population is assumed to be continuous, then we have :

$P(x_k = x_{k+1}) =$ .... for k = 1, 2, ..., $n-1$

(a) 1 (b) 0

(c) $\frac{1}{2}$ (d) 0.5

Q.11 The intervals $(-\infty, x_1), (x_1, x_2), \ldots\ldots (x_n, \infty)$ are known as the (..........).

(a) Central block (b) Subblock

(c) Sample block (d) Average block

Q.12 The study of sampling distribution of order statistic and coverages, is useful for a study of (..........). inference.

(a) Non-parametric (b) Parametric

(c) Sequential (d) Stochastic

Q.13 Let $x_1, x_2, \ldots\ldots, x_n$ be the ordered statistics from a continuous distribution F(x). Then the distribution of $F(x_1), F(x_2), \ldots\ldots, F(x_n)$ is known as (.........) and is noted as D(1, 1, ... 1, : 1).

(a) Multivariate distribution (b) Dirichlet distribution

(c) Truncated distribution (d) Sampling distribution

Q.14 The distribution of the largest sample value x_n is given as :

$g(x_n) = \text{n } (\ldots\ldots\ldots.)\, f(x_n)\, dx_n$

(a) $\{ F(x_n) \}^{n-1}$ (b) $\{ F(x_n) \}^{n}$

(c) $\{ F(x_1) \}^{n-1}$ (d) $\{ F(x_1) \}^{n}$

Q.15 Let $x_1 \leq x_2 \leq x_3 \leq \ldots\ldots \leq x_n$ be an ordered sample of size n then the range $R = (x_n - x_1)$. Then generalizing this concept, the difference between m^{th} value from either side denoted by R_m is known as (..........) and is written as :

$R_m = (x_{n-m+1} - x_m)$

(a) $\left(\frac{m}{2}\right)^{th}$ range (b) m^{th} range

(c) $(m+1)^{\text{th}}$ range (d) $(m-1)^{\text{th}}$ range

Q.16 In Q.(15), the mth range will be defined as :

$$R_m = (\ldots\ldots\ldots.)$$

(a) $(x_{n-m+1} - x_m)$ (b) $(x_{n-m} - x_m)$

(c) $(x_{m+1} - x_m)$ (d) $\frac{1}{2}(x_m + x_{m+1})$

Q.17 In Q.(15), the mth mid range will be defined as :

$$M_m = \frac{1}{2}\ (\ldots\ldots\ldots.).$$

(a) $(x_{n-m} + x_m)$ (b) $(x_{n-m+1} + x_m)$

(c) $(x_n + x_m)$ (d) $(x_m + x_{m+1})$

9.4 Key Answers

Write True/False

Q.	Ans.	Q.	Ans.	Q.	Ans.	Q.	Ans.
1.	True	9.	True	17.	True	25.	True
2.	True	10.	True	18.	True	26.	True
3.	True	11.	True	19.	True	27.	True
4.	True	12.	True	20.	True	28.	True
5.	True	13.	True	21.	True	29.	True
6.	True	14.	True	22.	True	30.	True
7.	True	15.	True	23.	True	31.	True
8.	True	16.	True	24.	True	32.	True

Fill in the Blanks

Q.	Ans.
1.	$(1-x)^{q-1}$
2.	$\left\lvert\left(\sum_{i=1}^{k+1} n_i\right)\right.$
3.	Dirichlet distribution
4.	$\dfrac{n_i}{\sum_{i=1}^{k+1} n_i}$
5.	$\left(\sum_{i=1}^{k+1} n_i\right)$
6.	$\left(\sum_{i=1}^{k+1} n_i - n_i\right)$
7.	$\left(\sum_{i=1}^{k+1} n_i\right), \left(\sum_{i=1}^{k+1} n_i + 1\right)$
8.	Order statistic
9.	i^{th} order statistic
10.	0
11.	Sample block
12.	Non parametric
13.	Dirichlet distribution
14.	$F(x_n)$
15.	m^{th} range
16.	$x_{n-m+1} - x_m$
17.	$x_{n-m+1} + x_m$

Multiple Choice Questions and Answers

Q.	Ans.	Q.	Ans.	Q.	Ans.	Q.	Ans.
1.	(a)	6.	(a)	11.	(c)	16.	(a)
2.	(b)	7.	(a)	12.	(a)	17.	(b)
3.	(c)	8.	(b)	13.	(b)		
4.	(a)	9.	(a)	14.	(a)		
5.	(b)	10.	(b)	15.	(b)		

❑❑❑

Bibliography

1. Alexander M. Mood, Franklin A. Gryabill and Duane C. Bose : (1974) Introduction to the Theory of Statistics. McGraw. - Hill.
2. Agarwal, B.L. (1999) : Basic Statistics. Wiley Eastern Ltd., New Delhi.
3. Fisher, R.A. (1925) : Theory of Statistical Estimation, Proceedings of Combridge Philosophical Society.
4. Gupta, S.C. and Kapoor, V.K. (1983) : Fundamental of Mathematical Statistics. Kushwaha, K.S. Sultan Chand and Sons, New Delhi.
5. Goon, A.M., Gupta, M.K. and Dasgupta, B. (1980) : An Outlines of Statistical Theory. The World Press Pvt. Ltd., Kolkata.
6. Kushwaha, K.S. and Kumar, Rajesh (2009) : The Theory of Sample Survey and Statistical Decisions, New India Publishing Agency, New Delhi.
7. Kushwaha, K.S. and Kumar, Rajesh (2009) : Basic Concept in Statistics. New India Publishing Agency, New Delhi.
8. Kendall, M.G. and Stuart, A. (1961) : The Advance Theory of Statistics, Vol.-2. Charles Griffin and Company, Ltd. London.
9. Murray, R. Spiegel (1961) : Schaum's Outline of Theory and Problems of Statistics. Schaum Publishing Company, New Yark.
10. Raw, C.R. (1973) : Linear Statistical Inference. John Wiley.
11. Steel, R.G.D. and Torrie, J.H. (1980) : Principles and Procedures of Statistics, A Biometrical Approach. McGraw Hill Kogakusha, Ltd. International Student Edition.
12. Searle, S.R. (1971) : Linear Models. John Wiley.
13. Saxena, H.C. and Surendran, P.U. (1990) : Statistical Inference. S. Chand & Company Ltd. New Delhi.
14. Wald, A. (1947) : Sequential Analysis. John Wiley.

❑❑❑

FSC
www.fsc.org
MIX
Papier | Fördert
gute Waldnutzung
FSC® C083411

Zeitfracht Medien GmbH
Ferdinand-Jühlke-Straße 7
99095 Erfurt, Deutschland
produktsicherheit@kolibri360.de